ALLE · ZEIT · WACH · 1842

Environmental Physiology and Biochemistry of Insects

Edited by
Klaus H. Hoffmann

With 78 Figures

Springer-Verlag
Berlin Heidelberg New York Tokyo 1985

Professor Dr. KLAUS H. HOFFMANN

Allgemeine Zoologie (Biologie I)
Universität Ulm
Oberer Eselsberg
7900 Ulm, FRG

ISBN 3-540-13762-9 Springer-Verlag Berlin Heidelberg New York Tokyo
ISBN 0-387-13762-9 Springer-Verlag New York Heidelberg Berlin Tokyo

Library of Congress Cataloging in Publication Data. Main entry under title: Environmental physiology and biochemistry of insects. 1. Insects–Physiology. 2. Insects–Ecology. I. Hoffman, Klaus H. (Klaus Hubert), 1946-. QL495.E58 1984 595.7'01 84-14182

Offsetprinting: Beltz Offsetdruck, Hemsbach/Bergstraße, Bookbinding: J Schäffer OHG, Grünstadt.

2131/3130-543210

Preface

Of all the zoological classes the insects are the most numerous in species and the most varied in structure. Estimates of the number of species vary from 1 to 10 million, and 10^{18} individuals are estimated to be alive at any given moment. In their evolution, insects are relatively ancient and, therefore, they have proved to be a phenomenally successful biological design which has survived unchanged in its basic winged form during the last 300 m. y. Insects were the first small animals to colonize the land with full success. Their small size opened many more ecological niches to them and permitted a greater diversification than the vertebrates. What is it about this design that has made insects so successful in habitats stretching from arid deserts to the Arctic and Antarctic and from freshwater brooks to hot springs and salines? Is it due to the adaptability of their behavior, physiology, and biochemistry to changing environmental conditions?

Three features of insects are of particular importance in determining their physiological relationship with the environment: their small size, as mentioned above, the impermeability and rigidity of their exoskeleton, and their poikilothermy. Of course, as with any other animals, the insects' success in its environment depends on its ability to maintain its internal state within certain tolerable limits of temperature, osmotic pressure, pH or oxygen concentration (homoeostasis). The mechanisms utilized to control rates of metabolic functions under varying environmental conditions clearly illustrate a key interplay between time and the adaptive strategy. The first strategy is that of the adaptation of life over a period of millions of years (genotypic adaptation). The second concerns the periodic adaptations to the annual cycle or seasons, to the monthly cycle governed by the movement of the moon, and to the 24 h day/night cycle. Such adaptive processes generally require several days or weeks for completion and are customarily referred to as acclimatization if the organism is responding to environmental parameters within its natural ecosystem or as acclimation if the organism is responding to a single environmental factor in a laboratory situation. The third time scale is a very short one and covers the rapid response to acute changes in the immediate environment of the animals.

Knowledge of the physiology and biochemistry of insects has developed extensively over the last 20 years. Reasons for this increased interest in insect physiology and biochemistry are the growing realization that insects can be useful not only as organisms to act as a model system for experimental studies of general principles, but also as an economic model. Since in our view the functions of insects are only really meaningful when studied in the context of the environment with which the organisms are interacting, we shall try to present examples of the major strategies of adaptation of selected physiological and biochemical functions (e.g., growth and development, aerobic and anaerobic energy metabolism, salt and water exchange, respiration, communication and defense) to varying conditions of temperature, light, humidity, salt concentrations, oxygen tension, or food supply. Beyond that, since hormones are involved in the operation of all systems of the insect body, we shall try to give some evidence of our hypotheses that the insect endocrine system is a mediator between environmental factors, such as temperature or light and the (sub)cellular responses, such as protein, fat, and carbohydrate metabolism.

In several chapters previously unpublished data are presented in order to give proof of these new hypotheses.

In general, recent summaries or reviews are cited, but not always the articles on which they are based. Consequently, the present book cannot be a complete coverage of the literature on environmental physiology of insects, but an introduction to it. It should prove useful not only to researches of the Insecta, but also to teachers and graduate students.

This book would not have been possible without the helpfulness of many colleagues. Above all I should like to mention my former teacher, Prof. Dr. H. Remmert, for introducing me into insect physiology research. The editor and contributors also wish to thank Dr. D. Czeschlik, Springer-Verlag, for his sustained interest in the presentation of this book. Our students and co-workers deserve special thanks for certain ideas and results discussed in this book. Finally, we are deeply obliged to the Deutsche Forschungsgemeinschaft (DFG) for the establishment of the "Schwerpunktprogramm: Physiologische Mechanismen Ökologischer Anpassungen bei Tieren". Most of the authors are (were) members of this research program and their work was generously supported by the DFG.

Ulm, October 1984 Klaus H. Hoffmann

Contents

List of Contributors

Dr. Walter Behrens, Allgemeine Zoologie (Biologie I), Universität Ulm, Oberer Eselsberg, 7900 Ulm, FRG

Priv. Doz. Dr. Gerd Gäde, Institut für Zoologie IV, Universität Düsseldorf, Universitätsstraße 1, 4000 Düsseldorf, FRG

Priv. Doz. Dr. Frank Hevert, Institut für Allgemeine und Spezielle Zoologie, Universität Giessen, Stephanstraße 24, 6300 Giessen, FRG

Prof. Dr. Klaus H. Hoffmann, Allgemeine Zoologie (Biologie I), Universität Ulm, Oberer Eselsberg, 7900 Ulm, FRG

Dr. Paul Kestler, Zoologisches Institut, Universität Göttingen, Berliner Straße 28, 3400 Göttingen, FRG

Dr. Hans Toni Ratte, Lehrstuhl für Biologie V, Rheinisch-Westfälische Technische Hochschule Aachen, Kopernikusstraße 16, 5100 Aachen, FRG

Prof. Dr. Rolf Ziegler, Institut für Tierphysiologie, Freie Universität Berlin, Fachbereich 23, Grunewaldstraße 34, 1000 Berlin 41, FRG

Chapter 1

Metabolic and Enzyme Adaptation to Temperature

Klaus H. Hoffmann[1]

Contents

1 Introduction

Temperature determines life of organisms more than many other environmental factors. The cause of this temperature sensitivity lies in the fundamental fact that all organisms are built up by chemical compounds and that all processes of life are made by chemical reactions which follow the laws of thermodynamics. The temperature acts not only upon the rate of all chemical reactions according the Arrhenius equation, but also causes conformational transitions of proteins, phase transitions of lipids, changes in the structure of water, etc. (Alexandrov 1977).

The temperature range for biological activity is relatively small both because of the specific properties of the biomolecules and because of the temperature coeffi-

1 Allgemeine Zoologie (Biologie I), Universität Ulm, Oberer Eselsberg, 7900 Ulm, FRG

Environmental Physiology and Biochemistry of Insects (ed. by K. H. Hoffmann)

cients of chemical reactions. The relationship between temperature (T) and rate (R) (velocity) of a given reaction (process) is usually defined by its temperature coefficient, Q_{10}. This is the factor by which a rate is speeded up, if temperature increases by 10°C. Generally Q_{10}-values of "normal" chemical reactions and biological processes range between 2 and 4. In other words, without any regulatory mechanisms deviations of only some degrees of centigrade from suitable temperatures would slow down mandatory metabolic, physiological, and behavioral processes to dangerous inactivity or speed up metabolic rates to an extent so that prevailing of food and oxygen would become impossible (Hochachka and Somero 1973). Longer periods below or above a critical temperature cause irreversible destruction of their bioactive structure and function and lead to a desorganization of metabolic processes, if their temperature coefficients differ slightly.

In order to circumvent these difficulties organisms have developed specific adaptations during the course of their evolution to become more or less independent of ambient temperatures. Two particularly prevalent types of adaptations are homeothermy and poikilothermy. True homeothermy is found only in mammals and birds. These organisms use metabolic heat to maintain a relatively high, optimal constant body temperature, whereas the body temperature of poikilotherms corresponds close to that of the external environment. Nevertheless, various species of poikilotherms, including some insects (Sec. 5, this Chap.), are able to maintain the temperature of parts of their body more or less constant while active. Thus, they have been designated as heterotherms. Two other terms, ectothermy and endothermy, refer to the source of heat used to maintain temperatures above or below ambient. Ectothermic organisms gain heat from the environment, while endotherms gain it from their metabolic processes.

Most insects are essentially ectothermic organisms with body temperatures close to ambient. Within their habitats they have to tolerate a considerable temperature range (Bursell 1974a). Two main problems to master are: (1) overcoming longer lasting seasonal adversary temperatures (e.g., low winter temperatures); (2) maintaining an active state during the short-term daily and local fluctuations of ambient temperature (microclimate). The latter is mainly a problem for terrestrial rather than aquatic insects. Both, adaptations to seasonal temperature changes and to daily or local temperature fluctuations make insects survive and function at widely different habitat temperatures. Short-term temperature changes are mastered by altering many aspects of their physiology and biochemistry in a manner that often compensates for changes within a species-specific temperature range. If cooling or overheating surpasses the adaptive capacities of active life, some insects move to a state of concealed life (Chap. 3). They retain, however, the ability to return to normal life upon reestablishment of acceptable temperature conditions.

The adaptive mechanisms can be divided into two types: modificational and genotypic (Precht et al. 1973). Modificational (nongenetic, phenotypic, physiological, capacity) adaptation exists when in an organism the dependence of experimental temperature (ET) is influenced by the previous adaptation temperature (AT). In contrast, genotypic (genetic) adaptations become apparent without any provocative action of a temperature fluctuation.

Both, modificational and genotypic adaptations involve all levels of biological organization ranging from molecular processes (this Chap.) to population dynamics (Chap. 2).

2 Temperature Dependence of Metabolism and Activity

Oxygen consumption is often taken as a measure of the overall metabolic rat of an animal. Its measurement has been employed more than any other experimental parameter to monitor changes in insects' metabolism associated with temperature (for a summary of data see Keister and Buck 1974). Figure 1.1 depicts the usual relationship between resting metabolic rate (per unit weight) and body temperature in insects. Of course, a relationship like that is found only in those situations where body temperature approximates the ambient temperature. The relationship between metabolic rate and body (ambient) temperature has Q_{10}'s of 2-3 and is usually only approximately exponential. It does not truely follow the exponential Arrhenius function with temperature coefficient, Q_{10}, being constant over all temperature ranges. When, however, metabolic range is not a strictly exponential function of temperature, then Q_{10} varies with the particular range of temperature considered. When Q_{10} changes with temperature it typically does so continuously, that means there is a steady drop in Q_{10} with increasing ambient temperature:

$$Q_{10} = \left(\frac{R_2}{R_1}\right)^{10/(T_2-T_1)} \quad \text{van't Hoff equation,}$$

where R_2 is the rate at any temperature T_2 (in °C) and R_1 is the rate at any lower temperature T_1. The formula indicates that the Q_{10} value per se is temperature dependent.

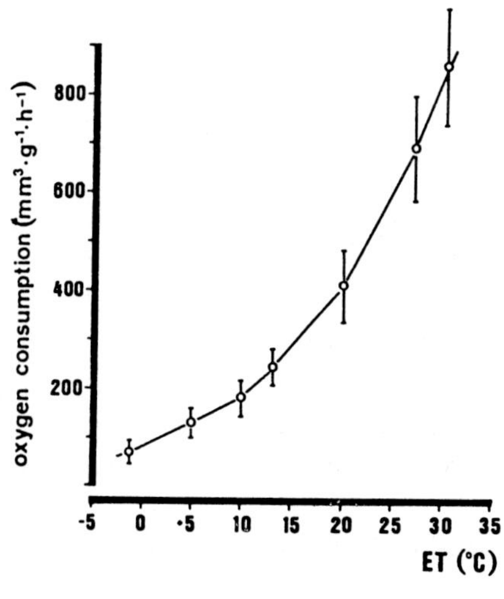

Fig. 1.1. Oxygen consumption of male adult crickets, *Gryllus bimaculatus*, at several experimental temperatures. *Vertical bars* represent SD. (After Hoffmann 1973)

In last instar larvae of the house cricket, *Acheta domesticus*, Q_{10} for oxygen consumption is about 2 in the normal temperature range 20°–40°C (Roe et al. 1980). Standard metabolic rate of male blowflies *Calliphora erythrocephala*, as it was measured by oxygen consumption of the whole animal and of isolated flight muscle tissue, is temperature dependent over the range 10°–30°C with Q_{10} values of 1.34–2.90 (Tribe and Bowler 1968). In the cockroach *Periplaneta americana*, the level of metabolic rate depends strongly on the acute experimental temperature (Welbers 1976) only during the activity phase. In the resting metabolism of *Periplaneta* an influence of the temperature pretreatment has been shown (see p. 12, this Chap.). Curves similar to that in Fig. 1.1 have also been obtained for single enzymatic reactions. It would, however, be unjustified to refer the overall respiration-temperature curve directly to enzymic activity. In a system in which overall gas exchange almost certainly involves a complex of sequential and parallel reactions, it is illusory to apply an analysis based on assumed single rate-limiting or "master" reactions (Keister and Buck 1974).

Many other physiological processes in insects (e.g., motoric activity, heart rate, or electric activities) also generally show a nearly exponential thermal dependency. In mosquitoes *Aedes sollicitans*, fed on a single meal of sugar, the rate of triglyceride synthesis exactly followed the Arrhenius equation over the entire range from 10°–30°C (van Handel 1966). In starving mosquitoes, the logarithm of the time in which 50% of the calories were used, again followed the Arrhenius relation with Q_{10}'s of 2.1–2.3 (van Handel 1973). Even processes such as development and reproduction (Chap. 2) or behavioral activities, which are based on the interaction of many single reactions like those described above, show similar temperature coefficients.

Summarizing, the temperature coefficient, Q_{10}, can be used to describe the thermal sensitivity of many quantitiable physiological rate functions in insects. One could assume that this conflicts with the ideas of thermal adaptation. However, thermal adaptation of insects does not seem to be a rare phenomenon, as will be evidenced in the following sections.

3 The Relationship of Time and Adaptive Strategy

The relative independence of live processes from changes in temperature is achieved through a variety of adaptive mechanisms that have evolved in many poikilotherms during the course of evolution. To fully appreciate these phenomena it is necessary to clearly discriminate between the effects of acute (experimental temperature, ET) and chronic (adaptation temperature, AT) exposure to temperature changes. As has been discussed by Hochachka and Somero (1973) and Hazel and Prosser (1974), such compensations to temperature may occur over at least three distinct time-course periods:

1. instantaneous temperature compensation (acute temperature change);
2. thermal compensation associated with a period of adaptation (chronic temperature change, nongenetic adaptation);
3. evolutionary rate compensation (chronic temperature change, genetic adaptation).

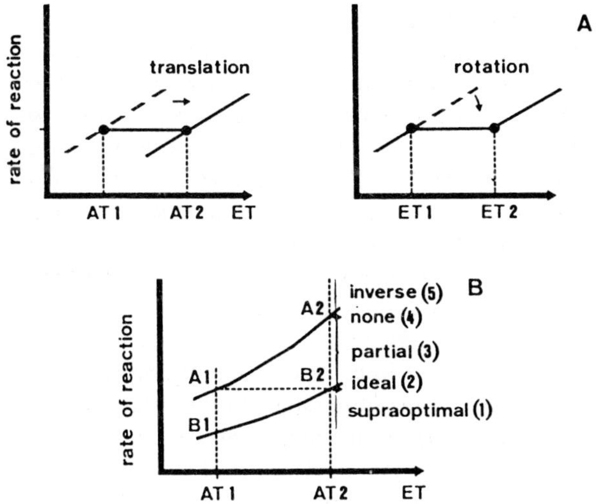

Fig. 1.2 A. The two main types of response to changes of temperature usually termed "adaptive" in ectotherms. Prosser (1958) refers in the first case to a translation of the rate in reaction/temperature curve, and in the second to a rotation. (After Wieser 1973). **B** Types of capacity adaptation (1–5) according to Precht (1949). Adaptation temperature for curve A1-A2: AT1, for curve B1-B2: AT2. (After Precht et al. 1973)

The first generalizations regarding a temperature independence of physiological processes in poikilotherms are according to Precht (1949) and Prosser (1958). After Prosser the reaction rate (R) responds to two distinct ways (Fig. 1.2A): The one shows a plateau of temperature independence when animals are acclimatized (abscissa: AT = ET), whereas the plateau of the other occurs in the course of acute temperature changes (abscissa: ET). Prosser refers in the first case to a translation of the R-T curve (up or to left with cold, down or to right with cold) and in the latter to a rotation (clockwise – Q_{10} reduced in cold, counterclockwise – Q_{10} increased in cold). It is irrelevant, however, whether plateaus of this kind occur in the acclimatized or in nonacclimatized animals, since such cases only differ with respect to the responsible mechanisms and not in principle (Wieser 1973). Typically the R-T curve is steeply in the lower temperature range (Q_{10} $\sim$ 2) and then flattens out (Q_{10} in the middle range $\sim$ 1). Therefore, it is perhaps generally significant that high Q_{10} values are so often found at the lower end of biological temperature ranges. Here many ectothermic organisms are obliged to speed up their metabolism as much as possible. It seems that temperature-insensitive phases frequently occur around the mean temperature expected in the usual environment of the animals. Figure 1.2B presents the type classification of temperature adaptations by Precht (1949) (supraoptimal, ideal, partial, none, and inverse compensation).

Most fundamental research on temperature adaptation has been done with ectothermic vertebrates and intertidal mollusks. Insects, as behavioral thermoregulators (see p. 29, this Chap.) were presumed for a long time to posses little capacity for metabolic compensation against thermal fluctuations. However, since insects have radiated successfully into a wide variety of habitats with different microclimates, important temperature adaptations at the physiological and biochemical level should also have been evolved during evolution in these animals.

3.1 Instantaneous Temperature Compensation

Metabolic adaptations to acute (short-term) temperature changes are well-illustrated for intertidal animals, but not restricted to this group of organisms. Some insect species subject to less drastic fluctuations in their thermal environment have also succeeded in compensating acute temperature fluctuations over at least a small temperature range (Keister and Buck 1974). Some strains of *Drosophila* pupae show an approximately independent respiration rate at 18°–22°C and 27°–30°C, on either side of the rearing temperature of 25°C. A dung beetle shows no respiratory increase between 12°–24°C and an ant none between 20°–34°C. A plateau of temperature independence, or even a minimum, also occurs in the temperature-respiration curve of some Carabidae (Schmidt 1956). In the same insect species the transpiration rate remained constant within a certain narrow temperature range. In the antarctic beetle *Hydromedion sparsutum* activities of the digestive enzymes α-amylase and proteases are temperature compensated between 5°–15°C (Q_{10} ~1.3); (Haderspeck and Hoffmann 1983). The rate of glycogen synthesis in the mosquito *Aedes sollicitans* increases with temperatures from 10°–22.5°C, but is independent of temperatures between 22.5°–30°C (van Handel 1966).

In a second group of insects, plateaus in the R-T plot are found at temperatures considerably lower than the rearing temperature and at which locomotion may be stopped or restricted (e.g., respiration rate in larvae of *Phormia* between 10° and 15°C (Keister and Buck 1974) or in *Drosophila* pupae at 15°C). Such regions of constancy in respiration-temperature plots are difficult to explain with respect to an unitary respiring system, but can be visualized as the resultant of a temporary progressive slowing of the metabolic rate of one reaction system superimposed on continuing increase in rate of a second process (Fig. 1.3). That temperature independence of whole-animal metabolism may result from properties of mitochondria is postulated by the low Q_{10} values reported to succinate and pyruvate oxidation (measured at low substrate concentrations; standard metabolic rates) in mitochondrial preparations from the desert locust, *Schistocerca gregaria* (Newell 1967). In addition, α-glycerophosphate- and ADP-stimulated respiration of blowfly sarcosomes, *Calliphora erythrocephala*, have a temperature independent plateau at intermedial temperatures (Davison 1971). In contrast, in blowfly mitochondria Tribe and Bowler (1968) and Danks and Tribe (1979) found no strong evidence for a plateau of temperature independence in respiration. Fleming and Miquel (1983) have observed different effects of temperature on the metabolic rate of young and old *Drosophila.* The Q_{10} over the entire temperature range was almost 1.5 times as great for the adult flies (Q_{10} = 2.8) as it was for the young flies (Q_{10} = 1.8), but there was no difference in respiration rates at the temperature in which the flies were raised.

Meyer (1978) has pointed out that short high and low temperature pulses immediately increase the anaerobic metabolism of nonfeeding *Callitroga macellaria* larvae, resulting in a drastic increase in formation of polyols or polyolphosphates. Starvation appears to be an important factor that affects both the level and the relation to temperature of metabolism (see Chap. 4). Immediate changes in temperature also often have disturbing effects on animals. For example, a rise in temperature frequently causes a temporary, but drastic increase in motoric activity. Such overshoots are also

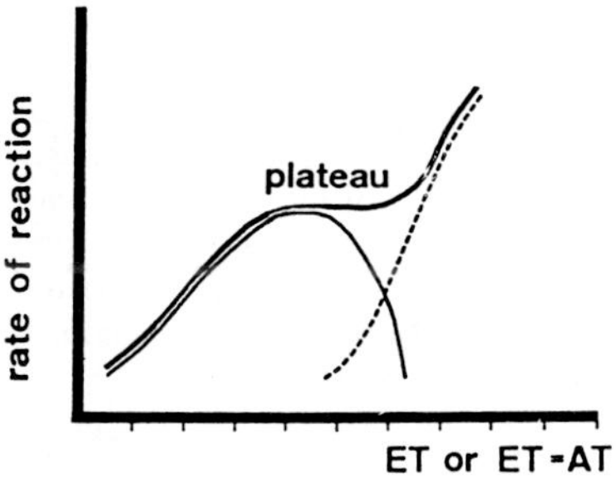

Fig. 1.3. Rate in reaction/temperature curve, where a plateau may be caused by the overlap of two separate R/T curves. (After Wieser 1973)

found in the speed of movements of insects (Precht et al. 1973). Additionally, rapid changes in temperature can have transitory effects on endogenous rhythm of insects in the form of acceleration or retardation.

What are the mechanisms for such temperature independence? A nearly instantaneous increase in enzymes' abilities to bind substrate as temperature is lowered might represent the most prevalent mechanism (Hochachka and Somero 1973). Although studies in this direction have not been accomplished, there are some lines of evidence that are consistent with the hypothesis that positive thermal modulation of substrate binding (that means an increase in apparent Michaelis constant, K_m, for substrate with increasing experimental temperature in a temperature range above the temperature of minimal K_m; see also p. 18, this Chap.) is important in effecting immediate metabolic compensation. For insects, however, only a few data are available that demonstrate the influence of experimental temperature on substrate binding, and these few data do not evaluate this concept. Within the habitat temperature range of three cricket species, *Gryllus campestris, G. bimaculatus,* and *Acheta domesticus,* K_m values of the glycolytic enzymes pyruvate kinase for phosphoenolpyruvate are hardly influenced by experimental temperature (Fig. 1.4A). A positive temperature modulation of enzyme-substrate affinity is avoided within their physiological temperature range (Hoffmann 1976a; Hoffmann and Marstatt 1977). Selection rather may have led to the development of enzyme species ("eurytolerant" enzymes) that are able to maintain their ligand-binding abilities at relative stable levels over the full range of habitat temperatures. An increase in app. K_m values at experimental temperatures beyond the habitat temperature, as can be observed from Fig. 1.4A, may render their metabolic rate less sensitive to extreme situations.

Temperature also influences the rate of enzymatically catalyzed reactions by determining the proportion of molecules in a given population that possess sufficient energy (energy of activation, E_a; enthalpy of activation, $\Delta H^{\neq}$) to react and form an activated complex. A temperature dependence of the barrier of free energy of activation, therefore, is proposed as another possible molecular basis for spontaneous temperature compensation. A biphasic nature of Arrhenius plots (ln rate vs 1/T), and that means different E_a values at different experimental temperatures, are a common phenomenon in membrane-bound enzymes and perhaps reflect changes in the membrane lipid component (for details see Sec. 3.2.6, this Chap.). Recent results of Wood and Nordin (1980) on mitochondrial respiration of the blowfly *Protophormia terranovae* and the housefly, *Musca domestica,* show that for both insect species, ADP re-

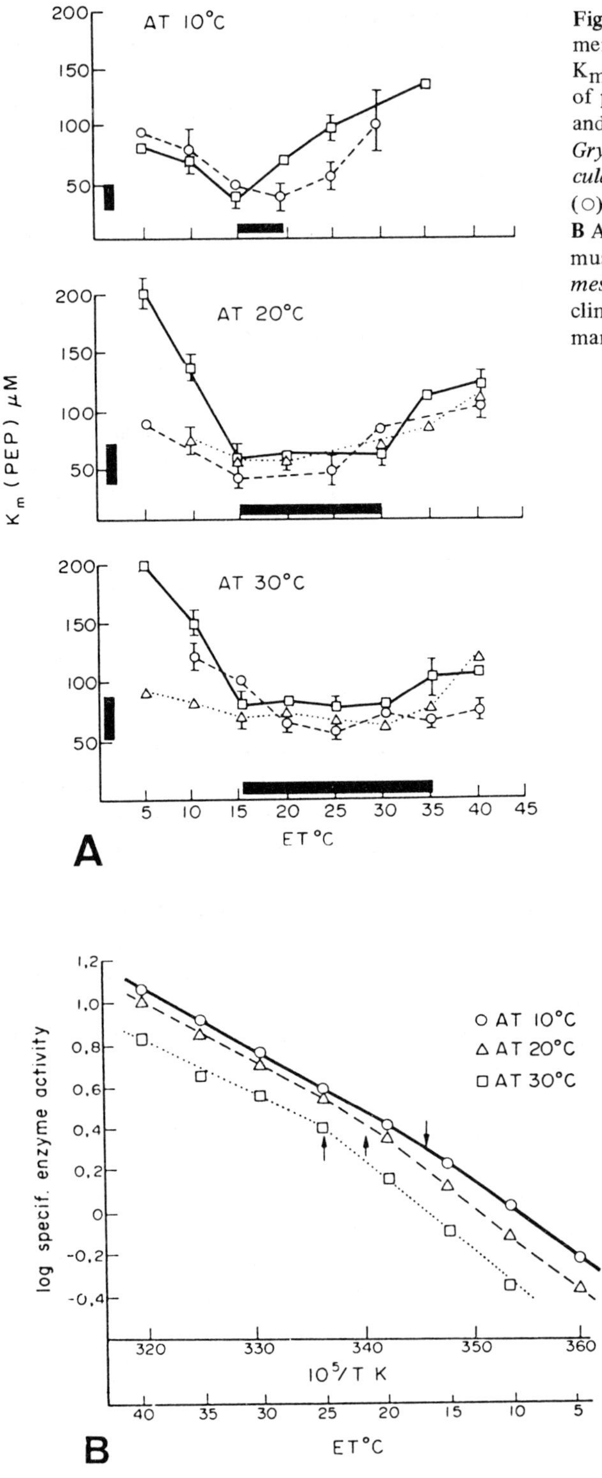

Fig. 1.4. A The influence of experimental temperature (*ET*) on the app. K_m of phosphoenolpyruvate (*PEP*) of pyruvate kinase (*PK*) from muscle and fat body of three cricket species: *Gryllus campestris* (□), *Gryllus bimaculatus* (△), and *Acheta domesticus* (○). *AT* acclimation temperatures. **B** Arrhenius plots of PK activity from muscle and fat body of *Acheta domesticus* after periods at various acclimation temperatures. (After Hoffmann and Marstatt 1977)

quiring step(s) in oxidative phosphorylation become rate limiting below an ET of 11.5°C. The observed increase in E_a below 11.5°C may slow depletion of energy reserves when their mobility is impaired due to a temperature decrease. Breaks in Arrhenius plots have also been described for cytoplasmic enzymes, e.g., for pyruvate kinase of Trichoptera and Plecoptera larvae, of the blue alder leaf beetle, *Agelastica alni*, of the flour beetle, *Tenebrio molitor* (Hoffmann 1976a), and of various cricket species (Fig. 1.4B). These findings indicate that cytoplasmic enzymes can also exist in at least two temperature dependent (per se induced) conformational states. An Arrhenius plot of phosphofructokinase from the cold-hardy gall fly larva, *Eurosta solidaginis*, (see also Sec. 4, this Chap.) shows an activation energy of 19,800 cal/mole (82.8 kJ/mole) and a very high Q_{10} of 3.64 beetween 10° and 0°C (Storey 1982). In addition, kinetic properties of this enzyme are strongly negatively modulated by low temperature.

A third mechanism for maintenance of temperature independent catalytic function is referable to the effect of experimental temperature on the interactions between enzymes and various modulating metabolites (Hazel and Prosser 1974). In the house cricket, *Acheta domesticus*, pyruvate kinase activity is stabilized by fructose-1, 6-bisphosphate against inactivation by elevated temperatures (Hoffmann 1975). In gall fly larvae, *Eurosta solidaginis*, the effects of physiological levels of AMP on the activity of phosphofructokinase (PFK) are drastically lessened at low temperatures. In addition, two compounds, which are normally not effectors of PFK, gylcerol-3-phosphate and sorbitol, both decrease enzyme affinity for fructose-6-phosphate at low temperature. PFK in cold-hardy *Eurosta solidaginis* is an excellent example of the exploitative strategy used by these larvae with respect to temperature changes. Rather than producing compensatory effects to allow continued, normal glycolytic flux at all temperatures, temperature-modulator effects on the enzyme enhance the direct flux of carbohydrates into sorbitol (a cryoprotectant in this animal species; see p. 26 at low temperatures.

A temperature dependent conformation change resulting in an interconversion between two forms of an enzyme, each exhibiting distinctly different kinetic properties, has been reported as an additional mechanism of immediate compensation, also for insect enzymes. The activity of pyruvate kinase from *Protophormia terranovae* displays cooperativity with respect to phosphoenolpyruvate at 20°C (Hill coefficient, n_H = 1.5), but cooperativity appears negligible at 0°C (n_H = 1.0) (Fig.1.5). Thus, while the maximum velocity decreases by a factor of 16 at saturating substrate concentrations, at physiological levels of PEP the decrease is only about sevenfold (Wood et al. 1977).

From above results it appears that the picture on immediate temperature compensation is rather complex and that the temperature relationship of any single enzymic reaction can probably not be used as an argument for or against the occurrence of temperature compensation of whole animal metabolism.

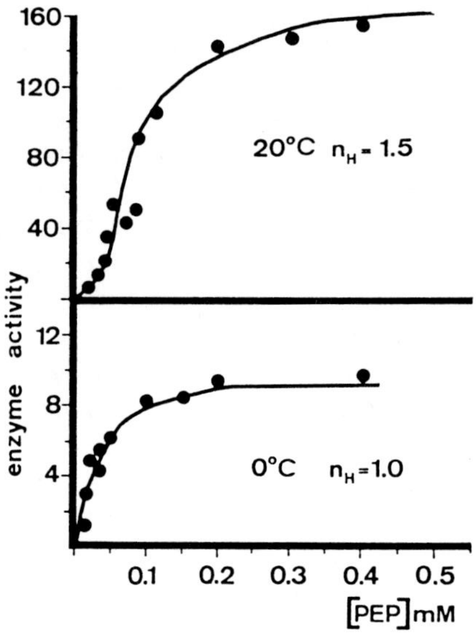

Fig. 1.5. Substrate saturation graphs of pyruvate kinase from *Protophormia terranovae* at 20° and 0°C. n_H Hill coefficient; *PEP* phosphoenolpyruvate. (After Wood et al. 1977)

3.2 Thermal Compensation Associated with a Period of Adaptation

This nongenetic type of adaptation exists when the experimental temperature-dependence curve of an organism is influenced by the adaptation temperature, and has been investigated most intensively in insects. Such compensations occur within the life span (season) of an individual of the species and generally require several days or weeks for completion. Physiological compensations associated with a period of adaptation to constant temperature conditions within the normal temperature range of life have been reported for all levels of biological organization in insects. The extent of those capacity adaptations and even their type may differ in different parts of the AT-range.

3.2.1 Activity and Performance

Rate acclimation at the locomotory level in insects has recently been studied, e.g., by Anderson and Mutchmor (1971) and Kenagy and Stevenson (1982). Typically, the range of temperature compatible with adequate motion function is shifted downward by cold acclimation and upward by warm acclimation. Usually 2-3 days were necessary for acclimation after a temperature transfer. In the German cockroach, *Blatella germanica*, rates of acclimation depended on size and range of the temperature change. In the cold-intolerant flour beetle, *Tribolium confusum*, about 17 days were required for locomotory acclimation, whereas in the cold-tolerant housefly, *Musca domestica*, locomotory acclimation to cold took approximately 40 h only. In the latter case a lag of about 20 h occurred after the change in temperature before any locomotory acclimation changes could be detected. The lag may be due to metabolic acclimations that must occur before locomotory acclimation can be expressed. However, the lag may also be of importance in reducing the effect of very short-term environmental temperature changes. However, it should be remembered that motile, terrestrial insects can simply avoid unfavorable temperatures, a behavior that could substitute for compensations. Tenebrionid beetles of the sagebrush steppe in eastern Washington, U.S.A. are active within the same 10°–30°C range of body temperature at all seasons. In order to perform this activity pattern, the beetles are strongly day-active in early spring and fall, mostly active during twilight and at night in the midsummer, and active bimodally (morning and evening) between these seasons.

The unusual event of an inverse temperature compensation has been found by Gunn and Hopf (1942) for the activity of the beetle, *Ptinus tectus*.

Because "fitness" of an animal is dependent on energy intake and food is generally a limiting factor (see Chap. 4), several authors have investigated the effects of temperature on both the ingestion rate and absorption efficiency in insects. Examples for these were first given by Andrewartha and Birch (1960). In the Mediterranean field cricket, *Gryllus bimaculatus*, food consumption increases with increasing AT between 13° and 34°C, whereas production of excrements is constant in the temperature range of 20°–38°C (Hoffmann 1974). Both sexes of the crickets ingest food in considerable amounts even at 10°–13°C, however, nearly no feces are produced under those temperature conditions. Consequently, the ratio Q of food consumption to feces production increases to values of 7–39 at 10°–13°C as compared to Q = 4–6 at 20°–38°C. The basic stimulus for insect feeding might be the need for energy (Applebaum et al. 1961). This statement, of course, refers to a close relation between the temperature dependence of food digestion (see p. 14) and food utilization. In addition, the rate at which food reserves are depleted is reflected in the rate at which oxygen is consumed, and this is also known to be closely related to temperature.

3.2.2 Oxygen Consumption

As already stated on p. 3, with respect to effects of temperature on respiration of insects, modifications due to rate changes in experimental temperature and past temperature history (AT) of the insects must be taken into account. Insects, in general, are considered to have rather little ability to compensate whole animal and tissue respiratory rates (Keister and Buck 1974). When acclima(tiza)tion in oxygen consumption (in resting metabolism) does occur, the effect is usually a single shift in the position of the respiration-temperature curve rather than a change in shape (Fig. 1.6A, B; Precht's type 3 of partial compensation). Consequently, insects acclimated to a low temperature respire at any given temperature more than warm adapted individuals. Meyer and Schaub (1973) have confirmed these results with experiments on oxygen consumption of larval stages of Calliphoridae. After 1–3 weeks at 10°, 16°, and 26°C cockroaches, *Periplaneta americana,* also showed acclimation: with those being kept at the lowest temperature respired with higher rates at any given temperature than the warm adapted ones. Nymphs acclimated to a greater degree than adults, and small adults more than large ones (Dehnel and Segal 1956). In contrast to these results, however, Welbers (1976) has not found any temperature compensation in oxygen consumption of *Periplaneta* nymphal stages during the activity phase. However, in rest-

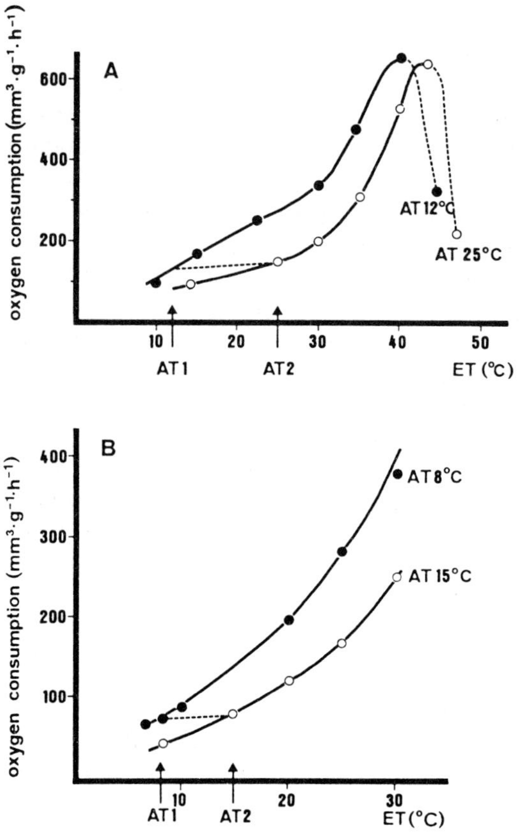

Fig. 1.6. A Oxygen consumption of *Melasoma populi* at various experimental temperatures. ●, animals acclimated to 12°C; ○, animals acclimated to 25°C. The *dotted line AT1-AT2* represents the type of adaptation (see Fig. 1.2B). **B** Oxygen consumption of *Leptinotarsa decemlineata* at various experimental temperatures. ●, animals acclimated to 8°C; ○, animals acclimated to 15°C. The *dotted line AT1-AT2* represents the type of adaptation (see Fig. 1.2B). (After Marzusch 1952)

ing metabolism of these animals an influence of the temperature pretreatment was shown. In all experimental temperatures the metabolism of animals pretreated at 25°C was higher than in animals adapted to warmer or colder temperatures. Similarly, the author has not observed a temperature compensation in the metabolism of pink bollworm larvae, *Pectinophora gossypiella* (Welbers 1975b). Data of Grigo and Topp (1980) on oxygen consumption of three species of rove beetles, *Atheta fungi, Philonthus carbonarius,* and *Lathrimaeum atrocephalum,* adapted to 6°, 16°, and 26°C and at different photoperiods in diapause and nondiapause indicate that in *Philonthus carbonarius* seasonal influences are responsible for a changing response to temperatures. The oxygen consumption of *Atheta fungi*, an eutropic cosmopolitan insect, was only influenced by the adaptation temperature. In the stenotropic *Lathrimaeum atrocephalum*, the oxygen uptake was not dependent on the AT, and its mean metabolic rates were significantly higher in nondiapause (winter) than in diapause. Hunter (1968) has studied the effect of temperature on oxygen consumption in several species of *Drosophila* from different thermal habitats. She found that stenothermal flies showed no adaptive response to temperature, but that eurythermal flies did show an adaptation response to temperature. Similar differences have been observed by Marzusch (1952) for stenothermal and eurythermal beetles. For whole animal metabolism rate Precht's type 5 of inverse compensation has also been reported, especially at low ambient temperatures (Sømme 1968).

Measurement of oxygen uptake by isolated tissues or mitochondria preparations from *Calliphora erythrocephala* and *Sarcophaga bullata* have revealed comparable acclimation patterns and temperature dependence as observed with the intact animal (Spencer-Davis and Tribe 1969; Danks and Tribe 1979). In contrast to these results, in *Periplaneta americana* respiratory compensation against thermal alteration exhibited by the entire insect is not completely reflected at the level of different tissues. Although living female roaches exhibit a total lack of cold adaptation (15°C), the coxal muscle of nymphs, males and females demonstrates an excellent capacity for thermal acclimation in the rate of endogenous respiration (Das and Singh 1974). While the fat body of the nymphal and adult male cockroaches also failed to show any thermal acclimation in the rate of endogenous oxygen consumption, this tissue of females exhibited warm acclimation (35°C) of the rate process.

A situation, converse of that normally observed in acclimatized living animals, was found by Davison (1971) in blowfly *(Calliphora erythrocephala)* sarcosomes. Sarcosomes from 34°C acclimatized flies were capable of much higher rates of ADP-stimulated respiration than those from a lower acclimatization temperature.

3.2.3 Cellular Metabolism

One of the experimental approaches used extensively to answer the question for biochemical and biophysical alterations underlying the process of acclimation and acclimatization has been the measurement of enzyme activities. In general, enzymes associated with pathways of energy production, i.e., enzymes involved in the generation and utilization of ATP (such as succinate dehydrogenase, malate dehydrogenase, cytochrome oxidase, alkaline glycerophosphatase, mitochondrial adenosine triphosphatase, glutamate-aspartate transaminase, glucose-6-phosphate dehydrogenase), have proven to increased activities in cold-acclimated individuals (Marzusch 1952; Mutchmor 1967; Burr and Hunter 1970; Hunter and Cediel 1970; Sømme 1972). This is entirely in line with the whole animal data which show that metabolic rate is frequently elevated between one- and twofold at a given test temperature in cold acclimated individuals. An elevated metabolic rate implies increased rates of ATP generation and utilization. Higher activity rates and lower temperature coefficients of muscle apyrase (adenosine triphosphatase) in cold- than in warm-acclimated insects (*Periplaneta americana*), as they have been observed by Thiessen and Mutchmor (1967), were accompanied by an increase in the number of mitochondria in muscles from cold-acclimated individuals.

An alternative response to cold acclimation has been shown by Anderson and Mutchmor (1971) and Das and Das (1982a), namely, an increased thermosensitivity of Mg^{2+}-activated ATPase from *Tribolium* and *Musca,* and LDH and SDH activity in the fat body of *Periplaneta americana*, in the low temperature range.

The data of Burr and Hunter (1970) on temperature acclimation of glutamate-aspartate transaminase activity in various species of *Drosophila* support the hypothesis that eurythermal species (*D. melanogaster, D. immigrans*) have greater capacity for adaptation than stenothermal species (*D. pseudoobscura, D. willistoni*) (see also p. 12). However, data on temperature adaptation in respiratory enzymes of various *Drosophila* species (Hunter and Cediel 1970), although showing evidence for a general temperature adaptation in *Drosophila*, do not support this hypothesis.

Sømme (1972) demonstrated that within one and the same insect species, *Ephestia kuehniella*, various enzymes are affected in different ways by acclimation. In larvae following acclimation at 20° or 6°C, the activities of glucose-6-phosphate dehydrogenase and catalase were higher in cold (CA) - than in warm-acclimated (WA) individuals, whereas acetylcholine esterase activity was higher in WA than in CA larvae. Almost all the patterns of thermal acclimation described by Precht (1949) and Prosser (1958) are exhibited by the three enzymes LDH, SDH, and aldolase in the coxal muscle of *Periplaneta americana* (Das and Das 1982b). While aldolase shows a "translation" (Prosser's pattern II) and a "partial" (Precht's type 3) compensation in males, but a noncompensation in females, the LDH activity reveals a "translation-with-rotation" (Prosser's pattern IV A) and a partial acclimation in females, but a "rotational" (Prosser's pattern III) acclimation (Precht's type 5 at lower temperatures and type 3 at higher thermal range) in males. SDH activity has a greater capacity for cold adaptation than for warm adaptation in males, but females show a lack of thermal acclimation of this enzyme. The biological significance and the molecular mechanisms underlying such sex specificity of response are inexplicable at this time (see also p. 20).

Nevertheless, from the aforementioned results it can be concluded that temperature acclimation in insects may affect both the rate of carbon flow through a given pathway and the contribution of various metabolic pathways to the total metabolism. In general,

1. magnitude of compensation for glycolytic enzymes is less then that characteristic of electron transport and Krebs-cycle enzymes, and
2. adaptation to cold frequently shifts metabolism towards anabolism, accompanied by a marked activation of the pentose shunt enzymes (Moreau 1975).

3.2.4 Digestive Function

Only a few insect species have been investigated so far with respect to thermal adaptation of their digestive enzymes. Applebaum et al. (1964) have demonstrated an almost 100% increase in midgut proteolytic activity of 13°C adapted *Tenebrio molitor* larvae over 33°C adapted insects, when the assay was conducted at an intermediate temperature of 23°C. No homoeostatic tendency has been noted with respect to amylase activity. The absence of any compensation of amylase activity may be connected with the hypothesis that decreased activity at lower temperatures would result in reduced carbohydrate utilization, owing to the need for more frequent feeding in order to converse continuously a high level of energy substrate. Increased food consumption at lower temperature may be termed as behavioral compensation. Janković-Hladni et al. (1976) have reported an inverse compensation (Precht's type 5) of the midgut amylotic activity in *Tenebrio molitor* adults. In addition, males were more sensitive to temperature changes than females. Partial compensation of midgut amylase and protease activities was exhibited by *Morimus funereus* larvae

(Ivanović et al. 1976). In American cockroaches, *Periplaneta americana*, various thermal acclimatory responses of salivary amylase are evident (Singh and Das 1977). Nymphs demonstrate a translational and males a rotational pattern, while females exhibit a small inverse compensation. Almost all the patterns of compensatory modulation of maximal catalytic activities are exhibited by gastric amylase, gastric protease, and lipase in male and female cockroaches acclimated to 16° and 32°C (Das and Das 1982b). Gastric protease of males shows translational, but only partial compensation. A significant warm acclimation, but no cold acclimation in both sexes, was found for gastric amylotic activity, whereas gastric protease of females and gastric lipase of both sexes are characterized by the lack of an adaptive compensation to temperature. The sex differences in digestive enzymes agree with differences in respiratory rate (see p. 13) and in the activities of metabolic enzymes (see p. 14). However, further studies are necessary in order to reveal the modifying role of the nutritional status of the insects on the thermal acclimation of digestive enzymes.

3.2.5 Changes in Body Composition

Only quite fragmentary information is available on biochemical changes in tissue composition of insects acclimated to different temperatures. The results confirm the assumption that the aforementioned enzymatic reorganization during thermal acclimation is reflected in alterations of biochemical tissue composition. One possible way of modifying the intensity of metabolism in a cell consists of a change in the free water content, and thus, the size of the reaction volume. Although in Precht et al. (1973) an association of cold acclimation with diminution of free water content of tissues in ectotherms is indicated, an increase in water content, rather than a decrease, seems to characterize the acclimation of most insects to low temperature (Hoffmann 1973; Singh and Das 1980).

Singh and Das (1980) have presented an investigation on the quantitative strategy of compensatory alterations in the chemical composition of a storage tissue like fat body and a highly active tissue like coxal muscle of nymphal and adult cockroaches, *Periplaneta americana* (Fig. 1.7). The most important biochemical alteration accompanying cold acclimation is the remarkably high degree of protein accumulation in the fat body. The increase in protein content of coxal muscle during cold acclimation is far less pronounced than that of fat body. The glycogen reserve has also been shown to increase in fat body and coxal muscle of cold acclimated animals compared to warm adapted ones. Sex-correlated differences in biochemical tissue composition of adults (protein, RNA, glycogen) as well as in the extent of compensatory changes during acclimation (protein and RNA) may be due to hormonal changes associated with reproduction in the two sexes (see p. 19). The significance of change in the concentration of amino acids in thermal acclimation is still unresolved. A diminution in the free amino acid/DNA ratio due to cold acclimation has been recorded in tissues of nymphal and adult female cockroaches, but not in males. No qualitative change occurs in the free amino acid spectrum of hemolymph and tissues of this insect species, although Anders et al. (1964) found a diminution of glutamic acid content (in spite of an increase of the total free amino acid level) in *Drosophila melanogaster* due to a reduction of the rearing temperature. In hemolymph and tissues of

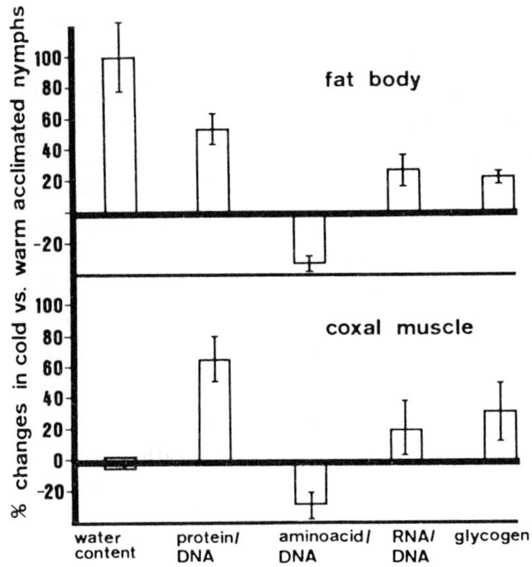

Fig. 1.7. Percentage changes in the chemical composition of fat body and coxal muscle of 15°C-acclimated *Periplaneta americana* nymphs over 35°C-acclimated animals. *Vertical bars* represent SD. (After Singh and Das 1980)

Mediterranean field crickets, *Gryllus bimaculatus*, concentration of total ninhydrin-positive substances significantly increased with a decrease in rearing temperature (Hoffmann 1973).

Whether the steady state concentration of certain key metabolites, such as the free amino acids glycine and histidine in cyprinid fish (Knapp and Wieser 1980), could act as "metabolic thermometers" also in insects, has to be proven in further experiments. Lipid composition of ectotherms is discussed as another indicator for changes in environmental temperature.

3.2.6 Lipid Metabolism and Composition

Environmental temperature has a pronounced effect on both the general level of lipid metabolism and the lipid composition of ectotherms. In many insects the total content of neutral fats and lipids increases with decreasing rearing temperature (Hoffmann 1973; Precht et al. 1973). A relationship between environmental temperature and fatty acid composition has been suggested for a wide variety of insect species (for a summary see Fast 1970; Downer and Kallapur 1981). The trend that emerges from these studies is that an inverse relationship exists between temperature and the degree of saturation of the total fatty acid complement. The rationale that underlies such investigations is that the lipid composition of a biological unit membrane will respond to temperature changes so as to maintain optimal membrane permeability and fluidity (Hazel and Prosser 1974). Thus, the most marked effects of temperature changes are likely to be expressed in membrane lipids (phospholipids). Danks and Tribe (1979) have analyzed the fatty acid composition of mitochondria phospholipids from blowfly (*Calliphora erythrocephala* and *Sarcophaga bullata*) flight muscles and found a significant increase in long chain unsaturated fatty acids (C16:2, C 18:3, C 20:3, C 20:4, C 20:5) at decreasing acclimation temperature. Kallapur et al. (1982) did not observe any apparent differences in the fatty acid composition of a total phospholipid extract of *Schistocerca gregaria* flight muscle mitochondria, but suggest that analysis of the fatty acid composition of individual phospholipid species might reveal temperature-related effects also in these insects. Further support for such a suggestion is provided by the highly characteristic fatty acid profile identified by individual phospholipids extracted from housefly larvae, *Musca domestica* (Bridges and Watts 1975), and by the demonstration in housefly larvae that the ratio of unsaturated to saturated fatty acids increased with decreasing rearing temperature in phosphatidylserine, phosphatidylcholine, and phosphatidylethanolamine, but that these changes were different to each of the phospholipid species (Robb et al. 1972).

Studies of Downer and Kallapur (1981) and Kallapur et al. (1982) provide evidence that homoeoviscous adaptation in membrane function may also result from changes in phospholipid composition and sterol content. Locusts held at 31°C showed lower levels of phosphatidylcholine and higher levels of phosphatidylethanolamine than 45°C acclimated insects. A trend towards increased cholesterol: phospholipid ratio was also observed at the higher temperature. Wide angle X-ray diffraction procedures corroborated the assumption of a higher lipid phase transition temperature in the 45°C acclimated samples. *Musca domestica* larvae might control the physical state of their phosphosphingolipids by varying the sphingosine composition (ratio of C_{16}: C_{14} sphingosines), possibly leading to an increased permeability of the membran at lower temperatures (Robb et al. 1972).

The fact that cold-acclimated organisms, in general, contain relatively more unsaturated fatty acids than warm-acclimated ones is probably related more to resistance than to capacity adaptation. Nevertheless, it might be of great significance for activity of enzymes which depend on bound lipids. Considering effects of temperature on the degree of saturation of fatty acids one has to note, of course, that such effects may interfere with effects of diet (Hoffmann and Stockmeier 1975; Geer and Perille 1977) or development in the life history of an insect (Clarke 1967).

3.2.7 Mechanisms of Thermal Acclimation on Enzyme Level

Acclimatory responses require a relatively long time to occur and are characterized by marked changes in the level of protein synthesis. This indicates that physiological acclimation probably involves feedback to the genetic material. Three main strategies seem to be involved in thermal acclimation of ectothermic enzymes (Hochachka and Somero 1973; Hazel and Prosser 1974):

1. Acclimated populations differ in the rate at which a given gene is being transcribed or alternately in the rate of degradation of the specific gene product. With this respect warm- and cold-acclimated animals differ quantitatively (different enzyme concentrations).
2. Warm- and cold-acclimated species will differ primiarily in the portion of the genotype being expressed. In other words, warm- and cold-acclimated populations are qualitatively different (specific iso(en)zymes).
3. Regulation of cellular processes caused by temperature induced alterations in membrane lipid composition (see also p. 16).

All three strategies are demonstrated to exist also in insects. That thermal acclimation may be mediated by temperature-induced alterations in specific enzyme concentrations is suggested for several digestive enzymes which exhibit compensatory changes in maximal velocity [e.g., proteolytic activity in *Tenebrio molitor* larvae (Applebaum et al. 1964) or salivary amylase in females and gastric protease in males of *Periplaneta americana* (Das and Das 1982b)]. A 25% increase in the number of mitochondria (proteins) was reported after cold acclimation in the housefly, *Musca domestica,* (see p. 13). This has been assumed to be partially responsible for the observed increase in muscle apyrase acitivty in this insect (Thiessen and Mutchmor 1967). However, as far as I know, in no insect system have quantitative changes in enzyme concentrations been unequivocally demonstrated.

In a few cases it has been shown that reduced environmental temperatures induce the synthesis or loss of isozymes. Kinetic data obtained for neural Na/K-ATPase in acclimated (16°C) and nonacclimated cockroaches, *Periplaneta americana*, by Piccione and Baust (1977) indicate the existence of thermally distinct isoenzymes.

Mills and Cochran (1967) have found four different ATPases in the thoracic muscle of *Periplaneta americana*, each showing a different temperature tolerance. ATPase IV lost 100% of its activity after exposure to low temperature (0°C) for 6 h, whereas ATPase II did not lose activity during the same low temperature exposure. The other two ATPases showed an intermediate loss of activity. As opposed to room temperature controls, dialyzed homogenates of cold stressed *Protophormia terranovae* contain one-third more activity of a phosphatase capable of catalyzing the hydrolysis of L-glycerol-3-phosphate (Wood et al. 1977). In the opinion of the authors this enzyme might consist of a mixture of thermal variant isozymes, one of which might be synthesized slowly during overwintering and losing its activity slowly when shifted to a higher temperature. The converse has been found by Storey et al. (1981). The overwintering gall fly larva, *Eurosta solidaginis*, did not show the production of new enzyme variants as a strategy in low temperature metabolic regulation.

At acclimation temperature an enhanced enzyme-substrate affinity provides for stabilization of an enzyme in given conditions (Alexandrov 1977). It has to be emphasized, however, that minimal K_m values appear to be restricted to isoenzymes, whereas other enzymes are characterized by a linear dependence of K_m on assay temperature over a wide temperatur range (see p. 7).

Environmentally induced alterations in membrane structure do not only maintain an appropriate membrane fluidity, but may also modulate the activity of membrane bound enzymes during thermal acclimation. After Hazel and Prosser (1974) the results indicate that changes in membrane composition regulate the activity of these enzymes rather than the quantity. The fluidity of the fatty acyl chains of membrane phospholipids may provide the motional freedom, allowing these membrane enzymes to realize conformational changes and movements associated with their activity.

3.2.8 Initiation of Acclimation Response

The findings discussed thus far make clear, how difficult it is to analyze the mechanisms of thermal acclimation, especially because a unified explanation can hardly be given. Most thermal acclimation data describe the types of physiological adaptations, but do not take into consideration, whether control systems, such as the (central) nervous system or hormones, play a role in acclimation. Hormones may influence synthesis of proteins by genes and, therefore, the concentration of enzymes, or they may have direct effects by stimulating or inhibiting enzyme activity or alternating membrane permeability.

In the regulation of the insect's response to temperature functions or actions of the nervous system are the perception of environmental temperature, the integration of this information with that derived from other sources, and the adjustment in the patterns of its own activity which will in their turn influence the activity of the endocrine glands and the patterns of muscular activity of the body (Clarke 1967). Little is known of compensatory changes of activity in nervous systems. Electrophysiological studies of the isolated nerve cord of *Periplaneta americana* have shown that units which respond differently to different temperatures exist in the nerve cord. One unit appears to be unaffected by temperatures at least between 10° and 20°C. In addition, isolated ganglia from insects kept at 22°C showed maximal frequency of on-going activity at 22°C; for those kept at 31°C the optimum was at 31°C (Kerkut and Taylor 1957). The amount of acetylcholine in the ventral nerve cord of *Periplaneta americana* increases with increasing temperature between 5° and 35°C; at 9°C, however, the amount is greater than at 15° or 25°C (Colhoun 1958). An Alaskan beetle, *Pterostichus brevicornis*, can live in winter temperatures at which the hemolymph is frozen (Baust 1972; see also p. 25). Extracellular recorded action potentials of the trochanter nerve of its hind leg showed block of on-going activity as follows: At -20°C, 0°C, and +20°C cold block occurred at -12, -8, and 0°C, respectively. Various lepidopteran insects exhibit a close correlation between their habitat (action) temperature and the temperature of maximal response in their electro-antennogram to pheromones (Tomescu et al. 1982; Bestmann and Dippold 1983).

The complexity of the interactions of neural activity, and endocrine, with temperature in influencing a physiological process can be seen in the factors that affect heart beat frequency in *Periplaneta americana* (Clarke 1967). The joint effects of hormonal and nervous influences can allow adjustments of their rates so that a harmonious response to temperature change can result.

The role of endocrines in metabolic compensation to thermal changes in poikilotherms is not yet finally resolved. The first documented case in insects was the report by Clarke (1966) on the plausible stimulation of protein synthesis at low temperature (15°C) in *Locusta migratoria* through neurosecretion, as evidenced by the exhaustion of corpora cardiaca. The results confirmed the findings of Lüscher (1965) that a metabolism-stimulating hormone is produced by the neurosecretory cells of the brain of a roach *Leucophaea maderae* and transferred to the corpora cardiaca, where it is stored or released, depending on the particulate temperature condition. Meanwhile several other authors have corroborated the function of insects' corpora cardiaca as a metabolic regulator on the one hand, as well as the temperature dependence of insects' neuroendocrine system on the other hand (e.g., Papillon et al. 1976; Peacock et al. 1977; Sharma et al. 1980). In most insect species studied so far, low temperatures inhibit the secretory activity of the neurosecretory cells. It is, therefore, clear that endocrine centers are very sensitive to fluctuations in environmental temperatures and unsuitable levels result in a prevention of growth and metamorphosis (see Sec. 2.5 in Chap. 2). This phenomenon probably plays an important part in saving the progeny from coming across adverse circumstances in which their survival would not be possible (see also Chap. 3). However, the mechanisms by which temperature (and also photoperiod) act through the neuroendocrine system to influence growth and reproduction are not fully understood. Data of Barker and Herman (1976) for the Monarch butterfly, *Danaus plexippus*, indicate that juvenile hormone is absent or ineffective at temperatures above or below the temperature optimum for reproduction. In females of *Schistocerca gregaria* a delay in the appearance of maximal hormonal levels (JH III, ecdysteroids) in the hemolymph at nonoptimal temperature conditions (28°C) appears to be sufficient for disturbing vitellogenesis (Porcheron et al. 1982). Similar observations were made in our laboratory for the Mediterranean field cricket, *Gryllus bimaculatus*, (Hoffmann et al. 1981; Behrens and Hoffmann 1983; Koch and Hoffmann 1984). In *Drosophila melanogaster,* Garen et al. (1977) have isolated a temperature-sensitive mutant, called ecd-1. Those ecd-1 individuals are normal at 20°C, but become deficient in ecdysteroids after a temperature shift to 29°C. The oogenesis defects which are observed in ecd-1 females depend on the stage at which the shift from the permissive temperature to the nonpermissive is performed (Audit-Lamour and Busson 1981). In the two last larval instars of *Schistocerca gregaria* a fall in diurnal rearing temperature from 33° to 28°C again induces lower circulating levels of JH III, which result in a lengthening of development and a slowing down of growth (Papillon et al. 1980).

The possible role of neurohormones in the process of acclimation and acclimatization of digestive enzymes has been extensively studied by Ivanović and co-workers (Ivanović et al. 1975, 1979, 1982). During the process of acclimatization in larvae of a phytophagous beetle *Morimus funereus* the neuroendocrine system shows cyclic changes in activity. The changes are also observed in the midgut proteolytic activity and in the hemolymph (proline, histidine concentrations) as well. It is suggested from these results that in *Morimus funereus* larvae neurohormones, i.e., the interrelationship of protocerebral neurosecretory cells themselves (membrane permeability and sites of hormone receptors) and their interrelationship with the neurohormone titer, play a crucial role in the complex mechanisms of acclimation and acclimatization in protein synthesis, energy metabolism, and excretion.

3.3 Physiological Adaptation to Cyclic Temperatures

Most experiments on metabolic adaptation processes have been carried out with constant acclimation temperatures. By contrast, the overwhelming majority of terrestrial insects lives in an environment that subjects them to changing ambient temperatures over short (irregular, diel) cycles and fluctuations. In alpine regions, for example, the most critical factor is temperature fluctuations during the animals' active

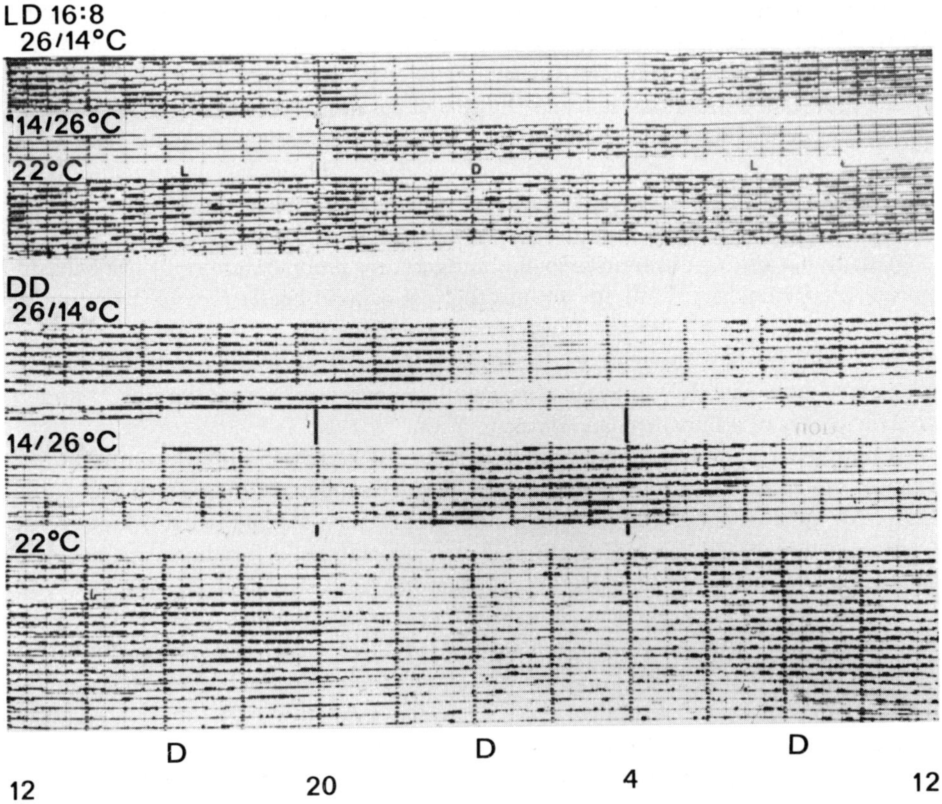

Fig. 1.8. Locomotoric activity of female adult crickets, *Gryllus bimaculatus*, under constant and daily alternating temperature conditions: ET 26/14°C, LD 16/8 h; ET 14/26°C, LD 16/8 h; ET 22°C, LD 16 /8 h; ET 26/14°C, constant darkness (*DD*); ET 14/26°C, constant darkness; ET 22°C, constant darkness. *L* light (4.00–20.00); *D* dark. (After Hoffmann 1980)

period – not heat or cold in themselves, but the way they vary rapidly from place to place and from time to time (Matthews 1976). Therefore, it may be asked whether rearing of eurythermal insect species at constant temperature is not "unbiological", in the sense that it does not allow any assertions to be made about the environmental requirements of the species (for details see Chap. 2).

Special adaptations to diurnally changing temperatures are suggested to exist in insects with daily activity rhythms. The daily light cycle, but also the daily temperature cycle, may act as a "Zeitgeber" and entrain circadian rhythms. Roberts (1962) demonstrated that a 24 h sinusoidal temperature cycle (22°–27°C; constant darkness) could entrain the locomotor rhythm in cockroaches (*Leucophaea maderae* and *Periplaneta americana*). Similar results have been shown in Hoffmann (1980) for female adult crickets, *Gryllus bimaculatus* (Fig. 1.8). It is to mention, however, that earlier reports, e.g., by Cloudsley-Thompson (1958), were conflicting or contradictory on this point. Since actions of temperature cycles as a Zeitgeber have been only observed during constant light or constant darkness, it may be supposed that at a simultaneous influence of light and temperature cycles effects of light changes will be domineering. In another cricket species, *Teleogryllus commodus*, temperature was postulated as an actual Zeitgeber for singing (Rence and Loher 1975). Temperature cycles may also entrain the eclosion rhythm of several insect species (Pavlidis et al. 1968; Pflüger 1973). Just as temperature cycle can entrain a circadian rhythm as well as a light cycle, a temperature pulse will cause a phase-shift in a manner comparable to a light pulse. For cockroaches *Periplaneta americana* this was first demonstrated by Bünning (1959).

A daily pattern of differences in the temperature preferendum (PT) has been reported by Remmert (1960) for an eurythermal carabid beetle *Cicindela campestris* in which the PT was 8°C lower at night than during the day.

Meats (1973) has shown that the fruit fly *Dacus tryoni*, reared at 25°C, could acclimate "immediately" to acute changes of temperature, so that the insects undergo daily cycles of acclimation and deacclimation.

A number of metabolic functions in insects are also known to show rhythmic diurnal changes (for a summary see in Rensing 1973 and Saunders 1976; see also Sec. 3.1 in Chap. 8). Most of these diurnal changes have only been studied in light/dark cycles. Consequently, effects of temperature cycles on metabolic functions are not well established. Welbers (1975b; 1976) has demonstrated a change in respiration of *Periplaneta americana* and *Pectinophora gossypiella* larvae with a temperature cycle. But he did not observe a metabolic rhythm of these larvae after a transfer into a constant temperature. Neville (1965) has shown that in a fluctuating environment with a light cycle and a temperature cycle the endocuticle of the desert locust, *Schistocerca gregaria*, was laid down in alternative "day" and "night" layers. The period of the cuticle rhythm was shown to be temperature compensated (Q_{10} = 1.04), although the quantity of material deposited was not (Q_{10} = 2.0).

From these results it again becomes obvious that increased rates of development at alternating temperature conditions, such as extensively discussed in Chap. 2, cannot be explained by an independence of metabolic rates (homoeostasis) from temperature changes. An actual hypothesis is that hormones (ecdysteroids, juvenile hormones, and neurohormones) are involved in the mediation of the effects of diurnally alternating temperatures on development rates of insects (Hoffmann 1980; Hoffmann et al. 1981; Behrens and Hoffmann 1982; Koch and Hoffmann 1984; Sec. 2.5 in Chap. 2).

3.4 Evolutionary Rate Compensation to Temperature

Thermal adaptation over evolutionary time periods is the process by which a species adapts to a specific thermal environment over many generations. Extrapolation of the metabolism-temperature curve of a temperate poikilotherm to polar temperatures (Arctic, Antarctic) would result in such a low metabolism that it is justifiable to ask whether this would be actually sufficient for maintenance of the animal (continuation of enzymic activity, circulation, excretion, muscle tonus, digestion, etc.). Therefore, several authors claim that the standard or resting metabolism of those organisms is elevated at low temperatures compared to that of temperate species measured at the same temperature (Young 1979; Clarke 1980). As a result organisms

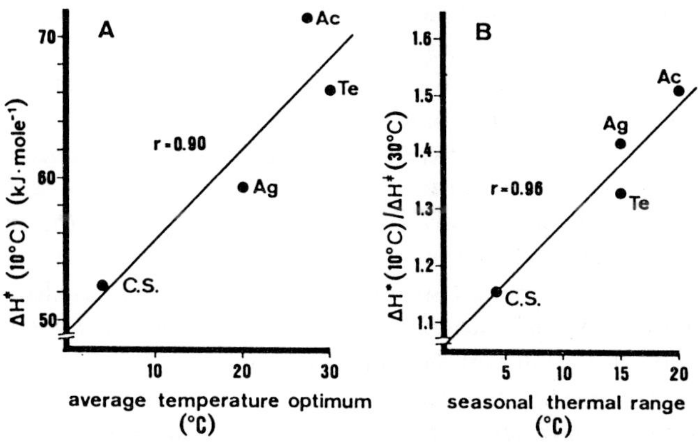

Fig. 1.9. A Correlation between enthalpy of activation $\Delta H^{\ddagger}$ of low temperature ($10^{\circ}C$) conformers (muscle pyruvate Kinases) and average optimal temperature at several insect species. *CS* "cold stenothermal" stream water larvae (Plecoptera, Trichoptera); ***Ag Agelastica alni*** (Coleoptera); *Te Tenebrio molitor* (Coleoptera); *Ac Acheta domesticus* (Orthoptera). **B** Correlation of enthalpy of activation $\Delta H^{\ddagger}$ of low ($10^{\circ}C$) and high ($30^{\circ}C$) temperature conformation state to each species' seasonal thermal range. (After Hoffmann 1976a)

from widely different latitudes often exhibit similar rates of physiological activity when measured and compared at their environmental temperatures.

For several polar invertebrates experimental evidence has been forthcoming in support of this hypothesis. Among these are aquatic and terrestrical micro-arthropods, like mites, but not terrestrial insects, except some Collembola. The Collembola represent one of the most important arthropods in the maritime Antarctic and are found in almost all terrestrial habitats in this zone. Studies by Strong et al. (1970) and Tilbrook and Block (1972) indicate some degree of cold adaptation in the respiration rate of the Collembola *Isotoma klovstadi* and *Cryptopygus antarcticus.*

Proponents of another view suggest that the standard metabolism of polar invertebrates is not elevated, since this strategy would divert energy from growth and reproduction and as such would be selectively disadvantageous (Block and Young 1978). Evidence to this has been gathered for example by Lee and Baust (1982) for the antarctic fly *Belgica antarctica.* A comparison of standard metabolism for polar and temperate chironomids also revealed no elevation of metabolic rate in polar forms. However, metabolic reactions of polar chironomid species exhibited lower activation energies (E_a, $\Delta G^{\ddagger}$) than temperate forms, indicating that their respiratory metabolism is relatively temperature independent. Data from several other authors tentatively indicate that activation energies may be correlated with natural habitat temperatures in general (McFarlane and McLusky 1972; Hoffmann 1976a). For example, a good correlation was found between the value of $\Delta H^{\ddagger}$ (enthalpy of activation) for homologous muscle pyruvate kinases of insects (Trichoptera and Plecoptera larvae, *Agelastica alni, Acheta domesticus, Tenebrio molitor*) and the average habitat temperature of the species (Fig. 1.9A). A good correlation also exists between the rate of $\Delta H^{\ddagger}$ ($10^{\circ}C$) to $\Delta H^{\ddagger}$ ($30^{\circ}C$) and the variability in body temperature experienced by the species (Fig. 1.9B). Variability in genetically determined protein primary structure is suggested to provide the basis and avenue for the natural selection of this adaptive molecular properties. However, the level of thermostability of the bulk of proteins in an organism seems to be absolutely independent of the temperature range (maximum) for animal life (Alexandrov 1977).

Other evidence for the existence of evolutionary temperature adaptation has been found in comparing the organic body constituents of flies (*Protophormia terranovae*) from Spitsbergen and from a laboratory stock (Hoffmann 1976b). High amounts of storage lipids are considered as characteristic for high arctic flies. Laboratory stock individuals on the contrary, do not accumulate such quantity of lipids, but have low molecular carbohydrates for their main organic constituent. This, of course, has great importance with regard to the caloric value of the individuals and their food quality in an ecosystem.

Two antarctic beetles *Hydromedion sparsutum* and *Perimylops antarcticus* are adapted to this environment in other respects (Block and Sømme 1983). Adults and larvae are found in small moss clumps, on which they obviously feed. The utilization of moss as a food resource, which is not usually consumed by phytophagous insects, permits these beetles to live in habitats that would otherwise be impossible to colonize.

4 Freezing Resistance and Freezing Tolerance

Survival at subzero temperatures is common to many insect species inhabiting the temperate and polar regions, as well as for many high altitude insect species. Where behavioral avoidance (e.g., migration, warm hibernation) is not possible, survival depends upon physiological and biochemical adaptations for cold-hardiness (for a literature review see Baust et al. 1982). Among insect species two general strategies for cold-hardiness are apparent. Freeze-tolerant species are capable of surviving ice formation in their extracellular body fluids, while freeze-susceptible species die if frozen and, therefore, must lower the freezing and supercooling points of their body fluids to temperatures below those to which they will be exposed during winter.

Insect cold-hardiness has often been equated with diapause. In most cases, however, this putative association is complicated and unclear (Horwath and Duman 1983a) (see also Chap. 3).

4.1 Freeze-Tolerant Species

Freeze-tolerant species survive extracellular ice formation, but adaptations control the amount of ice formed, protect the cells from dehydration and osmotic damage, and protect the cells from intracellular ice formation (Storey and Storey 1983). Of course, freezing tolerance in insects is a relative term, since survival in many species obviously depends on the cooling rate (Miller 1978) and decreases with both decreasing temperature and time subjected to the frozen state (Sømme 1982). So far no freezing-tolerant eggs have been found, and no freezing-tolerant species have been found among Collembola. The first adult insect species found to be freeze-tolerant was a small carabid beetle *Pterostichus brevicornis* that overwinters in decaying stumps and downed timber in interior Alaska (Baust and Miller 1970). Several individuals of this population tolerate temperatures up to -87°C without apparent damage. Meanwhile all other overwintering adult beetles so far studied in Alaska were found to be freezing-tolerant. All exhibited relatively high supercooling points (SCP; the temperature at which spontaneous freezing occurs during gradual cooling; for experimental procedure see Hoffmann and Krüll 1979) and produce high concentrations (several molar) of polyhydroxy alcohols (glycerol, sorbitol, threitol, mannitol) in response to cold (Miller 1982). The fact that all the freeze-tolerant beetles have high SCP (above -12°C) lends further support to the theory that nucleating agents (nucleating proteins) with their attendant high SCP are essential to the development of freeze tolerance. At least five other insect orders are also represented by adults that tolerate freezing (Lepidoptera, Diptera, Neuroptera, Homoptera, Hymenoptera), but in contrast to some adult beetles, all exhibit relatively low SCP (-5 to -10°C) associated with a tremendous build-up of glycerol. A similar coexistence of freezing tolerance and low SCP has been found for larvae of the alpine coleopteran species *Pytho deplanatus* (Ring 1982). However, the evidence for any importance of polyols in conferring freezing protection still remains largely circumstantial. Adult *Phyllodecta laticollis* beetles tolerate freezing to -42°C in winter, but show a shift in cold-hardiness in spring. Beetles having the same hemolymph osmolality and glycerol concentrations tolerate approximately 10°C lower temperature in spring than in winter. Whereas the cold-hardiness in winter seems to be due to the collegative properties of glycerol, the increased freezing tolerance in spring seems to be based on other mechanisms (van der Laak 1982). Asahina (1966) has postulated that some protoplasma change may be important in the genesis of freezing tolerance and Miller (1982) assumed some tissue changes at the membrane or molecular level (proteins, enzymes) to be of critical importance for the development of freezing tolerance. High concentrations of glycerol in the extracellular component may also afford frost tolerance by dehydrating the cells. Possibly when dehydrated or glycerinated cells finally do freeze, the ice crystals are small and less damaging than the large, jagged crystal lattice resulting during the freezing of solutions of low solute concentrations (Heinrich 1981).

Duman (1979) has reported the existence of proteins which produce a thermal hysteresis (a difference between the freezing and melting points) also in the hemolymph of a few freeze-tolerant species (for the role of hysteresis proteins in freezing-susceptible insects see the next section). Larvae of the beetle *Dendroides canadensis* accumulate protein antifreezes during winter. Juvenile hormone I treated animals significantly elevated the levels of hysteresis proteins (THP) in their hemolymph (Horwath and Duman 1983b). These results indicate a direct hormonal mechanism involved with seasonal control of THP production. The function of THP in freeze-tolerant insects is uncertain since these species should not usually need a supercooling agent. One possibility is that THP are useful as supercooling agents during those times (e.g., during the early fall, midwinter thaws, or late spring) when insects are not freeze-tolerant. Another possible function of THP is to act as a cryoprotectant to protect against freeze damage, and a third possible function is that these substances are also involved in water balance and for the ability to rest dessication during overwintering. A last possibility is that THP function as storage proteins.

From the standpoint of energetics it is most efficient to freeze at as high temperature as possible, since freezing causes a sharp drop in metabolic rate. In several freezing tolerant beetles anaerobic metabolism seems to be an important part of overwintering (Conradi-Larsen and Sømme 1973) (see also Chap. 5).

In recent studies, Storey and co-workers have used the freeze-tolerant gall fly larva, *Eurosta solidaginis*, as a model system to investigate extensively low temperature control of polyol synthesis, metabolism in the frozen state, bound water as an adaptation for cold-hardiness, and the possible role of bound water and of the free vs bound levels of intracellular metabolites in controlling low-temperature metabolism (for a summary of results see Storey 1983). The gall fly larva overwinters as a third instar inside stem galls on goldenrod and can endure temperatures as low as -40°C. Extracellular freezing occurs whenever temperatures fall below -8 to -10°C. Two cryoprotectant polyols, glycerol and sorbitol, are accumulated during cold acclimation, and lesser accumulations of trehalose and proline may also have cryoprotective functions. The regulation of polyol synthesis in *Eurosta* is a good example of the exploitation of temperature change to alter metabolic make-up (Storey and Storey 1983). Northern and southern populations utilize quantitative distinct adaptive strategies (Rojas et al. 1983) resulting in 3–4 times greater concentrations of glycerol in northern specimens (Baust and Lee 1982). The glycerol production is anticipatory, occurring before the first cold exposures, whereas sorbitol is triggered by the first exposure to frost. The carbon source for polyol synthesis is glycogen. The regulation of glycogen metabolism appears to exhibit temperature change, temperature effects on enzyme kinetics, and temperature/modulator interactions with enzymes producing the alterations in metabolic flux, which lead to differential polyol synthesis. For instance, temperature/modulator effects on phosphofructokinase appear to be the major factor controlling carbon flow into glycerol synthesis or, alternatively, into the pathway of sorbitol synthesis (Storey 1982). The aerobic metabolism with continued polyol synthesis is maintained up to temperatures as low as about -10°C, until extracellular freezing occurs. Below -10°C mitochondria appear to be metabolically inactive. Evidence for this includes the cessation of polyol, sugar, and amino acid accumulation at this temperature, a drop in the adenylate energy charge of the animals, and a build up of lactate at -30°C (Storey et al. 1981) (see also Chap. 5). A restrictive effect of high bound water content may be one factor causing the strong depression of metabolic activity at temperatures below -10°C. Bound water content of the larvae increased significantly at low temperatures (Fig. 1.10) and is attributable to changes in water binding by both low molecular weight (polyols) and high molecular weight (proteins, glycogen) subcellular components. In addition, bound water may have a more subtle effect in determining the relative pool sizes of bound vs free metabolites in the cell. An increase in the content of bound metabolites with low temperature may restrict metabolism by limiting the availability of substrates and effectors of enzyme reactions. High intracellular pH, a consequence of low temperature effects on intracellular buffers, may also affect a decrease in metabolic activity resulting in dormancy during the winter (Storey and Storey 1983) (see also Sec. 5 in Chap. 3).

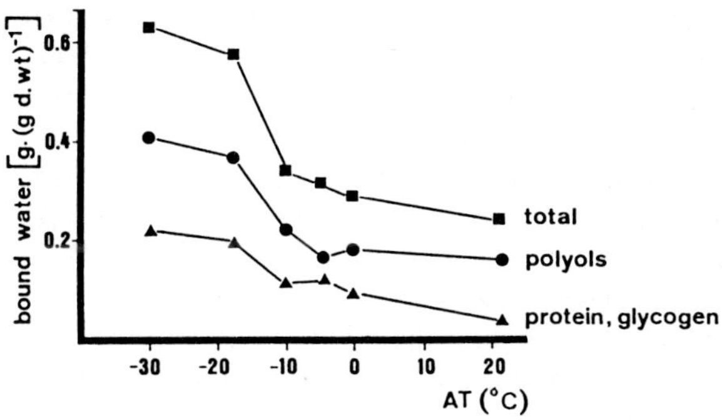

Fig. 1.10. Effect of acclimation temperature on the amount of water bound by soluble subcellular components of *Eurosta solidaginis* larvae. Polyols, low molecular weight components; protein, glycogen; high molecular weight components. (After Storey 1983)

4.2 Freezing-Susceptible Species

Freeze-sensitive species cannot survive the formation of internal ice and prevent this occurring by several mechanisms:

1. depression of the supercooling point of their body fluids;
2. elimination of ice nucleation sites in the body;
3. accumulation of high concentrations of polyols as cryoprotectants, and
4. accumulation of antifreeze proteins in their hemolymph.

It is generally assumed that the critical parameter for freezing-susceptible insects is the temperature to which they will supercool before spontaneous nucleation occurs. This SCP generally agrees with the lower lethal temperature of these species. Dependence on supercooling as a way of survival at low temperatures has evolved in all stages of insects. In general, lower mean SCP (e.g., as low as -51.3^{o}C in eggs of the tortricid *Zeiraphera diniana;* Sømme 1982) are found in insect eggs than in other stages. Very low mean SCP (around -30°C) have also been found in overwintering larvae. Relatively few pupae have been studied and it appears that their supercooling ability is generally poorer. Supercooling in freezing-intolerant adult insects has in particular been studied in a large number of beetles. Very low SCP have been found for several species of arctic bark beetles, overwintering in the trees. In the ice-avoidance strategy supercooling capacities are extended in the presence of cryprotectants. Several studies have implicated glycerol as an antifreeze also in freeze-susceptible species (Duman et al. 1982). Other polyols with probable antifreeze functions include sorbitol, mannitol, arabitol, and ribotol. Sugars, such as glucose and fructose, may also be antifreeze agents. Low SCP, and therefore, low freezing points would protect insects that overwinter in moist habitats from inoculative freezing across the cuticle. Although it is generally assumed that the large increase in polyol concentrations in winter functions primiarily to prevent freezing, these solutes may perform other functions as well. One possible additional function is as an energy source for the time immediately after the overwintering period. In addition, glycerol is known to stabilize enzymes at low temperatures, and thirdly glycerol demonstrated importance in avoidance of dessication damage in anhydrobiotic organisms. Little is known about the metabilic processes responsible for the regulation of glycerol accumulation in freezing-intolerant insects. Recently, it has been demonstrated that in the arctic blowfly *Protophormia terranovae* the source of glycerol is carbohydrate (Wood and Nordin 1976; Wood et al. 1977). In this species, polyol build-up during cold-stressing is not a consequence of a lack of terminal oxidative processes, but studies of the effects of low temperature in vitro on certain glycolytic enzymes demonstrated possible importance of glycerol-3-phosphate dehydrogenase, pyruvate kinase (see also p. 9, this Chap.) and a phosphatase in the regulation of glycerol accumulation.

Evidence for a relationship between supercooling and the feeding status of an animal has been corroborated by Sømme and Block (1982) for the antarctic Collembola *Cryptopygus antarcticus.* In general, the presence of food material in the gut increases the probability of freezing occurring in a supercooled animal because such material contains efficient nucleating agent.

The presence of THP in the hemolymph is apparently common among freeze-susceptible insect species [e.g., in *Mercantha contracta, Tenebrio molitor,* and *Coccinella novemnotate* (Coleoptera), *Oncopeltus fasciatus* (Hemiptera), and *Boreus westwoodi* (Mecoptera); Duman et al. 1982; Duman and Horwath 1983]. The role of THP in supercooling capacity species is rather unclear. Most of these insect species produce THP only during the winter, and coincident with the depression of the freezing point of the hemolymph by the THP the supercooling points of these insects were depressed. However, the correlation between the production of THP and the depression of SCP in winter does not prove that the THP are directly responsible for the lowered supercooling points. Recent studies by Zachariassen (cited in Duman and Horwath 1983) indicate that THP may function to stabilize the supercooled state. The major advantage of THP over small molecular weight antifreeze is that the THP do not generate the tremendous increases in osmotic pressure which accompany production of polyols or carbohydrates. Several insect THP have now been purified and their compositions analyzed (Patterson and Duman 1979; Schneppenheim and Theede 1980; Patterson et al. 1981; Patterson and Duman 1982; Hew et al. 1983). In contrast to the situation in fish THP, the insect THP lack a carbohydrate component, possess fairly normal levels of alanine, and have higher percentage of polar amino acids. However, the significance of these differences is not yet clear. Temperature and photoperiodic cues seem to control the annual cycle of THP production, thus, providing a failsafe system ensuring that THP are produced not too late in autumn and not lost in spring until the danger of freezing is past (Duman and Horwath 1983).

The question arises as to whether a freezing will also occur if cooling of an insect is stopped above its SCP, and if the insect is left at this temperature for increasing time intervals. Therefore, more work is needed to determine whether prolonged exposures to temperature close to the SCP may be harmful per se, even when freezing does not occur.

In conclusion, it can be stated that the strategies of freezing tolerance/avoidance in insects are heterogeneous and rarely lend themselves to uniform hypothesis. Four environmental factors, including temperature, state of hydratation, nutrient balance, and photoperiod appear critical to cold hardening.

5 Heat Tolerance and Endothermy

In contrast to the large differences in minimum temperature tolerated by active as well as inactive insects, there is relatively little variability in the maximum temperature (40°–50°C), which their tissues can tolerate (Heinrich 1981). Nevertheless, insects whose temperature depends to a great deal on that of the environment appear to have higher lethal temperatures when they live in hot environments rather than in cool ones. Microclimatological differences in particular environments, however, may reduce the value of such large-scale geographic comparisons. A long-legged beetle *Stenocarpa phalangium*, for example, lives in about 4°C cooler thermal environment than others with short legs active at the same time and place in the Namib Desert (Henwood 1975a).

Although insects are generally considered to be poikilotherms, there is considerable evidence that most species are capable of some degree of thermoregulation. Insects achieve thermoregulation either by varying the extent of heat exchange with the environment (anatomical adaptive feature; behavioral thermoregulation; ectothermic insects) or by generating metabolic heat (physiological thermoregulation; endothermic insects). The clear distinction between physiological and behavioral thermoregulation, however, is often difficult because many insects can be ectothermic or endothermic, depending upon the circumstances. The physiological sensitivities and behavioral responses of ectotherms (e.g., some butterflies, cicadas, beetles) to temperature have been the subject of several recent reviews (e.g., Heath 1970; Roberts 1974; Hoffmann 1978; May 1979; Reynolds 1979; Casey 1981; Kenagy and Stevenson 1982; Kingsolver and Watt 1983) and will not be taken into further consideration here. In endothermic insects (some butterflies, many moths, bees and wasps, some dragonflies and beetles, and a few katydids and flies) increases of body temperature are the result of heat produced by the flight muscles. These muscles are metabolically the most active tissue known (Heinrich 1974) (Chaps. 4, 5, and 6) and produce heat in all insects during flight, although the smaller fliers, which rapidly lose heat by convection in flight, are not necessarily endothermic. In large endothermic fliers, the muscle temperature range for flight is commonly narrow, and those insects are unable to fly until their wing vibration rate reaches the value that is observed during flight. This high wingbeat frequency, however, is achieved only when thoracic temperature is 32°C or higher. Therefore, a preflight warm-up becomes necessary to achieve sufficient high thorax temperatures (Fig. 1.11A). The heat production during preflight warm-up represents a variation of the flight behavior itself (Heinrich 1981). The weight-relative rates of heat production seem to be similar in both types of insect flight muscles, the myogenic and the neurogenic muscles.

During warm-up insects produce heat by contracting the flight muscles largely against each other (shivering) rather than against the wings (Kammer 1981). All the energy of muscle contraction is dissipated as heat, and temperature of the well-insulated thorax may rise as much as 10°C/min, whereas temperature of the poorly insulated abdomen remains near ambient (Fig. 1.11B). In moths preflight warm-up increases O_2 consumption some 2.3 times, and fat is the principal fuel (Hanegan and Heath 1970). The rate of warm-up increases with increasing ambient temperature. However, in other insect species the warm-up rate is suggested to be independent of

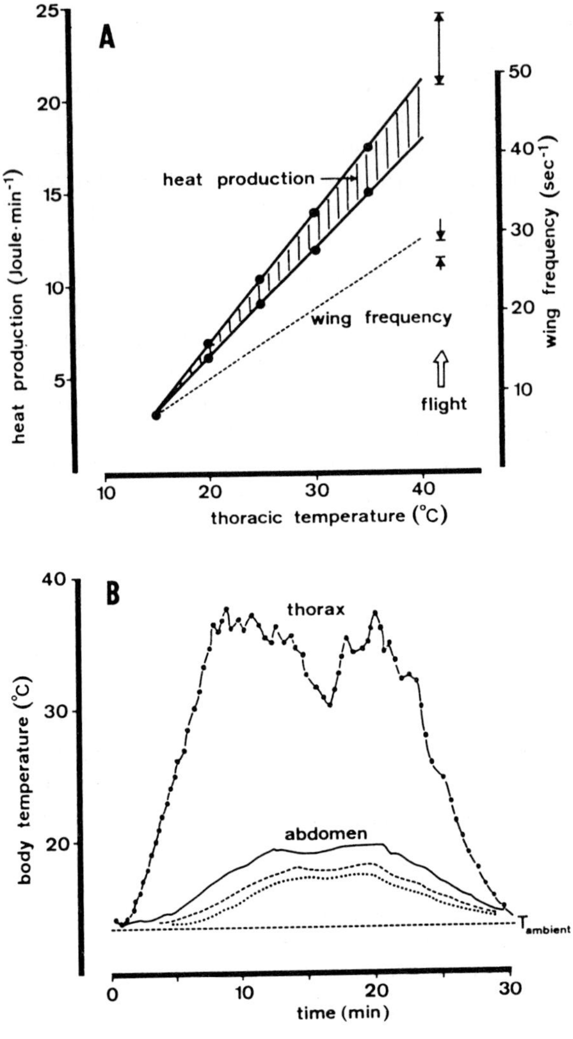

Fig. 1.11. A Wing vibration rate and calculated heat production during shivering thermogenesis in *Manduca sexta* in relation to thoracic temperature. *Arrows* at 41°C indicate average wing frequency and heat production of a 1.6 g moth during flight. (After Heinrich 1974). **B** Body temperature of bumblebee queens, *Bombus vasnesenskii*, during preflight warm-up, flight activity, crawling activity, and rest.———, ventral abdomen; – – – . . . . . , dorsal abdomen; air temperature 13.5°C. (After Heinrich 1975)

ambient temperature (May 1979). Little is known about possible acclimation in endothermic insects. Kammer (1981) tested warm-up contractions of Monarch butterfly wing muscles after acclimation to 5° or 23°C. At a mean temperature of 15°C, 80% of the animals acclimated to 5°C showed shivering contractions, but only 40% of the AT 23°C individuals. In animals kept at the higher temperature warm-up commenced later. During preflight warm-up the temperature of the thoracic nerve cord is more important than that of the muscles. Heating the nerve cord with a thermode in quiescent moths can stimulate the animals to fly without warm-up behavior. Warm-up behavior has been modified for several other functions. In foraging bees, for example, the interflight intervals are often only fractions of seconds in duration and preflight warm-up during these intervals at low ambient temperature functions to maintain and stabilize an already elevated thoracic temperature (Heinrich 1981). The heat production in order to warm up nests in honeybees, bumblebees, and other social insects which are vigorous fliers, involves the same heat-generating mechanisms. Fanning for nest temperature regulation is also a variation of flight or warm-up behavior.

Thermoregulation during flight potentially involves regulation of heat production and/or regulation of heat loss. Broad-winged insects, such as dragonflies, glide primarily in hot weather; in *Tramea carolina* the proportion of time spent gliding increases with ambient temperature from 20°–35°C (May 1979). The major physiological mechanism for regulation of body temperature in flight, however, seems to be the transfer of excess heat by the dorsal blood vessel to the cooler abdomen (Fig. 11B). Again the thoracic ganglia were predicted to control the heart rate and, hence, the dissipation of heat from the flight motor (Prosser and Nelson 1981). However, since most neural control centers in insects are located in the head, head temperature is also of particular interest. Data from Hegel and Casey (1982) show that in the sphinx moth, *Manduca sexta*, head and thorax temperatures were tightly coupled during preflight warm-up over a wide range of ambient temperatures, and that the observed head temperatures are the result of active heat transfer from the thorax via the hemolymph circulation to the abdomen. All insects in which heat dissipation mechanisms for flight have been implicated so far – sphinx moths, bumblebees, some dragonflies, and large beetles – also have preflight warm-up.

Little is known about the biochemistry of temperature-related aspects of flight muscle function. Organisms that must rely on high-speed locomotion for survival generally require enzymes that are highly efficient. Therefore, many large flying insects that inevitably heat up to 40°C or more have their biochemical machinery adapted to operate at these high temperatures. The higher the temperature set-point, the greater is the endurance of high-rate aerobic activity at high ambient temperatures (Heinrich 1977).

Endothermy during terrestrial activity (walking and running) has been shown to exist only in large tropical beetles. In some African dung beetles (Scarabaeidae), rates of specific energy metabolism during activity are similar to those of active mammals of similar size at the same ambient temperature (Heath and Heath 1982).

In some insect species endogenous heat production can be independent of flight or preparation for flight. A specific heat-generating metabolic process in bumblebees was first demonstrated by Newsholme and co-workers (1972). In resting bumblebee flight muscles, the glycolytic enzyme phosphofructokinase and the gluconeogenic enzyme fructose-1,6-bisphosphatase are simultaneously active and catalyze a cycle between fructose-6-phosphate and fructose-1,6-bisphosphate. Such a cycle produces continuous hydrolysis of ATP, with the release of energy as heat, which would help to maintain the thoracic temperature during rest periods at a level adequate for flight. The extent of substrate cycling was shown to be greater at low than at high ambient temperatures (Clark et al. 1973). When flight is commenced substrate cycling in bumblebees is no longer required and is suppressed. The rate of substrate cycling could be regulated by changes in the sarcoplasmic Ca^{2+}-concentration associated with the contractile process. Another glycolytic substrate cycle, the glucose/glucose-6-phosphate cycle has been described previously by Surholt and Newsholme (1983) for muscle and fat body of the hawk moth, *Acherontia atropos*. The marked increase in the cycling rate between glucose and glucose-6-phosphate upon flight contrasts with the findings of Clark et al. (1973) in the bumblebees, and suggests that the glucose/glucose-6-phosphate cycle in the hawk moth is unlikely to be utilized primarily for heat production. Morgan and Bartholomew (1982) have reported an enhanced heat production and regulation of body temperature in a large neotropical scarab beetle *Megasoma elephas* during exposure to decreasing ambient temperature, which was also not associated with any overt motoric activity. Energy metabolism and body temperature in these animals are conspicuously oscillatory with a given cycle in oxygen consumption peaking before the corresponding cycle in body temperature. The particular molecular mechanisms underlying this phenomenon, however, are yet unclear.

The advantages of temperature regulation in insects have been great (Heinrich 1981). In some insects all the major activities (feeding, mating, dispersal, oviposition) are associated with flight, and body temperature regulation is necessary to them to remain active. In other insect species, thermoregulation is restricted to specific activities, where it confers reproductive advantages, promoting success in the outcome of scramble and contest competition, in predator avoidance, or in resistance to viral infections.

6 Conclusions

Some insect species have evolved the ability to live in areas of the earth's greatest temperature extremes. In spite of some generalized conclusions in former literature regarding the absence of thermal metabolic acclimation in insects, a proper perusal of the literature reveals many well-documented cases of capacity and resistance acclimation to temperature, besides seasonal metabolic acclimatization, exhibited by insects at organismal, cellular, and even molecular level (Das and Singh 1972). From the aforementioned results it appears that the picture on metabolic temperature compensation is rather complex. The temperature relationship of a single enzymic reaction, for example, cannot be used as an argument for or against the occurrence of temperature compensation of whole animal metabolism. Nevertheless, adaptation processes render some ectothermic insects possible to remain active over moderate ranges of temperature.

In large flying (endothermic) insects the largest heat loads may be metabolically rather than environmentally induced, and show little or no acclimation (Heinrich 1981).

Acknowledgements. I am grateful to Dr. T. Mustafa, Odense, Dr. K.B. Storey, Ottawa, and Dr. H.T. Ratte, Aachen for critically reading the manuscript. Original research of our group described in this review was supported by the Deutsche Forschungsgemeinschaft.

Chapter 2

Temperature and Insect Development

Hans T. Ratte[1]

Contents

1 Introduction

The temperature dependence of insect development has been frequently investigated. Temperature affects both the time of development as well as fecundity; consequently, the appearance and dynamics of insect populations in the field are dictated by ambient temperature. For this reason there has been considerable interest for a long time in the temperature relationships of development time and fecundity mainly of filthy, annoying, and disease-transmitting insects in order to apply it in predicting the outbreak-time and the dynamics of pests. To do this, some empirical "rules" or "laws" about the temperature relation of developmental time have been derived which are employed with some success in pest control (Sec. 4). Although derived empirically with respect only to the time of development (this is only one variable from many others), these rules have greatly influenced the current concepts of temperature action and have been adopted by textbooks of environmental physiology and ecology. It has been suggested that the developmental process has the characteristic

1 Lehrstuhl für Biologie V, Rheinisch-Westfälische Technische Hochschule Aachen, Kopernikusstraße 16, 5100 Aachen, FRG

Environmental Physiology and Biochemistry of Insects (ed. by K. H. Hoffmann)

of a complex chemical reaction, the velocity of which is determined by the temperature dependence of the slowest step ("master reaction", "rate determining process") (p. 9 in Chap. 1).

However, many results from field observations and laboratory experiments in which the insects have been subjected to fluctuating temperatures cannot be brought into line with the current concept. Moreover, they are crucial in understanding the physiological basis of temperature action. While this may not be of major importance for the aforementioned applied purposes, from viewpoint of an environmental physiologist investigating mechanisms of adaptation, this is very unsatisfactory. A too simplistic physiological concept of the temperature action might also lead to erroneous insights into the insect's capabilities of adapting at a variety of different thermal environments.

In order to give a survey over the state-of-the-art as well as to overcome the difficulties, this chapter intends:

1. to exemplify the temperature relations of main variables resulting from development, such as time of development, adult longevity, adult size, and fecundity (Sec. 3);
2. to illustrate the current concepts of temperature action and to discuss their validity (Sect. 4);
3. to present an improved concept of temperature action, which unifies the conflicting results and offers a fundament for better understanding how insects might have been adapted to various thermal environments.

Before doing so, we have first to consider that development is restriceted only to a relatively narrow temperature range: the favorable range for development.

2 Favorable Temperature Range for Development

The temperature range in which insects can survive is much wider than the range for normal activity and nonarrested development. The latter, also called the "favorable range for development", will be treated in this chapter, whereas types of arrested development, such as diapause and torpidity are considered in Chap. 3. The favorable range for development has often been demonstrated by mortality data together with rearing experiments under laboratory conditions.

Constant temperatures. Generally, critically low or high temperatures cause a sharp increase in mortality leading to more or less U-shaped mortality vs temperature curves. Figure 2.1 shows that the form of the curves and the range of minimum mortality are species specific. The eggs of the nun moth, *Lymantria monacha*, from temperate climates show less than 50% mortality over a relatively wide range of lower temperature from 5° to 26°C. In contrast, the oriental fruit fly, *Dacus dorsalis*, survives within a relatively wide range of higher temperatures (from 15.5° to 35°C), whereas the development of the Douglas-fir tussock moth, *Orgyia pseudotsugata*, from California is limited to the small range from 16.5° to 28°C. The favorable

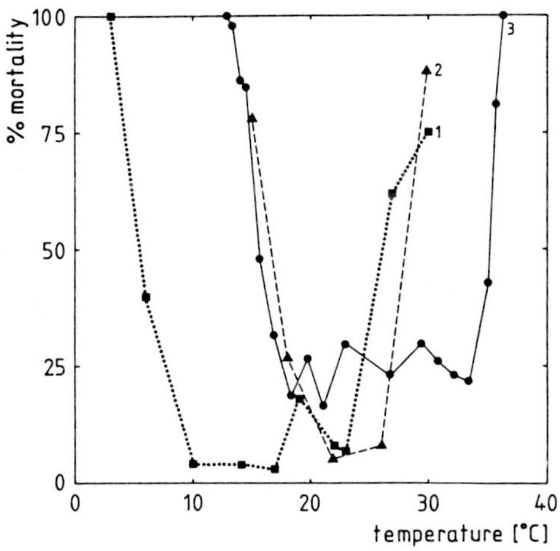

Fig.2.1. Favorable range for uninhibited development exemplified by mortality curves; *1* eggs of *Lymantria monacha*, temperate climate (after Zwölfer 1935); *2* larval instars of *Orgyia pseudotsugata*. California (after Beckwith 1982); *3* eggs of *Dacus dorsalis*, subtropical climate. (After Messenger and Flitters 1958)

range for development may be different for egg, larval, or pupal instars (e.g., Hammond et al. 1979).

Fluctuating temperatures. Some insect populations when reared under fluctuating temperatures show a mortality different to that under a comparable constant temperature regime. In the Mediterranean field cricket, *Gryllus bimaculatus*, only 18% of the eggs hatch successfully at a constant temperature of 20°C, whereas under some daily fluctuating temperature regimes round 20°C, the hatch success increases to 49–65% (Behrens et al. 1983). Similar results have been obtained in other species and for other developmental instars (Trpiš and Horsfall 1969; Siddiqui et al. 1973; Pfadt et al. 1975; Hofsvang 1976; Yeargan 1983). In some cases, mortality increases slightly with increasing temperature amplitudes (Siddiqui and Barlow 1973).

Not only may the mortality be lower under fluctuating temperatures, but also the favorable range for development may be expanded, especially in the lower temperature range. In any case, the period of lower temperature can fall outside the lower temperature limit. But even under temperature regimes showing a daily mean temperature below the lower constant-temperature limit, some insects are able to complete their development while they fail to develop under the mean comparable constant temperature (eggs of *Tribolium confusum*, Lin et al. 1954; eggs of *Dacus dorsalis*, Messenger and Flitters 1958; larvae of *Gryllus bimaculatus*, Behrens et al. 1983).

In contrast, the expansion of the favorable range towards temperatures higher than the upper constant temperature limit seems to be rather limited. A temperature elevation above this upper limit may cause the death of the entire experimental population even though the daily mean temperature does not exceed the upper temperature limit (Huffaker 1944; Champlain and Butler 1967; Butler and Lopez 1980). Under these conditions, the mortality also depends on the duration of the exposure to critical high temperatures.

Possible causes suggested for the increase in mortality at the extremes of the favorable range are that metabolic reactions differing slightly in their temperature dependence become disorganized or that the supply of food and oxygen is inadequate for the increased metabolic rate (Huffaker 1944; Hochachka and Somero 1973).

3 Developmental Variables as Affected by Temperature

3.1 Time of Development

Temperature effects on developmental time have been already amply reviewed and discussed by Andrewartha and Birch (1954), Bursell (1964), Laudien (1973) and for the embryonic development by Howe (1967). Some statistical problems are considered by Messenger and Flitters (1958) and Sharpe et al. (1977). In this section I will present only a few typical examples.

Constant temperatures. Typically, the time of development of a larval instar or of the total development decreases exponentially with increasing temperature as exemplified by the embryonic development of the oriental fruit fly, *Dacus dorsalis* (Fig. 2.2). In almost all insects, at lower temperatures the development time decreases steeply with increasing temperature: at medium temperatures the slope of the curve becomes smaller until it reaches zero at the so-called optimal temperature for development, which lies usually close to the upper thermal limit. The curve of developmental rate (reciprocal developmental time) takes a more or less sigmoidal course including an inflexion point mostly between 20° and 25°C, as well as a maximum typically between 25° and 35°C.

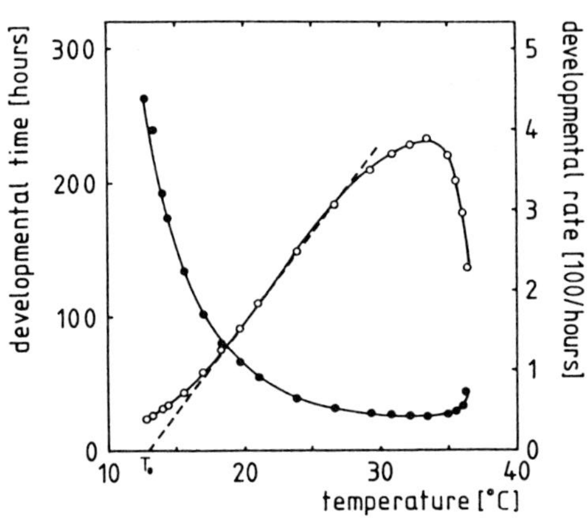

Fig. 2.2. Time of development (*black circles*) and rate of development (100/hours; *open circles*) of the oriental fruit fly, *Dacus dorsalis*; the curve is fitted by a catenary function and by its reciprocal, respectively (*solid lines*); the middle linear part is also approximated by a straight line (*broken line*) leading to extrapolation to the developmental zero (T_0). (After Messenger and Flitters 1958)

Such kinds of relationships have been obtained very frequently and also reported for different developmental instars. At the same time, males have often been found to develop slightly faster than females of the same experimental population. The relationship between developmental time and temperature may be approximated by different exponential functions, one of which is the well-known catenary function derived by Janisch (1925):

$$t_d = \frac{t_{min}}{2}\left(a^{(T-T_{opt})} + a^{(T_{opt}-T)}\right) \quad (1)$$

where:

T_{opt}: optimum temperature
T: : experimental temperature
t_d : time of development
t_{min} : empirical minimum time of development at T_{opt}
a : species specific constant.

Both expanded versions of this equation and other exponential functions have been employed with arguments of improved flexibility and, thus, better fitting properties (Logan et al. 1976; Hilbert and Logan 1983). The majority of developmental equations is composed of two exponential terms, one being a positive, the other a negative function of temperature. The justification of these equations is not only seen in their fitting quality, but also in their assumed grounding in the kinetics of chemical reactions, velocity constants being an exponential function of temperature. The "exponential law" proposed by Janisch (1925) assumes that such kinetics are exhibited by both accelerating processes, these dominating below the temperature optimum, and retarding processes which dominate at temperatures above the optimum. The retardation in development is believed to be due to a metabolic disorganization (see Sec. 2).

The rate curve of the oriental fruit fly, *Dacus dorsalis* (Fig. 2.2), displays a wide, approximately linear, range around the inflexion point; this part of the curve is often fitted by a straight line. Consequently, the time of development follows a hyperbolic function of temperature over this range. Similar relationships between developmental rate and temperature have been found frequently and have led to the establishment of the "rule of thermal summing" and the classical method of "temperature summation" (Blunck 1914; Peairs 1914; Shelford 1927; Eq. 2).

$$TC = t_d(T - T_o) \tag{2}$$

where:

t_d : time of development
TC : thermal constant
T : experimental temperature
T_o : threshold temperature
$T-T_o$: effective temperature above T_o.

In other words, the time of development multiplied by the "effective temperature" ($T-T_o$) is a constant (thermal constant). "Thermal summing" implies the existence of a distinct threshold temperature (T_o), at which the developmental rate becomes zero (developmental zero; developmental threshold). This zero-development has to be regarded as theoretical because it can be extrapolated only from a linear rate relation and cannot be determined by rearing experiments: the threshold may fall outside the favorable range for development and also the infinite developmental time at temperatures near the threshold makes its measurement practically impossible. In *Dacus dorsalis*, the developmental threshold of 12.8°C results from extrapolation of the linear part of the rate curve (T_o: Fig. 2.2). Although the cessation of develop-

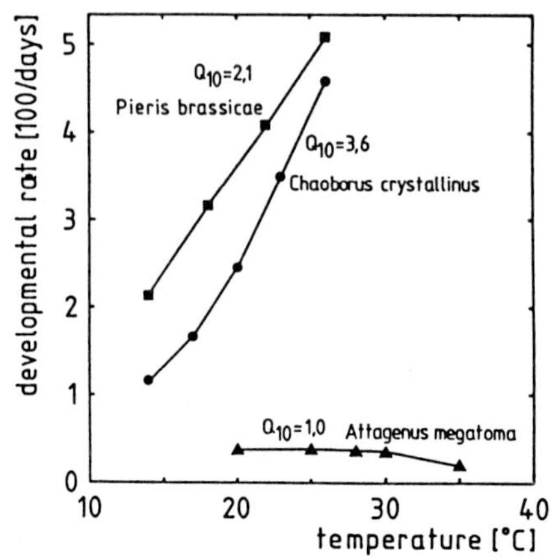

Fig. 2.3. Developmental rates of three insects exhibiting very different temperature coefficients (Q_{10}); *squares: Pieris brassicae* (LD 15:9; after Neumann and Heimbach 1975); *circles: Chaoborus crystallinus* (LD 12:12; after Ratte 1983); *triangles: Attagenus megatoma* (LD 0:24; after Baker 1983)

ment below critical temperatures can be observed easily, the concept of a real developmental zero has not been unequivocally accepted. When subjected to sub-threshold temperatures some insects develop faster after subsequent exposure to normal temperatures indicating that some development also takes place below the threshold (Hodson and Alrawy 1958; Eubank et al. 1973).

Finally, I should like to point to an important phenomenon of developmental rate curves: the degree of temperature dependence, expressed as slope of the rate curve or Q_{10}, shows a wide variation among various insects, which is illustrated in Fig. 2.3. The temperature coefficient may be about 2 (*Pieris brassicae*), as low as about 1 (*Attagenus megatoma*), or as high as 3.6 (*Chaoborus crystallinus*) and the developmental time of the carpet beetle, *Attagenus megatoma*, is nearly independent of temperature over a wide range: this shows that some insects do indeed possess the capability to develop independent of ambient temperature.

Fluctuating temperatures. The developmental time of insects reared under a fluctuating temperature regime has usually been found to differ from that under the corresponding constant temperature. This has also been claimed by earlier reviewers (Cloudsley-Thompson 1953; Howe 1967; Hagstrum and Hagstrum 1970; Beck 1983b). The majority of results is presented in Table 2.1. This listing contains only results from experiments conducted under diurnally fluctuating temperature regimes because presumably only these are relevant to the situation in the field. In order to illustrate the extent of acceleration or retardation of development caused by fluctuating temperatures, the percentage deviation from developmental time under comparable constant temperatures were calculated for each species (column R(eCT) of Table 2.1). It can be seen that considerably high accelerations may occur when the lower temperature falls below the developmental threshold (e.g., Lin et al. 1954; Messenger and Flitters 1959). The same has also been observed with nondaily alternating temperatures or single transfers from sub-threshold to favorable temperatures (Lin et al. 1954; Hodson and Alrawy 1958). Consequently, it has been concluded that under thermoperiodic conditions also below the threshold "some development" may take place: this is assumed to occur because the temperature requirements of some rate-limiting developmental steps are nevertheless satisfied (Beck 1983b).

Table 2.1. Development under fluctuating temperatures: percentage decrease (+%) and increase (-%) of developmental time as compared with results from constant temperature (R(eCT)) and with the predictions according to the rate summing concept (EXP(FT).)[a]

Species:	S	Php h:h	Thermoperiod Investigated range of eCT °C - °C	Range of daily temperature change °C	nr	Developmental time % deviation from R(eCT)	EXP(FT)	DRS	References
Coleoptera:									
Adalia bipunctata (L.)	L	16: 8	18.0	20, s	1	- 5	?	+a	Ellingsen 1969(*d*)
	P	16: 8	18.0	20, s	1	0	?	+a	
Anthonomus grandis (Boheman)	T	?	25.0 - 30.0	20, r	2	-29/ 2	?	+a	Fye et al. 1969(*c*)
Anthonomus thurberiae (Pearce)	T	?	25.0 - 30.0	20, r	2	-32/ 1	?	+a	
Hypera brunneipennis (Boheman)	E	15: 9	22.2	15. r	1	4	?	+a	Butler and Ritchie 1967(*c*)
	E	12:12	10.0 - 29.5	11.2, 16.6, s	3	-11/11	1/5	+	Madubunyi and Koehler 1974
	L	15: 9	16.7	8.3, r	1	0	0	+	Butler and Ritchie 1967(*c*)
	D	15: 9	16.7	8.3, r	1	21	21	–	
	J	12:12	10.0 - 29.5	11.2, 16.6, s	3	10/24	- 9/12	+	Madubunyi and Koehler 1974
Sitophilus oryzae (L.)	T	16: 8	25.0	10, s	1	- 1	?	+a	Hagstrum and Leach 1973
Tribolium castaneum (Herbst)	T	16: 8	25.0	10, s	1	10	?	+a	
Tribolium confusum (Duval)	E	?	15.4 - 19.0	3, 5, 7, 10, r	12	14/29	6/ 8	+b	Lin et al. 1954
	E	?	20.0 - 30.0	5, 10, r	13	-64/17	-68/41	-a	Mikulski 1936(*d*)
	J	?	22.5 - 30.0	5, 9.5, 10, r	10	-24/10	?	+a	
Trogoderma inclusum (Le Conte)	T	16: 8	25.0	10, s	1	10	?	+a	Hagstrum and Leach 1973
Diptera:									
Aedes aegypti (L.)	T	?	18.3 - 29.4	16.7, r	3	-18/ 6	-25/ 6	–	Headlee 1940
	T	0:24	15.5 - 30.0	5,5, 16.7, s	8	-43/27	-46/12	–	Headlee 1941(*c*)
	L	?	21.1 - 32.2	11.1, s	3	- 8/22	- 8/27	-a	Keirans and Fay
Aedes triseriatus (Say)	L	?	15.6 - 32.2	11.1, s	5	-22/20	-15/17	-a	1968(*d*)

Table 2.1 (cont.)

Species:	S	Php h:h	Thermoperiod Investigated range of eCT °C - °C	Range of daily temperature change °C	nr	Developmental time % deviation from R(eCT)	EXP(FT)	DRS	Reference
Anopheles atroparvus (van Thiel)	T	0:24	21.5 - 23.4	2.7 - 8.1, r	18	15/34	-11/11	+	Britz and Höhne 1955
Anopheles quadrimaculatus (Say)	T	?	20.0 - 27.1	7.9 - 13.6, s	14	- 6/13	- 4/14	+a	Huffaker 1944
Bradysia impatiens (Johannsen)	T	12:12	24.5	11.1, s	1	9	14	+a	Wilkinson and Daugherty 1969(*d*)
Calliphora erythrocephala (Meigen)	P	?	13.9 - 20.0	11, 18, 23, r	5	10/40	7/15	+	Ahmad 1936(*c*)
Ceratitis capitata (Wied.)	E	12:12	11.1 - 32.2	2, 6, 11.2, 17.8, s	17	-43/31	-41/15	–	Messenger and Flitters 1959
Chaoborus crystallinus (De Geer)	T	12:12	14.1 - 20.0	10, 12, s	8	-60/21	-45/39	–	Ratte 1979
	T	15:9	20.4	9, r	1	24	22	–	
	T	24:0	19.6 - 20.0	5, s	3	6/35	?	–	
Dacus cucurbitae (Coq.)	E	12:12	11.1 - 32.2	2, 6, 11.2, 17.8, s	17	-46/37	- 6/23	–	Messenger and Flitters 1959
Dacus dorsalis (Hendel)	E	12:12	11.1 - 32.2	2.6, 11.2, 17.8, s	17	-14/33	- 1/27	–	
Drosophila melanogaster (Meigen)	T	16:8	20.0 - 25.0	5, 10, 15, r	7	10/15	- 1/ 7	+	Siddiqui and Barlow 1972
Drosophila spec.	P	12:12	12.0 - 20.0	10, 16, s	5	10/37	-22/23	–	Remmert and Wünderling 1970
Exorista mella (Walker)	L	15:9	17.5 - 27.5	15, r	3	-25/14	- 5/ 4	+	Butler et al. 1968(*c*)
	P	15:9	17.5 - 27.5	15, r	3	15/21	11/16	–	
Hypoderma lineatum (de Villers)	T	0:24?	17.5 - 23.8	5, 10, r	6	3/15	-20/16	–	Pfadt et al. 1975
Musca autumnalis (De Geer)	T	?	11.8 - 31.9	5, 10, s	3	?	<10	+	Moon 1983
Muscina stabulans (Fallen)	P	?	13.9 - 21.7	13.4, 18, 25, 30, r	6	14/27	1/11	+	Ahmad 1936(*c*)
Syrphus corollae (Fabr.)	E	18: 6	18.0	20, s	1	7	?	+	Benestad 1970
	L	18: 6	18.0	20, s	1	-13	?	+	
	P	18: 6	18.0	20, s	1	5	?	+	

Table 2.1 (cont.)

Hymenoptera:									
Chelonus texanus (Cresson)	T	15: 9	19.4 - 28.2	15, 16, r	3	- 4/18	8/14	+	Butler 1966(*c*)
Diprion pini (L.)	T	?	20.0	10, r	1	3	?	+	Gösswald 1936
Praon exsoletum (Nees)	T	12:12?	10.5 - 26.7	11, s	5	-24/27	-17/21	-a	Messenger 1969(*c*)
Telenomus podisi (Ashmead)	T	12:12	18.0 - 26.7	12, 20, r	2	- 7/23	- 1/ 6	+b	Yeargan 1980
Trichogramma pretiosum (Riley)	T	14:10	18.4 - 32.2	16.7, s	6	-37/12	- 7/ 7	+a	Butler and Lopez 1980
Trisolcus euschisti (Ashmead)	T	12:12	18.0 - 27.0	10, 12, r	2	-24/ 7	?	?	Yeargan 1983
Lepidoptera:									
Anagasta kuehniella (Zeller)	T	16: 8	12.5 - 25.0	5, 10, 15, r	9	- 2/- 1	-15/- 3	+a	Siddiqui and Barlow 1973
Carpocapsa pomonella (L.)	T	?	17.2	14.4, r	1	-62	?	–	Headlee 1940
Danaus plexippus (L.)	T	12:12	22.0	10, ?	1	?	?	+	Zalucki 1982
Heliothis armiger (Hübner)	P	14:10	24.0	4, 8, r	2	- 2/ 8	- 1/15	+a	Foley 1981
	D	13:11	24.0	4, 8, r	2	- 4/ 4	?	+a	
Heliothis zea (Boddie)	E	14:10	21.1 - 35.0	11.1, 22.2, 27.8, s	14	-14/11	- 3/ 3	+	Eubank et al. 1973
Ostrinia nubilalis (Hübner)	E	?	19.3 - 25.7	11, r	4	-16/ 6	-26/16	–	Matteson and Decker 1965(*d*)
	L	?	22.1 - 25.7	11.1, 28.1,	4	0/14	- 1/14	+	
	P	?	22.1 - 25.7	11.1, 28.1,	4	- 3/14	- 6/14	+	
	T	0:24	15.0 - 22.5	5, r	3	< 10	< 10	+	Beck 1983(*a*)
Pectinophora gossypiella (Saunders)	E	15: 9	18.0 - 30.0	8, r	4	< 10	< 10	+	Welbers 1975(*a*)
	J	15: 9	18.0 - 30.0	8, r	5	- 3/ 9	- 4/ 9	+	
	T	12:12	23.3	18.4, r	1	4	?	+a	Philipp and Watson 1971
	T	11:13	28.3	17.3, r	1	-41	?	-a	
	T	10:14	32.3	8.4, r	1	3	?	+a	
Pieris brassicae (L.)	E	15:9	18.0 - 26.0	8, r	3	- 5/13	- 2/ 4	+	Neumann and Heimbach 1975
	L	15:9	18.0 - 26.0	8, r	3	- 3/ 9	0/ 7	+	
	P	15:9	18.0 - 26.0	8, r	3	- 3/ 2	- 7/ 4	+	

Table 2.1 (cont.)

Species:	S	Php h:h	Thermoperiod Investigated range of eCT °C - °C	Range of daily temperature change °C	nr	Developmental time % deviation from R(eCT)	EXP(FT)	DRS	Reference
Pseudaletia unipuncta (Haworth)	L	16: 8	17.0 - 24.5	9, 8, 14, 16, 24, r	5	- 5/39	- 3/ 5	+	Guppy 1969
Porosagrotis orthogonia (Morrison)	L1	?	9.0 - 27.3	14, 24, 29, r	18	-31/66	-37/30	–	Cook 1927(*c*)
Spilosoma lubricipeda (L.)	E	12:12	15.0 - 20.0	2 - 14, s	11	-20/50	-24/20	–	Baker 1971(*d*)
	D	12:12	12.0 - 20.0	2, 6, 10, 14, s	13	-11/48	-13/28	–	
Spodoptera exigua (Hübner)	E	15: 9	18.3 0 35.0	11.1, r	4	2/ 6	- 9/ 0	+	Butler 1966(*c*)
	J	15: 9	18.3 - 35.0	11.1, 15.0,	6	-11/ 8	-11/10	+	
Spodoptera littoralis (Boisduval)	T	12:12	17.5 - 30.0	10, 15, 20, r	4	-51/29	- 8/21	–	Sidibé and Laugé 1977
Spodoptera litura (F.)	J	16: 8	16.0 - 31.0	3, 6, 9, r	12	?	?	–	Miyashita 1971
Hemiptera:									
Acanthomia tomentosicollis (Reuter)	T	12:12	29.5	5, r	1	-12	-11	+	Egwuatu and Ajibola Taylor 1977
Anthocorus sibiricus (Reuter)	T	18: 6	18.0	20, s	1	3	?	+	Hofsvang 1976
Eurydema ornatum (L.)	E	?	20.0 - 25.1	9.7, 10.2, 13.1	5	- 4/30	2/23	–	Bonnemaison 1946
	L	?	20.1 - 25.1	17.4, r	5	- 5/11	- 3/ 2	+	
Leptopterna dolobrata (L.)	L	18: 6	16.0	9.6, r	1	20	20	–	Braune 1971
Lygus hesperus (Knight)	E	15: 9?	17.5 - 27.5	15, r	3	-12/ 4	- 7/ 1	+	Champlain and Butler 1967
Oncopeltus fasciatus (Dallas)	E	?	13.6 - 16.7	6, 7, 11, 12, r	12	- 5/37	- 8/28	–	Lin et al. 1954
	T	12:12	14.0 - 23.0	12, r, s	8	11/71	11/71	–	Kureck unpublished
Homoptera:									
Acyrtosiphon pisum (Harris)	T	16: 8	15.0 - 25.0	5, 10, 15, 20, r	10	7/12	- 1/ 9	+	Siddiqui et al. 1973

Table 2.1 (cont.)

Species	S	Php	Range	eCT	nr	a	b	DRS	Reference
Aonidiella aurantii (Maskel)	T	?	24.0	12, s	1	1	?	+	Munger and Cressman 1948(*c*)
Circulifer tenellus (Baker)	E	?	9.0 - 31.3	5.5, 11.1, 16.6 20.2, 27.8, r	25	1/67	0/27	–	Harries 1943(*c*)
Therioaphis maculata (Buckton)	T	12:12	8.0 - 32.2	11, 17, s	9	- 9/32	13/34	–	Messenger 1964
Orthoptera:									
Camnula pellucida (Scudder)	E	?	17.0 - 27.0	5, 10, 15, 20, 25, r	6	11/25	16/40	–	Parker 1930(*c*)
Gryllus bimaculatus (De Geer)	E	16: 8	13.0 - 34.0	12, 18, r	8	- 5/27	- 5/32	–	Hoffmann 1974
	J	16: 8	23.0 - 31.0	18, r	3	-28/26	6/9	+	Merkel 1977
	J	16: 8	20.0	12, r	1	?	-45	–	Behrens et al. 1983
Locusta migratoria migratorioides (R. & F.)	E	16: 8?	19.7 - 26.3	22, 27, r	3	37/57	4/ 7	+	Ahmad 1936(*c*)
Melanoplus mexicanus (Saussure)	E	?	17.0 - 27.0	5, 10, 15, 20, 25, r	6	21/29	21/40	–	Parker 1930(*c*)
	J	?	17.0 - 27.0	5, 10, 15, r	6	14/16	7/35	–	

[a] Deviations absolutely less than 15% are not seen as conflicting (+ sign in DRS-column), otherwise the results are signed –; ? : not determined; *a* % deviation estimated; *b* subzero development assumed; *S* developmental stage such as *E* (embryo), *L* (larvae), *P* (pupae), *D* (diapausing pupae), *J* (all juvenile stages), *T* (all stages, egg - adult); *Php, h:h*; photoperiod, hours of light: hours of darkness; *eCT* equivalent constant temperature, *r, s* rectangular, sinusoidal temperature regime; *nr* number of thermoperiods tested; for the sake of shortness some references which are cited by Hagstrum and Hagstrum (1970; marked with *c*) or by Hagstrum and Leach (1973; marked with *d*) are not included in the reference list.

In cases where in one species different thermoperiods have been tested within the entire favorable range for development, the following rule has often been observed: temperature fluctuations in the range below the inflexion point of the developmental rate curve may cause an acceleration in development, while the converse is true for thermoperiods alternating near the temperature optimum; there are only slight or no deviations in developmental time from constant temperature with fluctuations around medial temperatures of the favorable range. This rule was propounded quite early and can also be derived from the non-linearity of the developmental rate vs temperature relation established under constant temperature (for discussion see Sec. 4).

The study of the developmental rate under combinations of photo- and thermoperiods has usually been carried out such that the warmer phase of the thermoperiod coincided with the light phase – the situation encountered most commonly in terrestrial biotops. Effects of varying phase relationships between the thermo- and photoperiod have been rarely taken into consideration. In the aquatic midge, *Chaoborus crystallinus*, it has been demonstrated that at 20°C and with photoperiods of 12 h the time of development depends on the particular relationship between the sinusoidal temperature cycle and the light-dark cycle (Ratte 1979, 1983). This has not been observed in other insects (*Pectinophora:* Welbers 1975a; *Oncopeltus:* Kureck unpublished).

3.2 Adult Longevity

Constant temperatures. The survival period of adult insects is shorter at higher temperatures (King and Martin 1975; Butler and Foster 1979; Bari and Lange 1980; Moscardi et al. 1981). Very often, survival periods of females are significantly shorter than that of males. In the velvet bean caterpillar, *Anticassia gemmatalis,* unmated females live longer than mated ones at a given temperature (Moscardi et al. 1981).

Fluctuating temperatures. Analogous to the developmental time insects may either survive a shorter or longer adult period when living under a fluctuating temperature regime than under a equivalent constant temperature. The adult life span is shorter in the pink bollworm, *Pectinophora gossypiella* (Butler and Foster 1979), in the fruit fly, *Drosophila melanogaster* (Siddiqui and Barlow 1972), and in the milkweed bug, *Oncopeltus fasciatus* (Kureck unpublished). In contrast, the wax moth, *Galleria mellonella*, and the mosquito *Aedes sticticus* live longer under fluctuating temperatures (Destouches 1921; Trpiš and Horsfall 1969).

The temperature effects on longevity have been used to gain support for some theories about aging and dying in poikilotherms; these cannot be dealt with here, however, for references see Rockstein and Miquel (1973).

3.3 Size and Weight of the Adults

Temperature effects on adult size or weight of insects, although being the quantitative result of development, have not been so thoroughly investigated as the effects on developmental time, and "rules" or "laws" governing the size and temperature have as yet not been deduced. With respect to the adult weight, especially of females, one has to consider that the body weight may to a great extent contain the weight of the gonads; these may be either fully, partly, or not developed at the time of adult emergence (such as in Ephemeroptera, Plecoptera, some Hymenoptera, some Lepidoptera, and some Diptera); the temperature dependence of body weight, thus, may be partly explained by the temperature dependence of gonadal weight. In contrast, measurements of the length of the insect or of a sclerotized part of the body, such as head capsule width, reflect more closely the true size of the insect's integument and take no account of gonadal mass and its temperature dependence. Since, however, there is a relationship between size and weight, one may consider the body weights, especially of males, as a reflect variable of the true size of an insect. Generally, in females there is a close relationship between body mass and egg mass (Ratte 1979) as well as between body mass and number of ovarioles (Bennettova and Fraenkel 1981).

Constant temperatures. Different kinds of relation of adult size to temperature have been observed. First, some insects maintain a more or less constant size over a wide temperature range. The head capsule width and body length of the potato leafhopper, *Empoasca fabae*, does not change between 13° and 35°C (Fig. 2.4). Similar results have also been obtained after measuring instar weights immediately prior to the adult stage (pupae or prepupae), as well as recently eclosed adults: such was the case in the cabbage butterfly, *Pieris brassicae* (Heimbach 1973), the pink bollworm, *Pectinophora gossypiella* (Welbers 1975a), the Mediterranean olive fruit fly, *Dacus oleae* (Tsitsipis 1980), and the Mediterranean field cricket, *Gryllus bimaculatus* (Behrens et al. 1983).

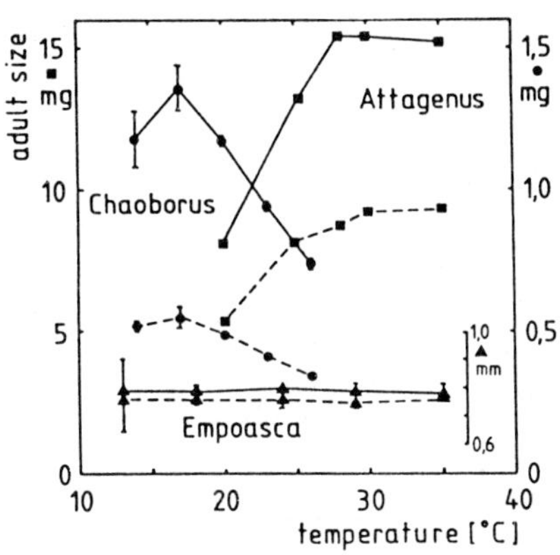

Fig. 2.4. Three types of body size relationships to temperature; *circles: Chaoborus crystallinus* (LD 12:12; after Ratte 1983); *squares: Attagenus megatoma* (LD 0:24; after Baker 1983); *triangles: Empoasca fabae* (LD 16:8; after Simonet and Pienkowski 1980); *broken lines:* males; *solid lines:* females

In some insects, temperature increase evokes lighter animals. In the phantom midge, *Chaoborus crystallinus*, above 14°C the dry weight of the adults decreases linearly with increasing temperature (Fig. 2.4). Similar results have been found in *Drosophila melanogaster* (Anders et al. 1964), in the armyworm, *Pseudaletia unipuncta* (Guppy 1969), in the parasitic fly, *Lixophaga diatraeae* (King and Martin 1975), and in the green cloverworm, *Platypena scabra* (Hammond et al. 1979).

An opposite relationship to temperature is shown by the weights of the carpet beetle, *Attagenus megatoma* (Fig. 2.4), and of the Douglas-fir tussock moth, *Orgyia pseudotsugata* (Beckwith 1982) as well as the head capsule width of the European corn borer, *Ostrinia nubilalis* (Beck 1983a). In these insects the body size or weight increases with rising temperature.

Fluctuating temperatures. If compared with equivalent constant temperatures, daily thermoperiods can produce either larger or smaller adults, although in some cases fluctuating temperatures do not markedly affect the adult size or weight (*Pieris brassicae*, Heimbach 1973). Under fluctuating temperatures, the adults are larger, e.g., in *Locusta migratoria* (Clarke 1967; Nicolas et al. 1979; Laugé and Launois 1980), in the pink bollworm, *Pectinophora gossypiella* (Welbers 1975a), and in the European corn borer, *Ostrinia nubilalis* (Beck 1983a). Smaller adults emerge in the noctuide *Spodoptera littoralis* (Sidibé and Laugé 1977) and in the aquatic midge *Chaoborus crystallinus.* In the latter species, the dry weight of the adults depends on the phase relationship between thermo- and photoperiod (Ratte 1979).

In summary, environmental temperature can have very different effects on the size or weight of different insects. Some insects can maintain their size or weight more or less independent of temperature, while others become smaller or larger with increasing temperature or temperature change (for the significance and possible physiological explanation, see Sec. 5).

3.4 Fecundity

Regardless of the timing of gonadal development (either occurring before or after metamorphosis), temperature affects the total number of eggs laid per female. The total number of eggs is also a function of the size of the females, even when gonadal development occurs after metamorphosis and egg production takes place over a longer period of time. The number of ovarioles depends on pre-imaginal temperature and on size (Cohet and David 1978; Bennettova and Fraenkel 1981).

Constant temperatures. The egg production of the aquatic midge *Chaoborus crystallinus* occurs prior to metamorphosis. Oviposition takes place immediately after adult eclosion and mating. Here the largest egg clutches, consisting of 350 eggs on average, are laid at 17°C (Fig. 2.5) whereas at the highest test temperature of 26°C, females lay 231 eggs on average. The relationships in some Simuliidae are rather similar (Colbo and Porter 1981). In *Chaoborus*, the favorable range for both egg production and larval development is identical. This appears not to be the case in insects undergoing postmetabolous gonadal development. In the spotted alfalfa aphid, *Therioaphis maculata*, and in the pink bollworm, *Pectinophora gossypiella*, the favorable range for egg production is conspicuously narrower than that for development (Messenger 1964; Welbers 1975a). Typically, in such insects the egg number vs temperature curves show a pronounced optimum (Fig. 2.5). The temperature at which the maximum of eggs is laid and the extension of the favorable temperature range for egg production are species specific and depend on the climate where the insects live. The Californian artichoke plume moth, *Platyptilia carduidactyla*, produces eggs between 10° and 28°C with maximum number at

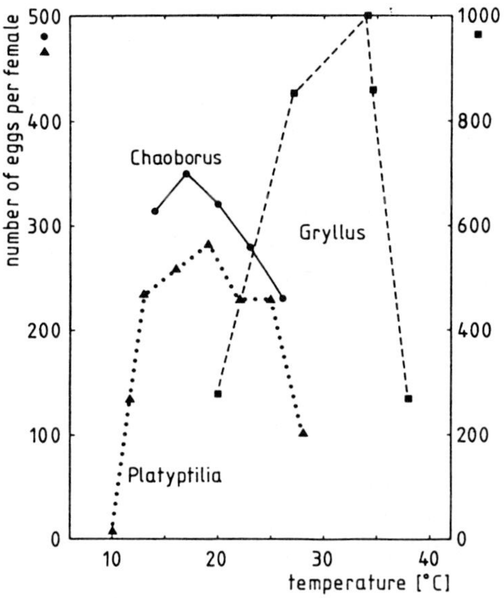

Fig. 2.5. Number of eggs per females as affected by constant temperature; *circles: Chaoborus crystallinus* (original, Ratte); *triangles: Platyptilia carduidactyla* (after Bari and Lange 1980); *squares: Gryllus bimaculatus* (after Hoffmann 1974)

19°C (Fig. 2.5), whereas the Mediterranean field cricket, *Gryllus bimaculatus*, lays eggs from 20° to 38°C (maximum number at 34°C). Additionally, close to the boundaries of the favorable range the total egg production of the entire experimental population may be not only affected by the decreased number of eggs per female, but also by a loss in fertility of an increasing percentage of females (Archer et al. 1980).

Fluctuating temperatures. Some characteristics of postmetabolous egg production resemble those of development itself. In other words, both the favorable range for egg production may be expanded and critical high temperatures during the warmer period may cause a drastic decrease in egg production. The first is exemplified by the spotted alfalfa aphid, *Therioaphis maculata*, which under fluctuating temperatures produces offspring between 8° and 32°C, whereas at equivalent constant temperatures eggs are only laid between 15° and 30°C (Fig. 2.6). Critical high temperatures decrease fecundity in the bug, *Lygus hesperus*, and in the fruit fly, *Drosophila melanogaster* (Strong and Sheldahl 1970; Siddiqui and Barlow 1972).

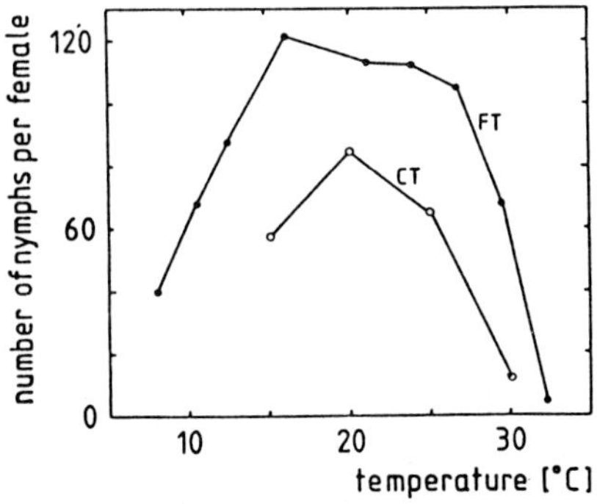

Fig. 2.6. The relationship between mean temperature and total fecundity of *Therioaphis maculata* (mean number of nymphs per female); *open circles*: fecundity occurring under constant temperature; *black circles:* fecundity under alternating temperatures. (After Messenger 1964)

Table 2.2. Effects of thermoperiod on fecundity (total number of eggs or of offspring)[a]

Species:	Php L:D	Thp °C/°C	hw:hc	eCT °C	P	Fecundity R(eCT)	R(FT)	Diff. %	Reference/Remarks
Coleoptera									
Tribolium castaneum (Herbst)	16: 8	30/20	12:12	25.0	s	97	124	27.8a	Hagstrum and Leach 1973 (total fecundity; a: not significant
Trogoderma inclusum (Le Conte)	16: 8	30/20	12:12	25.0	s	102	103	0	
Sitophilus oryzae (L.)	16: 8	30/20	12:12	25.0	s	80	77	0	
Diptera									
Drosophila melanogaster (Meigen)	16: 8	22.5/17.5	12:12	20.0	r	22791	31805	39.6	Siddiqui and Barlow 1972 (total fecundity)
		22/15		20.0	r	22791	30933	35.7	
		27.5/12.5		20.0	r	22791	15497	- 32.0	
		25/20		22.5	r	31342	36074	15.1	
		27.5/17.5		22.5	r	31342	18157	- 42.1	
		26/21		23.5	r	33000	33996	3.0	
		27.5/22.5		25.0	r	33922	25400	- 25.1	
Lepidoptera									
Anagasta kuehniella (Zeller)	16: 8	22.5/17.5	12:12	20.0	r	1758	1845	4.9	Siddiqui and Barlow 1973 (mean fecundity of 5 females)
		25/15		20.0	r	1758	1677	- 4.6	
		27.5/12.5		20.0	r	1758	960	- 45.0	
		25/20		22.5	r	1733	2032	17.3	
		27.5/17.5		22.5	r	1733	1679	- 3.1	
		27.5/22.5		25.0	r	1698	1594	- 6.1	
Pectinophora gossypiella (Saunders)	12:12	32.8/14.4	12:12	23.9	r	143	104	- 27.2	Philipp and Watson 1971 (total fecundity)
	13:11	36.7/19.4	13:11	28.3	r	159	108	- 32.1	
	14:10	36.7/27.8	14:11	32.1	r	21	163	776.0	

Table 2.2 (cont.)

	15: 9	22/14	12:12	18.0	r	0	31	-	Welbers 1975a
		26/18		22.0	r	87	120	37.9	(total fecundity)
		30/22		26.0	r	132	272	206.1	
		34/26		30.0	r	102	196	192.2	
		22/30		26.0	r	132	171	29.5	
Heteroptera									
Acyrtosiphon pisum	16:8	17.5/12.5	12:12	15.0	r	462	372	- 29.4	Siddiqui et al. 1973
(Harris)		20/10		15.0	r	462	384	- 26.8	
		27.5/12.5		15.0	r	462	396	- 24.2	(average fecundity of
		25/ 5		15.0	r	462	439	- 5.0	5 aphids)
		20/15		17.5	r	425	432	17.4	
		22.5/12.5		17.5	r	425	419	- 1.3	
		25/10		17.5	r	425	401	- 5.5	
		25/15		20.0	r	419	402	- 3.9	
		25/20		22.5	r	312	413	32.0	
		30/20		25.0	r	318	314	- 1.0	
		35/25							
Therioaphis maculata	12:12	28/20	12:12	24.0	r	16.5	35.0	112.0	Kindler and Staples 1970
(Buckton)		26/18		22.0	r	24.3	92.9	282.0	
		24/16		20.0	r	35.3	61.8	75.0	(mean number after 4
		19/11		15.0	r	26.5	25.9	- 2.3	days/5 apterous fem.)
		14/ 6		10.0	r	14.7	7.8	- 46.9	
	12:12	13.5/2.5	12:12	8.0	s	0	40	-	Messenger 1964
		16/ 5		10.5	s	0	68	-	
		18/ 7		12.5	s	0	88	-	(mean progeny per female)
		21.6/10.6		16.1	s	70	121	72.9	
		26.6/15.6		21.1	s	80	113	41.3	
		32.4/15.4		23.9	s	70	112	60.0	
		32.2/21.2		23.7	s	45	105	133.0	
		37.9/20.9		29.4	s	14	68	133.0	
		37.7/26.7		32.2	s	-	5	-	

Table 2.2 (cont.)

Species:	Php	Thp		eCT	P	Fecundity		Diff.	Reference/Remarks
	L:D	°C/°C	hw:hc	°C		R(eCT)	R(FT)	%	
Lygus hesperus	16: 8	35/12.8	12:12	23.9	s	150	242	61.0	Strong and Sheldahl 1970
(Knight)		37.7/12.8		25.3	s	165	30	- 82.0	(total fecundity)

[a] *Php* photoperiod; *Thp* thermoperiod range (min./max. temp); *hw, hc* hours at higher and lower temperature, respectively; *FT* results from fluctuating temperature; *eCT* results from equivalent constant temperature; *Diff %* percentage deviations from eCT; for further abbrevations see Table 2.1

In a number of "normal" thermoperiods around medium or lower temperatures, a tremendous increase in egg number or offspring per female has been frequently observed (Table 2.2). As compared with results from equivalent constant temperatures the increase may be as high as seven fold as reported for the pink bollworm (Philipp and Watson 1971; Table 2.2) and the Pitcher-plant mosquito, *Wyeomyia smithii* (Bradshaw 1980; not listed in Table 2.2). In contrast, thermoperiods may also evoke lower average fecundity (*Anagasta kuehniella; Acyrtosiphon pisum;* Table 2.2). The eggs laid under thermoperiodic conditions are smaller in *Gryllus bimaculatus* (Hoffmann 1974), whereas in the pink bollworm, no effects on egg size have been detected (Welbers 1975a).

In addition, egg number per female can depend on the phase relationship between photo- and thermoperiod and on the frequency of the temperature oscillations. When maintained at a fluctuating temperature around 26°C with the warmer period coinciding with the light (Table 2.2) females of the pink bollworm, *Pectinophora gossypiella,* lay 272 eggs on average; however, when living under the inverse combination of thermo- and photoperiod (warm period at dark time) females produce only 171 eggs (comparable constant temperature: 132 eggs per female). Also in the planktonic midge *Chaoborus crystallinus* females produce different egg numbers when reared under different phase relationships of photo- and thermoperiod (Ratte 1979). Under rapidly oscillating temperatures around 20°C of 2:2, 7:7, 8:8 h, crickets of *Gryllus bimaculatus* lay about twice the number of eggs than under equivalent diurnally fluctuating temperatures, and about four times the number than at a constant temperature of 20°C (Behrens et al. 1983). The pink bollworm, *Pectinophora gossypiella,* shows a higher oviposition rate during a longer adult survival period (Welbers 1975a).

4 Current Concepts of Temperature Action

The current accepted concepts are based on constant temperature results and mainly intend to explain the dependence of developmental rate on temperature. Other aspects of development, such as body size and fecundity, are not included.

The *rule of thermal summing,* in spite of both earlier and recent criticism (Huffaker 1944; Logan et al. 1976; Hilbert and Logan 1983), is widely employed in insect pest control and has acted as a stimulus for many laboratory experiments designed to determine thermal constants for many insect species. From the suggested thermal constant for any developmental instar of a given species the following predictions can be made: within the favorable range the temperature sum for the complete development should be the same and should be independent regardless of whether the temperature fluctuates around a lower or higher value. In other words, with a daily temperature cycle or irregularly fluctuating temperatures the time of development should exactly equal that under the equivalent constant temperature exhibiting the same temperature sum.

In this manner, the time of outbreak of insect pests can be predicted by temperature summing [day-degree summing, hour-degree summing; according to Eq. (3)]: if temperature rises above the theoretical threshold (T_0) and, therefore, begins to become "effective" one calculates the mean daily temperature, T(x) from an air temperature record obtained in the endangered area and adds this to the already accumulated sum of temperatures, TS(t) (for methods see Allen 1976).

$$TS(t) = \int_{t_o}^{t} T(x)\, dx \tag{3}$$

where:

t : time

t_o : time when the daily mean temperature rises above the threshold

T(x) : temperature at time period x

TS(t) : accumulated temperature sum at time t;
TS(t) = TC: condition for terminating development.

By means of the temperature forecast the day of outbreak can be easily calculated from the thermal constant (TC) obtained from rearing experiments and the already accumulated temperature sum TS(t). The applicability of day-degree models (also including the establishment of linear rate vs temperature relationships) has been shown repeatedly for a number of insects (recently again by Zalucki 1982; Engle and Barnes 1983; Ring et al. 1983). In other cases, however, field development does not agree satisfactorily with that predicted despite a linear relationship between developmental rate and temperature (Mishima 1976; Rock and Shaffer 1983).

Any discrepancy between results from fluctuating temperatures and corresponding constant temperatures can not be explained by the rule of thermal summing. Hence, it was suggested many years ago that this discrepancy could be due to the non-linearity of developmental rate vs temperature relationship (Kaufmann 1932; Ludwig and Cable 1983). For these cases temperature summing was thereafter replaced by the *concept of developmental rate summing* introduced by Kaufmann and based on the Janisch catenary function. Subsequently, rate summing was also employed with other nonlinear rate functions (e.g., Eubank et al. 1973).

Developmental rate vs temperature curves resemble those of metabolic rate or chemical reaction rates (see Sec. 2 in Chap. 1). This has led to the suggestion that development is a rather complex set of chemical reactions, the rate of which is determined by one "master reaction" or one rate-determining enzymatic process (Sharpe and de Michele 1977). Normal chemical reactions lead to the production of specific compounds, whereas the developmental reaction is assumed to accumulate "development" as a product. The accumulation speed depends on temperature and should show temperature coefficients similar to chemical reactions; indeed, this is often observed (Q_{10}, s of 2–3). In consequence of the chemical analogy, it is suggested that developmental rates behave additively. This means that under fluctuating temperatures the developmental rate is the same as under constant temperatures and, thus, results from summing rate proportions. Therefore, the developmental rate can also be predicted by simply adding rate proportions as derived from the constant temperature relationship (after Eq. 4): depending on the extent of temperature change, one has to sum at either daily, hourly, or minutely intervals the rate proportions, p(x), the value of which depends exactly on the temperature of the given time period, T(x), according to the relationship established from constant temperature results.

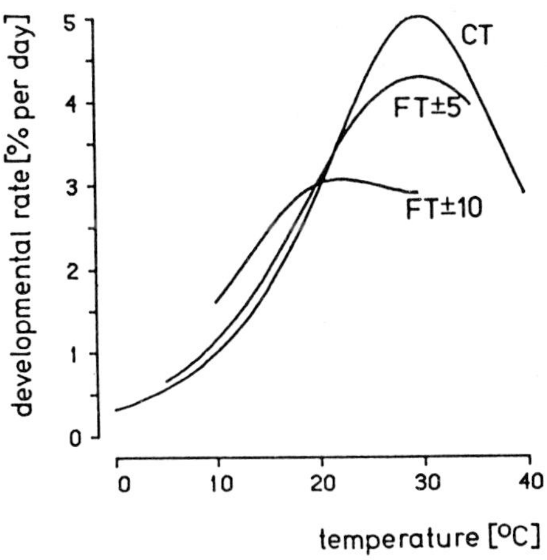

Fig. 2.7. Predicted curves for the developmental rate under rectangular fluctuating temperatures of ±5°C (FT ±5) and of ±10°C (FT ±10); the curves are calculated according to the developmental rate summing concept (Eq. 4) using the rate relation to constant temperature (CT); curve CT is a catenary function (Eq. 1) with the following coefficients: T_{opt} 30°C; t_{min} 20 days; a = 1.1214; percentage deviations are shown in Table 2.3

$$P(t) = \int_{t_o}^{t} p(x)\, dx \tag{4}$$

where:

t : time

t_o : starting time

p(x): rate proportion at time x according temperature T(x)

P(t) : accumulated development at time t.

The development is predicted to be completed, if P(t) reaches the value 1 (or 100%). Linear rate relations lead to the same expectations as predicted by the rule of thermal summing: identical rates under both constant and fluctuating temperatures. In contrast, nonlinear relations produce rates which may deviate from the constant temperature rate. To illustrate this the deviations from a developmental rate relation of the catenary type established for a "theoretical insect" have been calculated (Fig. 2.7). The insect was assumed to develop from egg to adult over a 20 day period at 30°C and over a 100 day period at 10°C (temperature coefficient, Q_{10} = 2.5). (Although only a theoretical insect, both this relationship of developmental rate as well as the following general considerations and conclusions may be regarded as representative of many real insect species, because it principally approximates the relationships found for many insects. Moreover, these considerations are also valid for chemical and metabolic rates as well as for growth rates should all be nonlinear under constant temperatures.)

The insect under consideration develops at constant temperatures according to curve CT (Fig. 2.7), under a rectangularly alternating temperature regime (amplitude 5°C) according to curve FT±5 and with amplitudes of 10°C according to curve FT±10, thus, showing in part considerable deviations from the rate found under corresponding constant temperatures. The percentage deviations resulting from these two thermoperiods are given in Table 2.3 together with values for two analogous sinusoidal temperature regimes. Additional values for two rate relations of different

Table 2.3. Percentage deviations of developmental rate from those under constant temperature (CT) as produced by different fluctuating temperatures of rectangular or sinusoidal shape (ranges of 10° and 20° C)[a]

Temperature range	Sinusoidal		Rectangular		Sinusoidal		Rectangular		Sinusoidal		Rectangular	
Q_{10}(CT)	2.0		2.0		2.5		2.5		3.0		3.0	
Amplitude °C	±5	±10	±5	±10	±5	±10	±5	±10	±5	±10	±5	±10
Mean temperature °C												
10	4.7	18.0	9.4	35.2	7.6	30.3	15.2	60.4	10.8	44.6	21.6	90.3
15	3.2	10.6	6.3	19.3	6.0	21.2	11.9	39.6	9.3	34.3	18.5	65.3
20	0.3	- 1.0	0.5	- 3.8	2.2	3.9	4.2	3.9	4.8	10.9	9.2	15.0
25	-3.4	-12.8	- 6.8	-24.8	-3.9	-15.2	- 7.8	-29.7	- 4.0	-16.0	- 8.1	-33.5
30	-5.3	-17.8	-10.4	-32.9	-7.5	-23.5	-14.4	-42.3	- 9.7	-28.7	-18.6	-50.4

[a] Values were calculated for three catenary relations exhibiting different temperature coefficients (2.0, 2.5, and 3.0; see also Fig. 2.7).

temperature coefficients (2.0 and 3.0, Table 2.3) show that the extent of deviation depends not only on the temperature program (shape, amplitude), but also on the temperature coefficient of the rate relation. The deviations follow the same general rules as those established also empirically (Sec. 3.1):

1. temperature changes around the inflexion point of the curve (linear range) cause neither a considerable retardation nor an acceleration in development;
2. temperature changes occurring to a greater part below the inflexion point (exponential range) cause more rapid development than the comparable constant temperature;
3. temperature changes above the inflexion point (optimum range) have a retarding effect.

In order to test the validity of the rate summing concept, we shall again consider Table 2.1 and compare the observed developmental times under fluctuating temperatures with the expected times calculated by rate summing (Eq. 4). Since many authors have neglected this, the developmental times expected for fluctuating temperatures have been calculated by using the original or interpolated constant temperature results. However, not all values necessary for these calculations could be drawn from the papers. In these cases, the relative deviations were estimated according to Table 2.3 and compared with the reported real deviations from constant temperature results (these are marked by an a in the DRS-column of Table 2.1).

A survey of Table 2.1 reveals that only in some cases do both the observed and expected deviations agree exactly. A certain amount of variation is probably due to the constant temperature results, these themselves exhibiting some variation. Although according to Connolly (1981), this probably does not cause very large errors in the predictions, the findings have been classified in agreement with the development rate summing concept only then, when they deviated less than absolutely 15% from the expectation (marked with + in the DRS-column of Table 2.1). According to this classification, the developmental rate summing concept might explain the results of 34 from 59 species (58%). Some cases (marked with b, Table 2.1) show agreement only because some development was also assumed to take place below the developmental threshold; this has also been proposed by some other workers conducting experiments with single temperature changes or with non-diurnal thermoperiods (Ludwig 1928; Ludwig and Cable 1933; Hodson and Alrawy 1958). In 25 species the developmental time of at least one instar showed more than 15% deviation from the expected value, although in some cases a part of the deviation indeed might be explained by the rate summing effect. In addition, field observations indicate a limited validity of the concept. Nonlinear relationships together with the developmental rate summing method have been employed in pest control with different degrees of success (Mellors and Bassow 1983; Moon 1983; Rock and Shaffer 1983).

Contrary to the rate summing concept, developmental rates depend on the phase relationship between photo- and thermoperiod, as it has been observed in the aquatic midge, *Chaoborus crystallinus* (Ratte 1979, 1983). According to the developmental rate summing concept this should not play any role. In a strict sense, also all findings of temporal programming of the adult emergence conflict with the concept. This concept is not able to explain how the developmental rate can be adjusted in such a manner that eclosion takes place only at distinct times of day, as is found in many insects (for references see Saunders 1982 and Sec. 3.3 in Chap. 1), or takes place only at distinct combinations of tide and day-time, as in the intertidal midge, *Clunio marinus* (Neumann 1976).

Results that cannot be accounted for by the concept as well as the increased fecundity are regarded as "true adaptations" of development to fluctuating temperatures. Such adaptations have been explained as follows (see also Hoffmann 1980):

1. An energetic advantage might result from a diurnal periodicity of activity and rest (especially of feeding and nonfeeding). At rest, the metabolism during the period of lower temperature requires comparably little energy. Therefore, if the food needs are satisfied completely during the activity period at higher temperature, the total respiration costs should be lower under fluctuating than under corresponding constant temperatures. This would lead to an energetic advantage which might be used up in producing additional body- or egg-substance (McLaren 1963). However, as far as I can see, as yet this postulate has not been confirmed by investigating experimentally the feeding activities of an insect together with the energy budget.
2. A temperature compensation for the metabolic rate (see Chap. 1) leading to higher metabolic rates and subsequent faster growth and shorter development as well as increased reproduction, could not be substantiated.
3. In order to explain the expanded favorable range for development or reproduction towards lower temperatures as observed under fluctuating temperatures, Laudien (1973) proposed the existence of a factor inhibiting development; this factor might then be destroyed by a short-term temperature increase above the critical threshold.
4. The neuro-endocrine system might be stimulated by thermoperiods leading to modified hormonal concentrations and subsequent stimulation of reproduction. Support for this comes from increased concentrations of ecdysteroids and juvenile hormones which correlate with the increased egg production in *Gryllus bimaculatus* under fluctuating temperatures (Hoffmann et al. 1981; Behrens et al. 1980; Behrens and Hoffmann 1983; Koch and Hoffmann 1984).
5. Metabolic reactions and/or developmental (probably neuro-endocrine) control reactions might be diurnally organized. In the aquatic midge, *Chaoborus crystallinus,* the diapause induction as well as the average weight gain per day depends on the temperature during different time periods of the day. This evokes developmental times, body weights, and diapause proportions showing a clear dependence on the phase relationship between photo- and thermoperiod (Ratte 1979, 1983).
6. Thermoperiods might act like photoperiods. Under conditions of constant illumination thermoperiods may act as Zeitgeber like photoperiods. In the parasitic wasp, *Nasonia vitripennis,* as well as in other species, in constant darkness thermoperiods cause an analogous diapause response as do photoperiods of the same length (for references see Saunders 1982). With constant light and temperature cycles around 20°C with warm periods of 8, 12, and 16 h/day *Chaoborus crystallinus* shows similar developmental times and body weights as at a constant tem-

perature of 20°C with photoperiods of 8, 12, and 16 h (Ratte 1979). It may be possible that in some experiments conducted under constant illumination thermoperiods caused the same effects as photoperiods.

The hypotheses (4) to (6) suggest the involvement of the neuro-endocrine system by which the effects might be brought about. At first sight, the partly applicable developmental rate summing concept and the additional hypotheses hardly seem to be able to be brought into line. How may a coherent developmental mechanism, which has to be postulated to be common to all insects, evoke such different results? In my opinion the variety of results can be explained and the conflicts overcome by introducing an improved concept of temperature action which distinguishes two different, but simultaneously acting, temperature effects on both mass processes and reactions of differentiation.

5 Towards an Improved Concept of Temperature Action

In order to explain the results and to name the different temperature effects it is necessary initially to distinguish strictly between two aspects of development: "growth", the quantitative aspect and "differentiation", the qualitative aspect, as already claimed by Wigglesworth (1959, 1964), Needham (1964), and Clarke (1967). Perhaps because some authors did not discriminate between the two aspects, then, as now, these are termed differently and used with some confusion (I am here following the terms of Needham, namely, that development = growth + differentiation). As exemplified in Fig. 2.8 we may consider solely the change of qualitative states with time, such as egg, larva, pupa, or imago; the duration of each instar may be measured by the time of development (Fig. 2.8a). On the other hand, we may determine simultaneously the mass characteristics of the process while measuring the size at different times and, thus, obtaining the growth curve (Fig. 2.8b). The measurable variables of both developmental aspects are defined in Fig. 2.8.

It is obvious from Fig. 2.8a that in a strict sense the developmental rate is a rate of differentiation (or rate of metamorphosis) and does not include any statement about growth or size of an insect. Insects that develop over exactly the same period may be of a quite different adult size. Nevertheless, it will be demonstrated in this section that the developmental rate may depend on both the growth rate and a neuro-endocrine process which sets a critical size for metamorphosis.

Also in gonadal development and egg production the quantitative and qualitative aspects are to be distinguished: gonads and egg masses grow in size with a typical velocity, whereas the number of eggs produced and egg size result from differentiation (egg maturation), which divides the total egg mass into single eggs of distinct size.

In previous sections evidence has been presented for both temperature effects on developmental time and on body size as well as on egg number and egg size, indicating that both processes, growth and differentiation, are targets of temperature action. The following sections intend to explain how temperature exerts its influence on the two processes. In doing this, we shall initially consider some fundamental

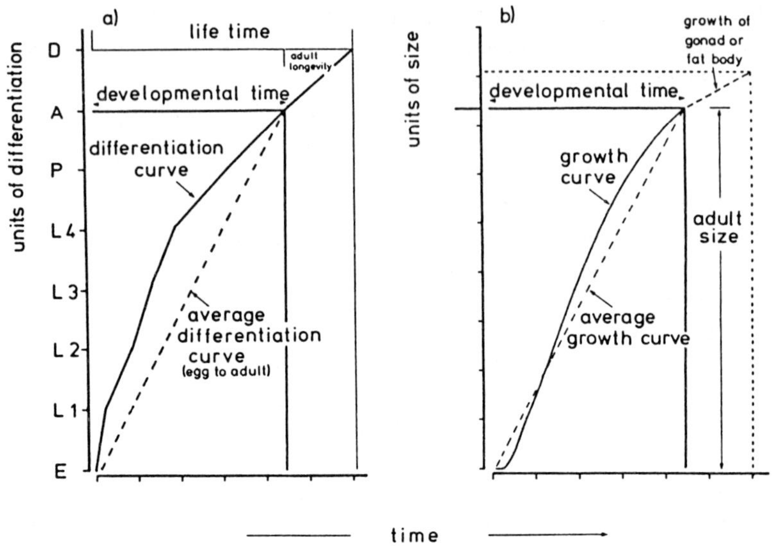

Fig. 2.8. **a** Qualitative aspect of development: during the time insects develop through different instars, such as egg (*E*), larva 1-larva 4 (*L1-L4*), pupa (*P*) and adult (*A; D*: death); variables of the qualitative aspect are time of development and rate of development represented by the slope of the average differentiation curve (which is obtained by setting "development to a given instar" equal to 1 (or 100%); **b** quantitative aspect of development: during the developmental time insects grow and emerge thereafter at a distinct species specific adult size; afterwards growth may take place during gonadal development and egg production; variables of this aspect are "adult size" as well as growth rate (true size increment per time unit) or "average size gain per time unit" (size devided by developmental time; slope of the average growth curve)

characteristics of growth, metamorphosis, and gonadal maturation. Thereafter, the "concept of dual temperature action" will explain some phenomena while some examples given in previous sections will be reconsidered.

5.1 Some Fundamentals of Growth, Metamorphosis, and Reproduction

Growth in insects is similar to other poikilotherms. Some deviations from the ideal course (presented in Fig. 2.8b) may occur because of losing weight or taking up water during molting. Depending on the timing of metamorphosis, holo- and hemimetabolous insects terminate the growth of their body by undergoing metamorphosis, as opposed to ametabolous insects and other nonarthropod poikilotherms. Therefore, in insects growth curves are very similar during the initial phase, but rather different during the final larval time (for examples see Bertalanffy 1957; Joosse and Veltcamp 1970; Roe et al. 1982). Provided that there is a satisfactory supply of material and energy (food, yolk, oxygen, etc.) during normal development the growth rate is maintained by the presence of the molting hormone (MH, also called ecdysone), which is also responsible for the moltings accompanying the enlargement of body. The action of MH seems to be more triggering than accelerating (Wigglesworth 1964). Depending on critical environmental conditions (e.g., temperature, photoperiod, food), growth may be slowed down to a minimum rate or even stopped, such as in diapause development (see Chap. 3).

From this, we have to expect two different types of temperature action on growth; first, an indirect effect mediated by the endocrine system (triggering) and second, a direct action of temperature on metabolic reactions and, therefore, growth rate, the velocity of which obeys indispensably the chemical mass law and which thus must behave strictly additively. Therefore, growth rates under fluctuating temperatures are to be explained by rate summing and can, thus, be predicted in the same way as performed for developmental rates (Sec. 4).

The *metamorphosis* of insects is mediated by endocrine mechanisms which have been investigated over the past decades with increasing intensity. Although there are as yet some uncertainties with regard to detail, precise concepts have evolved concerning the role of the brain, the function of the endocrine glands, such as the corpora allata and the prothoracic gland, and the action of the hormonal messengers, such as the juvenile hormone (JH) and the molting hormone (MH) (for reviews see Willis 1974; Steele 1976; Jungreis 1979; Chapman 1982). (Of course, the synthesis and degradation of hormones proceed via normal chemical reactions, but as opposed to normal growth reactions, hormones evoke quite another kind of effect while changing qualitative characteristics, such as the realization of the adult form during metamorphosis.)

With respect to the explanation of temperature effects on developmental rate, we have to ask how metamorphosis may be timed and how it is affected by temperature. Figure 2.8a offers two possibilities: (1) a critical size might be set which is to be achieved by the growing insect or (2) a critical time might be set independent of growth. The possibility I only wish to consider is that a critical size might be set for the onset of metamorphosis, because this is supported by recent findings. The other possibility, namely, that a critical time for metamorphosis might be set, has until now not been experimentally confirmed and would appear to be unlikely. If a critical size is set, the developmental rate depends on two different processes: the first sets the critical size, which then may be achieved by the second, the growth process. Newer findings in fact suggest that in some insects, if not in all, the timing of metamorphosis depends on attainment of a critical size before metamorphosis can take place (Nijhout 1979; 1981; Williams 1980; Woodring 1983). The critical size is also supported by experiments with starved larvae or with larvae that have been fed with diets of different quality. Here the developmental time was prolonged, whereas the final size or weight was little or not affected as compared with controls (Ratte 1979; Jones et al. 1981; Roe et al. 1982).

Classical experiments indicate that the critical size itself is set by the neuroendocrine system (Wigglesworth 1959; Ciemior et al. 1979; Safranek et al. 1980). By ectomization of corpora allata or by implantation of additional corpora allata, as well as by increasing the titer of JH, MH, or their analogs, one can make insects, say, "leave their growth curve" earlier or later than would be the case in normal development (Fig. 2.9a); this results in faster developing dwarfs and slower developing giants, respectively. In other words, the critical size for metamorphosis can be set higher or lower by changing the relative concentrations of JH and MH. The

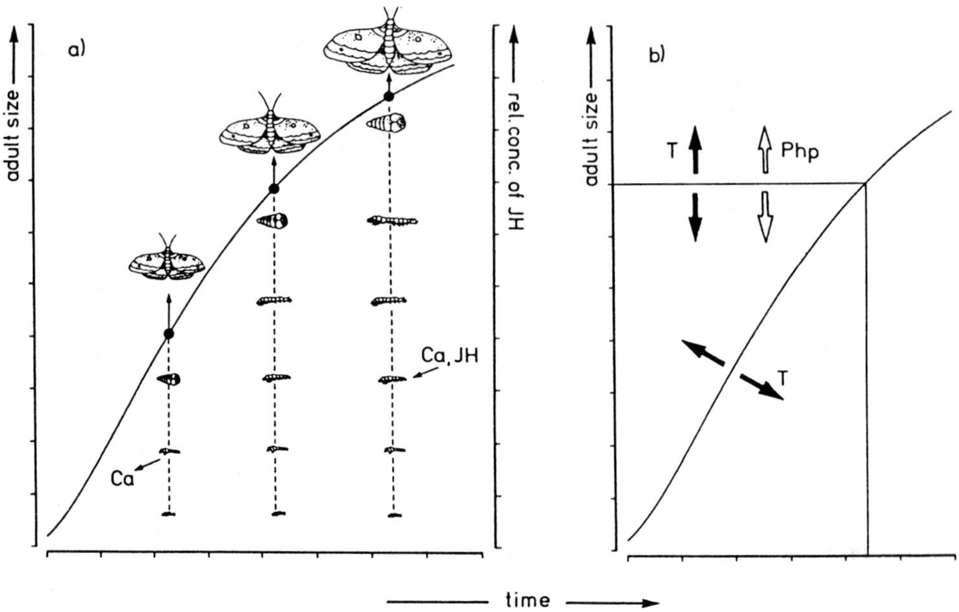

Fig. 2.9. a Principal changes of adult size and developmental time observed as a consequence of manipulating hormonal levels as well as of implantation and ectomization of corpora allata; **b** similar changes of developmental time and size as caused by temperature *(T; black arrow)* or by photoperiod *(Php; white arrow); Ca:* corpora allata; *JH:* juvenile hormone; for further explanations see text

growth curves suggest that the effects on size are not evoked by a slower or faster growth rate, but are in fact due to the modified hormonal conditions which corresponds to a modified critical size (Ciemior et al. 1979). This points to a considerably independent control of growth and metamorphosis and shows also that both growth and differentiation cannot only be distinguished from one another logically, but also experimentally.

The normal hormonal action, however, can only be maintained if there is at least a minimum supply of food. This is concluded from experiments where insects have been totally deprived of food; they emerged as undersized and sometimes sterile adults indicating that the "normal" hormonal action leading to "normal" size and "normal" ovarian development was affected by starvation [Fraenkel 1976; Weaver and Pratt 1981; Cymborowski et al. 1982; Khan et al. 1982b (see also Sec. 3 in Chap. 4)].

Gonadal growth and differentiation of eggs show close similarity to larval growth and differentiation. Not only do the growth curves of gonads and egg mass closely resemble those of larvae (e.g., Cohet and David 1978; Wagner et al. 1981; Porcheron et al. 1982), but also the same hormonal compounds are involved in controlling ovarian growth and maturation (for references see Engelmann 1970, 1979; Koeppe et al. 1980; Hoffmann et al. 1981; de Kort and Granger 1981; Redfern 1982). Both ovariole number and egg size are modified by manipulating JH concentrations (Deb and Chakravorty 1980). Repeated injections of ecdysteroids increase the ovarian weight and stimulate oviposition in *Gryllus bimaculatus* (Behrens and Hoffmann 1983). Similar effects have been shown by repeated applications of JH III onto female crickets (Koch and Hoffmann 1984).

Summarizing, the adult and egg size as well as ovariole number have been found to be under control of the endocrine system; changes in size or egg number are evoked by manipulation of hormonal levels.

5.2 Dual Temperature Action on Development and Reproduction, a Unifying Concept

It is remarkable that quite analogous changes in adult size as well as in egg size are caused by environmental factors, such as temperature, photoperiod, and thermoperiod. This strongly indicates that this kind of temperature effect is mediated by the neuro-endocrine system as shown by the following examples:

1. the adult weight of *Chaoborus crystallinus* depends on both different photoperiods at 20°C and several thermoperiods around 20°C (Ratte 1979). Typically, smaller adults emerge after shorter developmental times than larger ones. From an approximately equal average weight gain per day it was concluded that the growth rate was not markedly affected and the effects caused by a modified critical size. These findings, therefore, must be regarded as the result of the influence mediated by the neuro-endocrine system.
2. The Mediterranean field cricket, *Gryllus bimaculatus*, lays smaller eggs under thermoperiodic conditions than under corresponding constant temperatures (Hoffmann 1974). The tremendously increased egg number can be accounted for as being the result of a larger egg mass (caused by growth rate summing) as well as of a simultaneous reduction in egg size (effect on differentiation).

In the same species and also in others, direct support for a temperature sensitive endocrine mechanism comes from findings that histological changes and different hormonal titers are evoked by temperature change (Clarke 1966, 1967; Peacock et al. 1977; Hoffmann et al. 1981). As yet, however, the true nature of this temperature sensitivity remains unclear; but at least it must lead to modified concentrations of MH and JH.

The two different but simultaneously acting temperature effects are illustrated in Fig. 2.9b. Temperature increase speeds the growth rate, but may cause either larger or smaller adults. In some species an additional effect on the critical size is expected from photoperiodic action. From the rather different relations of size to temperature (see Sec. 3.3) one can conclude that the temperature sensitivity of the neuro-endocrine system is not uniform and seems to be species specific. In contrast to this, temperature effects on the growth rate are rather similar not only in insects but also in all poikilotherms (Needham 1964). Although there are differences in the absolute growth rate, the relationship between growth rate and temperature approximately follows a catenary function (Needham 1964; Hanski 1976). Therefore, the considerations made for the developmental rate (Sec. 4) are valid also for the growth rate including the same predictions as depicted in Fig. 2.7 and Table 2.3. The additive nature of growth rates has been demonstrated in larvae of the blowfly *Lucilia illustris* (Hanski 1977). The same was reported by Hoffmann (1974) for growth processes related to gonadal development and egg production: during the oviposition period, the crickets of *Gryllus bimaculatus* accumulate organic substance according to the rate summing concept when living under thermoperiodic conditions.

Summarizing, there is clear evidence to show that temperature does not affect the growth process alone, but also the control of metamorphosis and reproduction [whereas photoperiod seems to influence only the latter and not the growth rate (Needham 1964; Ratte 1979)]. The effects are mediated by the neuro-endocrine system and are recognized by the temperature-dependent sizes of adults and eggs.

5.3 Evidence for the Concept of Dual Temperature Action

Favorable range; sub-threshold development. In Sec. 3 evidence was presented that in several insects some development takes place below the developmental threshold. The present concept explains this by the combination of a relatively high critical size (or noncritical hormonal concentrations) and a simultaneous slow growth rate which hinders the growing insect in attaining its critical size over a definite experimental period. However, some enlargement of the body takes place, so that after subsequent transfer to a higher favorable temperature the critical size for metamorphosis can be reached earlier. In contrast, the threshold concept concerning the developmental rate in a strict sense postulates a sub-threshold rate of zero which leads to errors in predicting the developmental time according to the developmental rate summing concept. Correct predictions can be expected only when growth is fully stopped beyond a distinct growth threshold; this must not necessarily be identical with the theoretical zero point of development.

Varying temperature coefficients of the developmental rate. The consequences of the dual temperature action on both the adult size and the growth rate together with its effect on developmental rate is easily demonstrated in the cases of the three insects already presented (Fig. 2.3). In order to do this, the size of the adults or of the pupae was plotted against the developmental time (Fig. 2.10). The temperature de-

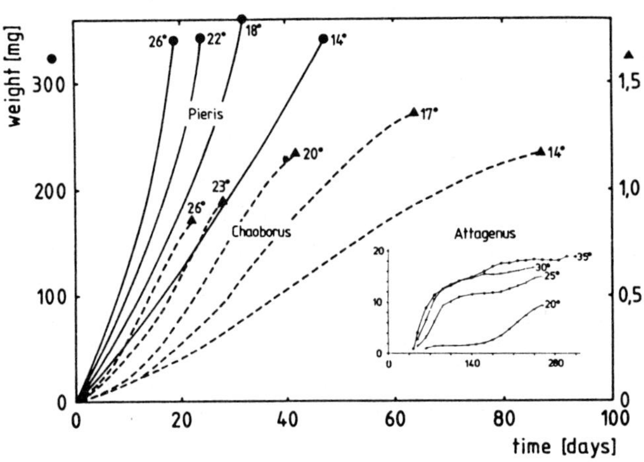

Fig. 2.10. Body weight plotted against developmental time at different constant temperatures, thus, explaining the varying temperature coefficients of the three insects of Fig. 2.3; in *Pieris* and *Chaoborus* growth curves are arbitrarily chosen, in *Attagenus* they have been redrawn from Baker (1983)

pendent developmental time of the cabbage butterfly, *Pieris brassicae*, maintaining a nearly constant body weight at different temperatures, exhibits a Q_{10} of 2.1; this can be perceived solely as growth rate effect because the weight shows only slight temperature dependence. In contrast, the adult size of *Chaoborus crystallinus* increases linearly with temperature between 17° and 26°C (Fig. 2.3). The relatively high Q_{10} of 3.6 is due to a decrease in the final size and an increase in the growth rate with increasing temperature. In the same species, the Q_{10} of the average weight gain per day approximating the growth rate is only 2.1, a value similar to that of *Pieris*. The carpet beetle, *Attagenus megatoma*, develops nearly independently of temperature over a wide range (Fig. 2.3). Growth curves presented by Baker (1983; included in Fig. 2.10) and size data explain this temperature independence as being caused by an interplay of different growth rates and final sizes.

By means of these examples, it has been clearly demonstrated that insects are indeed able to accelerate or to retard their development. This is achieved *not* by temperature compensation of the growth rate, but by modification of the critical size. Probably the adaptation of developmental time to the lower temperatures of the Arctic also occurs by alteration of the critical size which is comparatively low in many arctic insects (Remmert 1980). Slow-growing insects develop faster if only a small adult size is to be achieved.

Time of development under fluctuating temperature. Controlling developmental time by modification of the critical size and growth rate points to the fact that the developmental rate is not a true rate in a chemical sense as it is in the case of the growth rate. It should only be closely related to the latter and exhibit approximately the same characteristics when environmental factors do not considerably affect the adult size, as was shown for *Pieris*. In such cases summing of developmental rates is equivalent to summing of average growth rates and, thus, can be expected to predict the developmental rate under fluctuating temperatures with adequate success. This, however, will not succeed if the thermoperiod leads to considerably smaller or larger animals than arise under comparable constant temperatures. Unfortunately, tests of this hypothesis are not possible for the many findings that differ from the developmental rate summing concept because measurements of size have rarely been carried out. Evidence for the hypothesis, however, may be drawn from findings in *Chaoborus crystallinus* (Ratte 1979) and *Spodoptera littoralis* (Sidibé and Laugé 1977): conspicuously smaller animals emerge after shorter developmental periods under the influence of thermoperiods than under constant temperatures. Consequently, developmental rate summing here fails to be successful.

If thermoperiodic action leads to an increase in critical size, the effects on developmental time may be different: (1) the growth rate summing effect may not compensate for the increase in critical size which means that under thermoperiods the attainment of the critical size requires a longer time than under corresponding constant temperatures; (2) the growth rate summing effect may compensate for the larger size so that the critical size may be achieved during the same time or even during a shorter time than under comparable constant temperatures. Approximately the same developmental time together with an increase in adult size caused by thermoperiods have been reported for *Pectinophora gossypiella* (Welbers 1975a) and *Ostrinia nubilalis* (Beck 1983a), while Merkel (1977) observed even a shorter developmental time in larger crickets of *Gryllus bimaculatus.* However, in these cases it has not been investigated in detail, if the increased growth rates are in fact to explain solely by a rate summing effect. Other possible explanations of increased growth rates (such as temperature compensation, see Sec. 3.3 of Chap. 1) have not been as yet substantiated by experiments, but would be easily incorporated in the dual concept.

Summarizing, the concept of dual temperature action means that the rate summing method can be applied exclusively to true rates, such as the growth rate. The concept of developmental rate summing seems to be applicable only when the insects maintain an approximately constant size, independent of thermoperiod. In any case, the growth rate under fluctuating temperatures results from growth rate summing; this may lead to earlier or later attainment of the critical size if this rate is not linearly dependent on temperature.

Fluctuating temperature; increased fecundity. Also with gonadal development and egg production we have to consider not only that the egg mass may be increased by a growth rate summing effect, but also that steps of differentiation may be affected by temperature, modifying both the number of ovarioles and the size of single eggs as well as the time at which a distinct proportion of egg mass becomes a fully developed egg. It is plausible that the number of eggs produced per day depends on the accumulation rate of the egg mass, the number of ovarioles, and the final size of the single eggs. The conspicuous increase of fecundity due to an increased oviposition rate under fluctuating temperature has already been explained in a previous section using the example of the cricket *Gryllus bimaculatus.*

Photoperiodic action; thermoperiodic action. In several insects, long days lead to shorter developmental times (for references see Saunders 1982). The developmental time and body size of *Chaoborus crystallinus* living under various photoperiods at 20°C is similar to values obtained under constant light and corresponding thermoperiods around 20°C simulating long days with long periods of high temperature, etc. (Ratte 1979). This, together with analogous findings on diapause induction, is interpreted as being caused by Zeitgeber-effects due to the periodic nature of temperature cycles. Zeitgeber-processes may act on the photoperiodic clock leading to subsequent effects on metamorphosis. These are mediated by the endocrine system (Saunders 1982). The discussed effects are neglected by the developmental rate summing concept, but may be well incorporated in the concept of dual temperature action.

Temporal programming of adult emergence and oviposition. Adult eclosion and oviposition often show a diurnal rhythm, strongly indicating that there are gating processes which are controlled by the photoperiodic clock and mediated by the neuroendocrine system (for references see Saunders 1982). In development, this leads to rhythmical larval molting, puparium formation, and subsequent periodic emergence. In species with a diurnal rhythm, the developmental time is composed of one-day units, whereas in the intertidal chironomid, *Clunio marinus*, the developmental time consists of 15-day units. This clearly indicates that the developmental time of insects can be adjusted very precisely to various environmental situations, phenomena which are completely disregarded by the developmental rate summing concept, but are integrated in the new concept; modelling these phenomena certainly, however, requires an enormous amount of work.

In summary, the concept of dual temperature action is that developmental times, adult sizes, and fecundity are caused by both an indispensable temperature action on growth rates common to all processes of enlargement as well as an additional facultative temperature effect mediated by the neuro-endocrine system which seems very species specific.

6 Conclusions

This chapter has illustrated that in their development, insects may respond to temperature more plastically than it has often been assumed in the past; thus, insects are well equipped to master the problems arising from the "tyranny of thermodynamics". This plasticity often seems to be concealed when insect development is observed only under constant temperatures. Consequently, population growth under fluctuating environmental conditions may be faster or slower than under constant temperature. This was clearly shown by calculating the innate capacity for increase or population growth (e.g., Messenger 1964; Siddiqui and Barlow 1972, 1973).

This chapter has presented a concept of temperature action which in my opinion fulfills the three following conditions:

1. it explains the temperature dependence of both the qualitative and quantitative aspect of development;
2. it presents a perspective for establishing quantitative models of the temperature dependence of development; and
3. it offers an insight into how the development of insects might be adapted to a variety of different thermal environments.

With respect to the latter, it is probable that the neuro-endocrine control of metamorphosis and reproduction rather than growth is the major target of selective forces which are responsible for adapting species to their environment. This is not only obvious from the very different kinds of responses to ambient temperature, but also from the differences in types of development between species, such as from differences in timing of gonadal development (before or after metamorphosis), in relating final size to time of development as well as in temporal programming of the emergence. The precise adjustment of development to the home environment temperatures by use of a neuro-endocrine mechanism does not involve altering the complete growth and metabolism, but solely the modification of only a few (perhaps at least two) hormonal concentrations.

Future work should replace verbal modelling by quantitative models in order to predict the development and fecundity under fluctuating temperatures. According to the concept of dual temperature action this means establishing simultaneously the temperature relations of the growth rate and those of the final size (also those of the growth rate of the egg mass and of the size of single eggs) as well as the interrelation of the two variables.

It is probable that also in other arthropods an analogous concept of temperature action is applicable. The temperature dependence of development seems to be very similar in crustaceans and spiders. This, however, and some of the consideration of this chapter must remain, as yet, hypotheses. I hope to have provided here a stimulus for the verification of these ideas by further detailed experiments.

Acknowledgements. I wish to thank U. Capellmann and V. Fitsch for their technical assistance and Dr. H. Cooper-Schlüter for reading the English manuscript.

Chapter 3

Environmental Aspects of Insect Dormancy

Walter Behrens[1]

Contents

1 Allgemeine Zoologie (Biologie I), Universität Ulm, Oberer Eselsberg, 7900 Ulm, FRG
Present address: TAD Pharmazeutisches Werk GmbH, Heinz-Lohmann-Straße 5, 2190 Cuxhaven, FRG

Environmental Physiology and Biochemistry of Insects (ed. by K.H. Hoffmann)

1 Introduction

Temporary unfavorable environmental conditions, such as extreme temperatures and photoperiods, drought, or short food supply are normal events in nearly all biotopes which have been settled by insects. Only some tropical regions may provide constant favorable living conditions throughout the year.

To survive such unfavorable events insects have developed numerous mechanisms ranging from simply reducing metabolic rates with quick recovery (see Chaps. 1, 4, and 5) to complicated genetically programed diapause development. A nonobligatory developmental delay which is directly referable to the changing environmental conditions is called quiescence, whereas diapause is a phenomenon which may be an obligate or a facultative part of development. Diapause is always initiated by environmental stimuli long before the unfavorable conditions set in. It is characterized by very low metabolic rates and, for the most part, no morphological changes, organ development, or tissue differentiation can be observed during this time. Nevertheless, development is continued during diapause and can be demonstrated at least at the level of neurosecretory active cells in the nervous system (Beck 1980).

A large amount of information on insect diapause, on its induction and termination, has been collected by field studies as well as under laboratory conditions. A detailed classification into oligo-, para-, and eudiapause using the stimuli for the onset of diapause development (and also for the following postdiapause development) has been proposed by Müller (1970). Other classifications have been given by Mansingh (1971) and Thiele (1973). These concepts, however, have been criticized by Tauber and Tauber (1978) and Beck (1980) as unhelpful in describing all the different phenomena and aspects of insect dormancy.

Therefore, in the present paper the term diapause is used comprehensively. One has to take into account, for example, that under environmental conditions unfavorable situations can still persist although diapause development has already been completed. In such a case the status of diapause can be followed by a quiescence without any detectable transition.

Thus, it is difficult to decide – especially from field observations – whether the onset of postdiapause development after a time dormancy is due to diapause termination or to quiescence termination.

In the present paper this special problem will not be further considered. Environmental and to some extent, artificial factors enabling insects to complete their developmental cycles are discussed generally, special regard being given to the physiological and biochemical mechanisms and adaptations.

2 Diapause Stages in Insect Development

As mentioned above diapause enables insects to settle in geographical regions with temporary unfavorable environmental conditions. Thus, diapause is always correlated with markedly decreased metabolic rates. This way to withstand unfavorable conditions is a very economical one and may be considered as one of the outstanding abilities of insects which made them so successful in the evolutionary process.

Diapause may occur in every stage of the developmental cycle. However, within one species it is mostly genetically confined to a single and well-defined period. This period can be variable even among closely related species. Examples are given by Lees (1955): the tortricid butterfly *Cacoecia rosana* enters diapause as an embryo, *Cacoecia fumiferana* as a newly emerged larva. The silkmoth *Antheraea yamamai* has an embryonic diapause, whereas *Antheraea pernyi* diapauses as a pupa. The multivoltine syrphid *Epistrophe baltea* passes a (facultative) imaginal diapause, whereas the (univoltine) species *Epistrophe bifasciata* enters diapause at the end of the last larval instar.

These considerations indicate that diapause has been evolved independently and that the selective value is little affected by the stage in which it occurs. Müller (1970) has characterized the different forms and stages of dormancy as a modern phenomenon in phylogeny in which evolution proceeds rapidly.

2.1 Embryonic Diapause

Diapause within the egg membranes has been often referred to as egg diapause. However, this nomenclature should be avoided because it is misleading. Up to now no insect species is known to lay fertilized eggs in which embryogenesis only starts after a period of diapause. In fact, embryogenesis begins immediately after egg deposition. The earliest embryonic diapause seems to occur in the precleavage, the so-called fusion nucleus stage, e.g., of the crane fly, *Tipula simplex* (Hartmann and Hynes 1980). Others, like the butterfly *Notolophus thyellina* and the cricket *Homeogryllus japonica* diapause after blastoderm formation, while embryos of the silkworm, *Bombyx mori,* and the bug *Leptopterna dolobrata* start diapause slightly later in development when they have reached the so-called dumbbell-shaped form (Lees 1955; Braune 1980) (Fig. 3.1). The moth *Rhopobata naevana* diapauses when the somites have been formed and many Orthoptera when anatrepsis has been completed and the embryo is ready to revolve in the egg (e.g., *Melanoplus differentialis, Locusta migratoria*). Finally, some insect embryos diapause as completely differentiated larvae ready to hatch. This is true in the butterflies *Antheraea yamamai* and *Lymantria dispar,* in *Melanoplus bivittatus* (Orthoptera), in *Timarcha tenebricosa* (Coleoptera), and in *Aedes triseriatus* (Diptera) (Shroyer and Craig 1980).

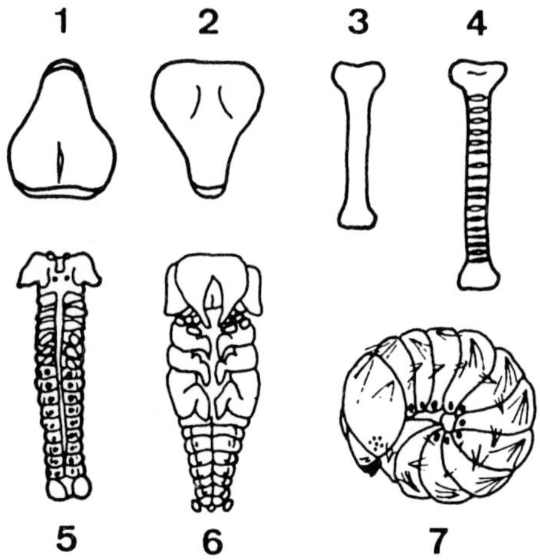

Fig. 3.1. Stages of embryonic diapause. **1** Pyriform embryo – several Lepidoptera; **2** reversed pyriform embryo – several Orthoptera; **3** dumbbell-shaped embryo – e.g., *Bombyx mori* and *Leptopterna dolobrata*; **4** segmented embryo – several Lepidoptera, e.g., *Rhopobata naevana*; **5** segmented embryo with appendages – several Lepidoptera; **6** segmented embryo with extremity buds – many Orthoptera, e.g., *Locusta migratoria;* **7** prelarval embryo – some Lepidoptera, e.g., *Lymantria dispar.* (After Umeya, redrawn from Lees 1955)

2.2 Larval and Pupal Diapause

Larval diapauses can occur at any stage from the newly hatched first instar to the mature nonfeeding phase of the final instar, the so-called prepupal stage. Both, larval and pupal diapauses are characterized by a reduced metabolic rate, reduction of body water content, and increased fat body reserves (see p. 85f). Although the growth is suppressed in diapausing larvae, they are not necessarily inactive and may move about and respond when suitably stimulated. However, in most of the species which diapause as larvae or pupae this is a stage of arrest which is preceded by a distinct prediapausing behavior. The aim of such behavior is to find suitable places for the time of dormancy, such as bark clefts or even specially built diapause chambers in the ground.

Prediapausing larvae of the southwestern corn borer, *Diatraea grandiosella*, descend at the end of their feeding period to the root zone of their host plant (corn or sorghum), excavate a cell in the stalk crown, prepare a plugged exit hole for the emergence of the adult and then molt from the dark spotted feeding larva to the immaculate diapause larva. After diapause termination the larva removes the plug from the exit hole and postdiapause development continues with pupation (Chippendale 1978a).

Larval stationary molts during diapause are also known for several other species, for example, the rice stem borer, *Chilo suppressalis* (Yagi 1981).

Larval or pupal diapause is normally restricted to one unfavorable environmental period, although in some insects prolonged diapauses have been observed. Ethniid moths can stay in larval diapause arrest for 2 years (Powell 1974). The wheat blossom midge, *Sitodiplosis mosellana*, is remarkable in that it can spend up to 12 winters in the soil, although nearly one-third of the diapausing larvae which entered the soil pupate after only one winter (Barnes 1952).

Dormant pupae of the moth *Rothschildia jorulla* emerge after 8 years and those of *Biston alpinus* after 6 or 7 years. In the walnut fly, *Rhagoletis completa*, pupal diapause can take up to 5 years (for review see Lees 1955; Boller and Prokopy 1976).

2.3 Adult Diapauses

Adult diapause is characterized as a state of reproductive delay particularly occurring in females. It may coincide with quiescence of the animal (the Colorado potato beetle, *Leptinotarsa decemlineata*) or may just be a functional pause of the reproductive organs (the water beetle *Dytiscus marginalis*). This type of diapause is most common among Coleoptera and Lepidoptera, but is also known from Diptera, Orthoptera, and Hemiptera.

In the Egyptian grosshopper, *Anacridium aegypticum*, the oocytes remain small after the onset of diapause and vitellogenesis will only occur after diapause termination (Geldiay 1967). By contrast, in the water beetle *Dytiscus marginalis,* the adult diapause is characterized by continuous vitellogenesis and oosorption. Oocytes with their accompanying nurse cells surrounded by follicle cells, continually pass down the ovarioles. However, when the oocyte is about 1-2 mm in diameter, it is resorbed by follicle cells which then form the corpus luteum. The same happens to the oocyte next in line in each ovariole (Joly 1945).

A reproductive diapause occurring in both sexes is reported for the monarch butterfly, *Danaus plexippus* (Herman 1981). In this species as well as in the red flour beetle, *Tribolium castaneum*, the reproductive delay in males is due to an inhibition of the accessory glands (Murad and Ahmad 1977).

An extraordinary form of adult diapause is found in the butterfly *Barbara colfaxiana.* This insect species is dormant as a fully developed (pharate) adult within the pupal cuticle (Sahota et al. 1982).

2.4 Dual Diapauses

Insects with developmental cycles taking more than 1 year may have one or more developmental delays. Such repeated dormancies may be facultative or obligate.

Two facultative diapauses are known to occur in the pitcher plant mosquito, *Wyeomyia smithii* (L_3 and L_4) (Lounibos and Bradshaw 1975), the coniferous budworm, *Choristoneura sp.* (L_2 and L_4) (Harvey 1967), and in *Dendrolimus pini* (L_2 and L_5).

The moth *Operophtera brumata* is said to undergo obligatory diapause as an embryo and as pupa, while the west Canadian grasshopper *Pardlophora apiculata* diapauses as embryo and as a late nymph (Pickfort 1953). The mosquito *Aedes sierrensis* exhibits an embryonal and a larval (L_4) diapause (Jordan 1980), whereas in the locust *Locusta migratoria* an embryonal and a reproductive diapause exist (Jones 1956, Poras et al. 1983).

3 Factors Effecting Diapause

The most important factor for diapause induction is the photoperiod, although other factors like temperature, food quality, moisture, the parental age, and intraspecific relations like crowding are known to modify the primer stimulus.

3.1 Light Effects

3.1.1 Light Intensity

It was found to be a general rule that the photoperiodic diapause reaction in insects is independent of intensity and total light energy, provided that the intensity exceeds a threshold value of about 1.0 ft-c $\hat{=}$ 10.8 lx. This limit may vary between different species (for review see Lees 1955). Such a low value means that daily fluctuations in light intensity due to clouds have no effect on photoperiod perception and that the effective day length includes the period of twilight, but that direct moonlight, which is usually within the range of 0.01-0.05 ft-c (0.1-0.5 lx), is uneffective (Beck 1980). However, the sensitivity to low light intensities may be influenced by the maximum intensity and by the time of day. Takeda and Masaki (1979) showed that the web-worm larvae *Hyphantria cunea* responded to stimuli of 0.5 lx after a 14 h treatment of 180-200 lx, but failed to respond even to 7.5 lx when the preceding intensity was 9000 lx. Furthermore, under natural conditions, the threshold for response was 1 lx at dawn, but 10 lx at dusk.

Several response thresholds for photoperiodic reactions are listed by Saunders (1974).

As a result of the high sensitivities, insects that live inside fruit and even the pupa of *Antheraea* inside its cocoon are affected by a photoperiod (Chapman 1969). It is possible that the mechanism of light perception in insects is temperature dependent (de Wilde 1962).

3.1.2 Light Quality

Insects can percept light qualities from the ultra-violet to the red end of the spectrum. Wavelenghts in the blue-green region have been found to be most effective in diapause induction. As early as 1933, Kogure studied the influence of light quality on the production of diapause and nondiapause eggs in the silkworm *Bombyx mori.* He found that wavelengths below 500 nm induced 89% diapause eggs, whereas orange and red light qualities induced 8% and 3%, respectively. Similar values were found in the pitcher plant mosquito, *Wyeomyia smithii* (Bradshaw and Phillips 1980). The authors suggested that two distinct pigments with different sensitivities (390-450 nm and 480-540 nm) influence the photoperiodic response. More spectral sensitivity data are listed by Saunders (1974).

3.1.3 The Site of Photoperception

Neither the compound eye nor the ocelli of insects seem to act as receptors of photoperiod. In larvae, the stemmata – at least in some species – may be involved, but most probably the brain and especially the cerebral neurosecretory cells react to the stimuli.

In the large white cabbage butterfly, *Pieris brassicae*, Claret (1966) demonstrated that light penetrates the larval head capsule and acts directly on the brain to regulate the secretion of prothoracicotropic hormone (PTTH) and the induction of pupal diapause.

In adult aphids (*Megoura viciae*), it was shown that only parts beneath the median line of the head capsule, most probably the neurosecretory cells, respond to photoperiods (Lees 1964; Steel and Lees 1977).

Pupae of the giant silkmoth, *Hyalophora cecropia*, reacted with nondiapause when treated with long photoperiods only in the anterior part. The same treatment restricted to the posterior part caused diapause, but this reaction could be abolished when another brain was implanted in this part.

3.1.4 Sensitive Stages and Day Times for Diapause Induction

Insect diapause results from the exposure of sensitive stages to distinct stimuli. This stage in the life history is usually fixed and is characteristic for each species. Furthermore, it is typical for a true diapause that the sensitive stage in the developmental cycle, as well as the onset of the diapause occur at a time when life conditions are still suitable. During sensitive and postsensitive stages the insects prepare themselves for the forthcoming metabolic demands during the unfavorable conditions in the diapause state (see p. 83ff). Nothing is known about the signals indicating that the preparations have finished and diapause can start. The time elapsing from diapause induction to its onset may be as long as a complete developmental cycle (*Bombyx mori*).

In some cases the diapause stage itself is the sensitive stage (Fig. 3.2).

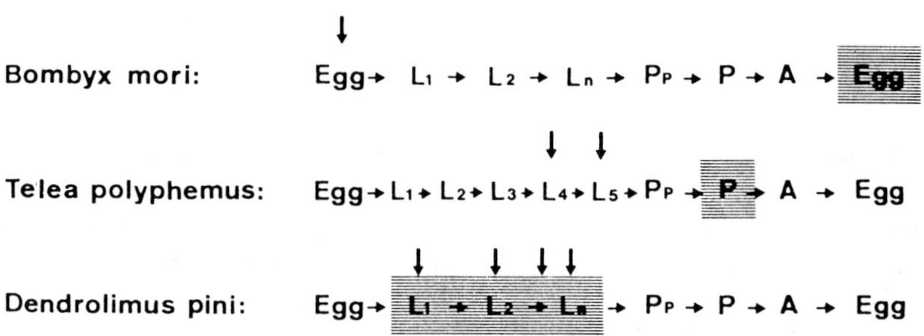

Fig. 3.2. Photoperiod sensitive (*arrows*) and diapausing (*shaded*) stages of some Lepidoptera. *L* larval instars; *Pp* prepupa; *P* Pupa; *A* adult. (After Highnam and Hill 1977)

In the silkworm, *Bomyx mori,* the embryo is most sensitive to photoperiod, although the early larval instars, too, show some sensitivity. In *Anteraea pernyi*, all larval instars are susceptible with markedly increasing sensitivity towards the latter ones, and the effects are cumulative. In many butterfly species, such as the silkmoths *Telea polyphemus* and *Philosamia cynthia*, the pupal diapause is inducible only by short photoperiods during the 4th and 5th instars. The tomato moth, *Diataraxia oleracea*, has been reported to have a sensitive period as short as 2 days at the end of the penultimate larval instar (Way and Hopkins 1950). On the other hand, diapause induction during a long lasting sensitive stage may be averted by a single misleading

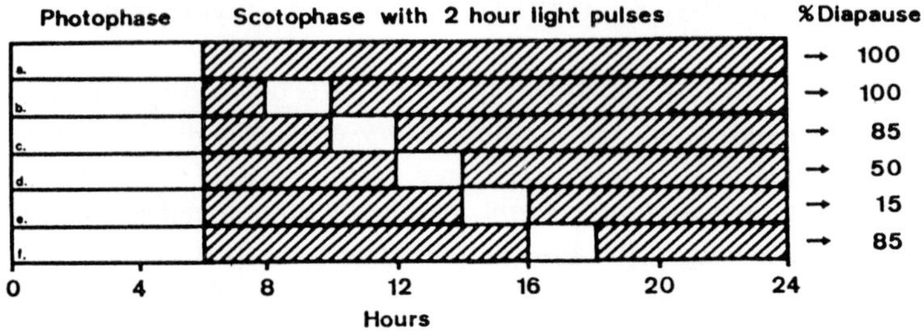

Fig. 3.3. Effects of 2 h light pulses during scotophase on diapause incidence in *Pieris brassicae* reared under short day conditions (L:D = 6:18). Night interruptions 14-16 h after the start of the photoperiod (*e*) are able to reverse the diapause inducing short day conditions.(After Highnam and Hill 1977)

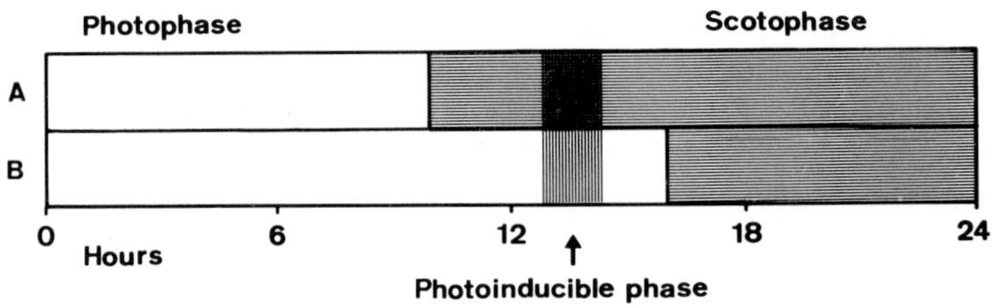

Fig. 3.4. A theoretical model for photoperiodic diapause induction. If during a sensitive developmental stage the photoinducible phase runs within the hours of darkness, diapause will be induced (**A**). Otherwise nondiapause reaction will be observed (**B**). (After Highnam and Hill 1977)

stimulus. Larvae of the ichneumonid wasp *Pimpla turionellae* are sensitive to photoperiod from L_3-L_5. Photophases below 11 h/day induced diapause, but a single photophase of more than 12 h/day operating in L_5 is enough to inhibit diapause (Schopf 1980).

Besides photosensitive stages during the life history of insects, special photo-inducible phases within the 24 h cycle of a day must be taken into account. When larvae of the large white cabbage butterfly, *Pieris brassicae*, are reared under short day conditions (L:D = 6:18), all subsequent pupae enter diapause. However, when the dark period is interrupted by 2 h of light at various times, the incidence of diapause is very much reduced, if the light interruption occurs 14 h after the beginning of the primary photoperiod (Fig. 3.3). The insects react as if they were in long day conditions. This suggests that there is a special photo-inducible phase at a particular time during the daily cycle. If photoperiod overlaps the photo-inducible phase, diapause is prevented, but if it is so short or out of phase, so that the photo-inducible phase lies in the dark period, diapause is induced (Fig. 3.4).

Similar observations have been made in the mosquito *Aedes atropalpus*, in which a 1 h light pulse between 7 and 9 h after the onset of a short day scotophase inhibited diapause (Beach and Craig 1979), or in the noctuid butterfly *Mamestra brassicae*, in which a 3 h light pulse reversed the long night effect during early scotophase (Furunishi et al. 1982).

3.1.5 Photophase – Scotophase Relations

Although it is commonly the scotophase (night length) which is measured by insects, reports in the literature generally describe the photophase (day length) in diapause studies (Beck 1980). The majority of species studied so far measure the absolute length of scoto- or photophase, respectively, and only a few exceptional species are probably able to detect increasing or decreasing scoto- or photoperiods. There is usually a species specific photoperiod above which the diapause response occurs in most or all of them.

For example, in the vetch aphid, *Megoura viciae* a 15 min decrease in photophase from 15.00 to 14.45 h switches some females from producing parthenogenetic offspring to produce sexually reproducing offspring. A further decrease to 14 h photophase causes the entire population to do so.

Because the diapause-inducing photophase – scotophase – ratio is not a precise value, but covers a species specific range, the critical photoperiod has been defined to be the elapsing time of day light under which 50% of the individuals of a distinct population enter diapause. In the example mentioned above the critical photoperiod is about 14.30 h/day (Saunders 1971). It is 12.25 h/day in the mosquito *Aedes dorsalis* (Mulligan 1980). More than 200 insect species with photoperiodic diapause induction have been listed by Saunders (1976).

The relationship between season and day length varies with latitude. Therefore, the critical photoperiod for diapause induction in a given species varies between the local pppulations from different latitudes. A well-studied example is the moth *Acronycta rumicis*, which exists in several local races in Russia, each with a different critical photoperiod (Table 3.1). In the northern population diapause is induced by rather long photoperiods because of the early onset of winter. Southern populations are induced to enter diapause at shorter photoperiods, thus allowing an additional generation in the mild climate at those latitudes. Similar observations on latitude-dependent critical photoperiods for larval diapause maintenance were made by Holzapfel and Bradshaw (1981) for the mosquito *Aedes triseriatus.*

Table 3.1. Differences in critical photoperiods and number of generations per year between several local races of *Acronycta rumicis.* (After Danilevsky 1965)

Local race	Latitude °N	Critical photoperiod h/day	Number of generations per year
Black Sea	43	14.30	3
Belgorod	50	16.30	2
Vitebsk	55	18.00	1–2
Leningrad	60	19.30	1

Diapausing insects are often divided into so-called long day species and short day species. The former develop without diapause when the sensitive stage falls within photoperiods longer than the critical day length, while the latter develop without

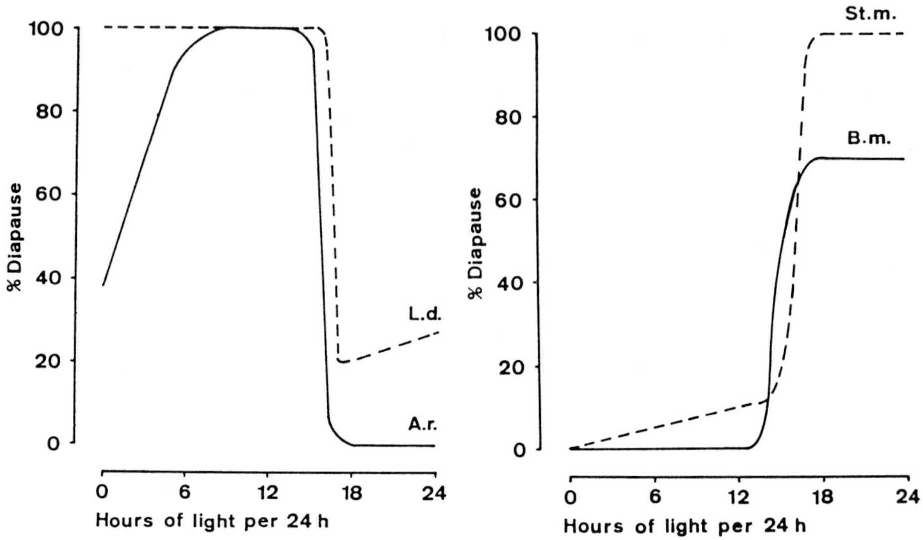

Fig. 3.5. Percentage of diapausing insects in relation to photoperiod. *Left panel*: long day type; L.d. – *Leptinotarsa decemlineata;* A.r. – *Acronycta rumicis. Right panel:* short day type; St.m. – *Stenocranus minutus;* B.m. – *Bombyx mori.* (Redrawn from Bligh et al. 1976)

diapause, when the sensitive stage falls within photoperiods shorter than the critical day length.

The majority of insects from the temperate regions belongs to the long day type (Fig. 3.5). Sensitive stages from these species experience short day conditions during fall and undergo a hibernating diapause (examples cited above).

A well-known example for a hibernating short day insect is the bivoltine strain of the commercial silkworm, *Bombyx mori.* The photosensitive period in this butterfly lasts from late embryogenesis to the first one or two larval stages. Under short photoperiods and low temperatures in spring, these individuals produce a summer generation of adults. The eggs laid by the summer generation are submitted to long photoperiods and high temperatures which lead to a second fall generation of adults producing diapause eggs. Further examples for short day insects are given in Fig. 3.5.

When insects have a photosensitive stage in spring or early summer, day length may induce an aestivating summer diapause. It is found in tropical insects as well as in those from temperate regions, and enables the animals to survive the dry and hot seasons with short food supply or to settle specific ecological niches (Masaki 1980).

The grasshopper *Oedipodia miniata* exhibits an aestival reproductive diapause during the dry summer in Israel (Pener and Orshan 1980). Crane fly embryos (*Tipula simplex*) aestivate in dry regions of California (Hartman and Hynes 1980). The noctuid moth *Diparopsis castanea* from tropical Africa aestivates as a pupa when the host plants stop growing because of drought (Lees 1955). An example of summer diapause in temperate regions is given by Okuda and Hodek (1983). The central Japanese population of the lady beetle, *Coccinella septempunctata bruckii*, aestivates, while the European population of *C.s. septempunctata* does not. In Europe, the southern population (e.g., 43°N) of the cabbage moth *Mamestra brassicae* displays hibernation and aestivation diapause in the pupa, but no aestivation can be found in the northern population (e.g., 48°N) (Grüner and Sauer 1984).

A detailed review with numerous examples on the widespread phenomenon of summer diapause is given by Masaki (1980).

3.2 Temperature Effects

Besides photoperiod temperature is the second most important factor in diapause induction. In temperate regions high temperatures, even when combined with short photoperiods, generally suppress diapause, whereas low temperatures enhance any tendency to enter diapause. However, the influence varies considerably between species, and several transition forms of temperature induced diapause are known.

3.2.1 Temperature Compensated Diapause Induction

The large white cabbage butterfly, *Pieris brassicae*, and the flesh fly *Sarcophaga argyrostoma* are good examples of temperature compensation in diapause induction.

In *Sarcophaga* diapause is induced by a photoperiod of about 13.30 h/day between 15° and 25°C, but at 20°C the maximum level of induction is only 80% and at 25°C very few larvae diapause (Fig. 3.6). Evidently, although the photoperiodic clock is fully temperature compensated, the control processes can be uncoupled at ecologically suitable temperatures.

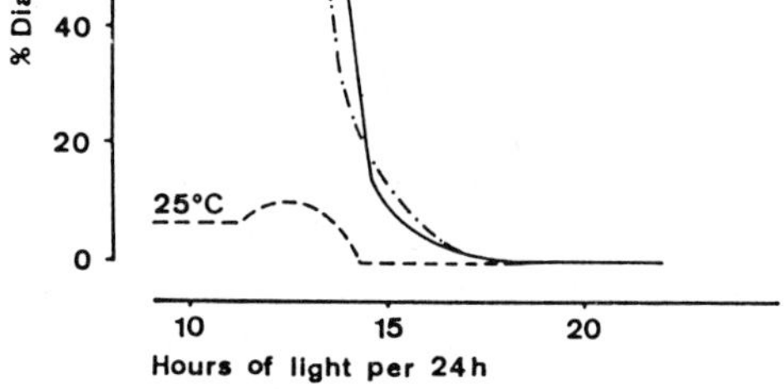

Fig. 3.6. Temperature compensated diapause induction in *Sarcophaga argyrostoma*. The critical photoperiod for all tested temperatures is about 13.5 h photoperiod per day, but the percentage of diapause incidence varies with the temperature. (After Saunders 1971)

3.2.2 Nontemperature Compensated Diapause Induction

A well-known example of nontemperature compensated diapause induction is the noctuid moth *Acronycta rumicis* (Fig. 3.7 and Table 3.1). In this butterfly the criti-

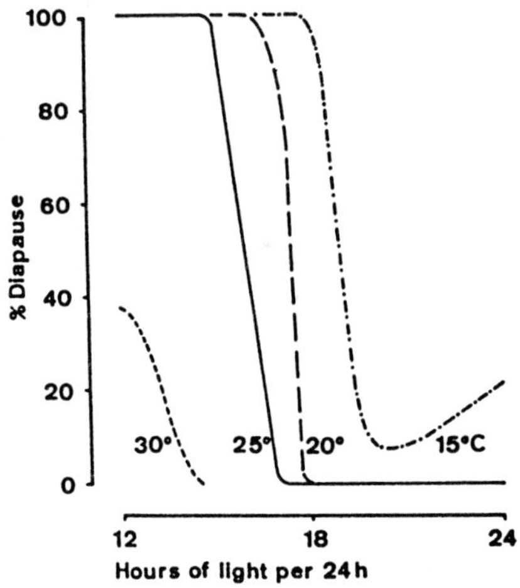

Fig. 3.7. Nontemperature compensated diapause induction in *Acronycta rumicis*. The critical photoperiod for diapause induction is strongly temperature dependent. (After Danilevsky 1965)

cal photoperiod for diapause induction varies from 16 h/day at 25°C to 19 h/day at 15°C (Danilevsky 1965).

3.2.3 Thermoperiod Diapause Induction

Only few investigations have been performed with combined photoperiod – thermoperiod cycles. Nevertheless they demonstrate clearly that low scotophase temperatures tend to increase the incidence of diapause (Beck 1983b).

The European corn borer, *Ostrinia nubilalis*, has a photoperiodically inducible diapause with a critical day length of 15 h at 26°C. The effect of thermoperiod on the induction of diapause has been tested at the critical day length, using a 21:31°C cycle (mean temperature of 26°C). The observed incidence of diapause was more than 90% when the lower temperature coincided with the scotophase, but a low incidence of 15% was found when the cryophase coincided with the photophase. This experiment clearly shows that the incidence of diapause is strongly influenced by the scotophase temperature (Beck 1983b).

To test the diapause-inducing competence of thermoperiods constant photoperiodic inputs (constant darkness or constant light) are necessary. All experiments reported on this topic so far found thermoperiodic induction of diapause to be possible only under constant darkness, but not under constant light. Under such artificial conditions, thermoperiodic response curves can be obtained analogous to the photoperiodic response curves (Beck 1983b).

3.2.4 Temperature-Induced Diapause

In some insect species diapause induction is thought to be only temperature-dependent and independent of photoperiod under natural conditions. The parasite *Cothonaspis boulardia* (Hymenoptera) from Guadeloupe exhibits a last larval instar diapause induced by a temperature of 17.5°C or below and independent of photoperiod (Claret and Carton 1980).

Eggs of the locust *Locustana pardalina* from arid South Africa react with embryonic diapause when submitted for a short period to high temperatures. The incidence of diapause is directly related to the exposure period (Matthee 1978).

3.3 Other Factors Influencing Diapause Induction

Beside the main diapause inducing factors, photoperiod and temperature, some others may influence or even induce diapause.

3.3.1 Dietary Conditions

Sub-surface feeding larvae of the cabbage root fly, *Erioischia sp.*, cannot respond to photoperiod, but enter diapause after short day lengths. The inducing signal may be the seasonally changing food composition (Hughes 1960). In the Colorado potato beetle, *Leptinotarsa decemlineata*, the age of the potato plants influences the response, although it is mainly induced to enter reproductive diapause by short day photoperiods. Feeding on senescent leaves induced diapause even under long day conditions (Fig. 3.8) (de Wilde 1962).

The diapause induction of the pink bollworm, *Pectinophora gossypiella,* also depends on the age of the diet. Old cotton bolls (> 20 days) exhibit high oil content which favors diapause incidence (Abul-Nasr et al. 1974a,b; Foster and Crowder 1980).

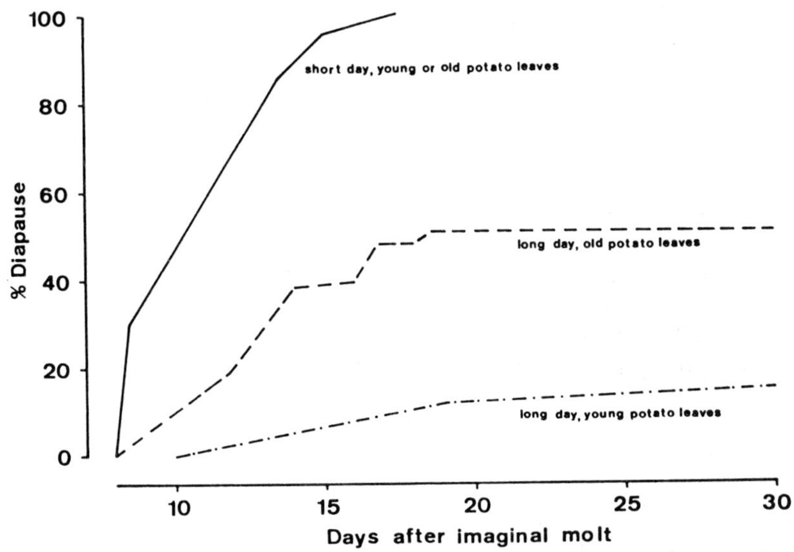

Fig. 3.8. Effects of photoperiod and food quality on diapause induction in adult female *Leptinotarsa decemlineata.* (After Danilevsky 1965)

In the cabbage moth, *Mamestra brassicae*, diapause incidence increases when the linoleic acid content of the diet decreases (Bracken 1982).

In the mosquito *Aedes triseriatus* and in the lacewing *Chrysopa mohave*, complete or partial absence of food stimulates the appearance of diapause (Clay and Vernard 1972). These examples clearly show the influence of nutrition on diapause incidence. It is probable that the influence of this factor has been underestimated in numerous earlier studies.

3.3.2 The Maternal Influence on Diapause Induction

In some species the physiological state of the parents is known to affect the incidence of diapause in the following generation. In the locust *Locustana pardalina* the proportion of diapause and nondiapause eggs is related to both the age and the phase of the mother. Phase solitaria females lay 100% diapausing eggs compared with only 42% by phase gregaria females. Old gregaria females lay more diapausing eggs than young. The physiological basis of the phenomenon in *Locustana* and of similar observations in other species (reviewed by Lees 1955) has not yet been elucidated. However, in the silkworm *Bombyx mori* (bivoltine strain), a diapause hormone (see p. 88f) from the mother determines the diapause incidence of the next generation embryos. The release of this hormone is induced by long day photoperiods experienced during the late embryonic stage of the mother. Consequently, developing embryos from the spring grow up to a summer generation of females laying nondiapause eggs, while embryos developing during summer mature to females forming the fall generation, which produces exclusively diapause eggs.

3.3.3 Crowding and Diapause Induction

As already seen on p. 71 some examples exist, showing that population densities influence diapause incidence. In the migratory locust, *Locusta migratoria*, the percentage of eggs entering diapause is directly related to the density of the adults (Tanaka 1982).

In the beetle *Trogoderma variabile* the larval density represents the main diapause regulating factor. When reared in groups the larvae pupate promptly, although they diapause when reared individually (Elbert 1979).

In the moth *Ephestia cautella* larval diapause induction is a function of the interaction between genotype and larval crowding. High larval densities stimulate diapause incidence (Hagstrum and Silhacek 1980).

3.3.4 Oxygen Supply

Embryonic diapauses in some insect species are thought to occur unevadably because the induction is merely genetically programed and needs no environmental stimulus, but depends on extraordinary development processes during morphogenesis. Such circumstances, for instance, are temporary oxygen-impermeable egg membranes, such as the chorion or the serosal cuticle, which are known for *Leptopterna dolobrata* (Heteroptera) (Braune 1971, 1980) and for *Philaenus spumarius* (Homoptera) (Witsack 1973).

In *Leptopterna dolobrata,* during embryonic prediapause development a cuticle which is secreted by the serosa is deposited before katatrepsis. This layer is impermeable to water and oxygen. The short oxygen supply leads to partial anoxia which primarily slows down morphogenesis and then stops it completely. The oxygen barrier was found to be constant and independent of the experimental temperature ($Q_{10} \approx 1$). Only temperatures $\leqslant 15^{\circ}C$ lasting for several days were able to restore oxygen and water permeability and thereby induce postdiapause development.

4 Diapause Maintenance and Induction of Postdiapause Development

Once diapause has been induced, development cannot be resumed at once by a return to favorable conditions. A minimum period in the dormant state must first elapse. This is what distinguishes diapause from mere quiescence, which can be terminated at any time by replacing favorable conditions. The relationship between diapause duration and environmental conditions is complex. Termination of diapause is often accelerated by some optimum temperatures but delayed by either higher or lower temperatures. Furthermore, water availability often plays an important role in the onset of postdiapause development (see p. 84).

4.1 Diapause Maintenance by Photoperiodic Conditions

As mentioned above photoperiod is the most important environmental stimulus for diapause induction (see p. 71ff), and some insects remain in diapause as long as the photoperiod persists beyond the critical day length, but terminate it when the photoperiod reaches the critical day length again. Two hibernating insects which follow this reaction scheme are the culicid *Wyeomyia smithii* and the neuropteran *Meleoma signoretti.* In other species the response to photoperiod may change during diapause, and examples are known in which the photoperiodic control of diapause maintenance is terminated near the time of winter solstice. Other insect species, which are induced to enter diapause under short day lengths, require long day lengths to terminate the diapause state (for review see Tauber and Tauber 1976).

4.2 Diapause Maintenance by Temperature Conditions

Laboratory studies indicate that some insect species have low temperature optima for diapause development during their hibernal diapause. One factor that is common to most of these species is that warm conditions during late summer and early autumn ensure diapause maintenance, becuase during these periods warm conditions slow the rate of diapause development. By this mechanism diapause is maintained until winter temperatures drop below the threshold for growth. This adaptation precludes untimely postdiapause development during the unfavorable season.

This generally accepted opinion on diapause maintenance has been recently contradicted by Tauber et al. (1982). Working with the hibernating alfalfa leafminer, *Agromyza frontella* (Diptera), these authors found that diapause induction can be reversed by favorable high autumn temperatures. This is an example of low temperature as a primary factor maintaining autumnal hibernal diapause in a species from the temperate region.

In summer diapausing species the optimum temperature for diapause development can be higher or lower than the usually occurring temperatures. Insect species react individually and no general statement can be made.

Tropical insects which may experience seasonal temperature changes but little day length changes are thought to rely mainly on temperature as a diapause regulating factor. An example of this category of insects is the flesh fly *Sarcophaga spilogaster*, in which diapause is induced and maintained by cool temperatures (Denlinger 1974).

4.3 Diapause – Postdiapause Transition

It has been generally accepted that diapause is a state which has to be terminated by environmental stimuli. This is often referred to as "diapause breaking". Chilling was thought to be one of the most important factors, because under laboratory conditions in numerous insect species a low temperature treatment hastens diapause termination. On the other hand, many hibernating species terminate their diapause more quickly under warm than under cold conditions. Furthermore, during the course of diapause there is generally a decrease in diapause intensity as well as a continual alteration in the animals' response to environmental stimuli (Tauber and Tauber 1976). Therefore, the present consensus is that diapause is a dynamic rather than a static state, during which the physiological conditions of the animal change. With few exceptions, no naturally occurring specific stimuli could be found to actively terminate diapause (Tauber and Tauber 1976). Nevertheless, at least in some insects, there is a gradual transition from the diapause to the postdiapause period under natural conditions (Masaki 1956).

Furthermore, field studies have revealed much longer periods of dormancy than experiments in which diapausing insects were collected outdoors and then studied under laboratory conditions. This led to the assumption that in most hibernating insects from temperate regions diapause development has come to an end by the time of winter solstice and is followed by a time of metabolic quiescence until in spring or early summer the postdiapause development starts (Müller 1976). This, too, indicates that under natural conditions no diapause terminating stimulus is needed.

4.4 Induction of Postdiapause Development by Artificial Stimuli

Some interesting artificial methods to induce postdiapause development have been listed by Lees (1955) and Beck (1980). Although none of them occurs under natural conditions and the mechanisms are as yet not understood, it may be noteworthy to discuss some of them. The findings show how complex the physiological basis for induction of postdiapause development is.

Diapausing embryos of *Bombyx mori* can be induced to start postdiapause development by electric shocks, brushing, or immersion in warm dilute hydrochloric and sulfuric acid, while temperature shocks are less effective. Diapausing larvae of the fly *Liparia lucens* react on pricking and singing with postdiapause pupation. Diapausing larvae of the sugar beet webworm, *Loxostege sticticalis*, and embryos of the grasshopper *Melanoplus differentialis* as well as pupae of the flesh fly *Sarcophaga crassipalpis* and the tobacco hornworm, *Manduca sexta*, react to several fat solvent vapors, such as xylol, hexane, toluene, carbon tetrachloride, and others (Denlinger et al. 1980). Water uptake after such treatments has been thought to be the primer step of postdiapause development. Urea and some other compounds have been reported to accelerate the termination of embryonic diapause in the cricket *Acheta commodus*.

Nevertheless, the response to such artificial stimuli is species specific. Thus, some diapausing insects are highly resistent or even withstand all of them, as is the case for the cherry fly, *Rhagoletis cerasi*.

4.5 Postdiapause Development

Temperatures above the threshold value for morphogenesis are the outstanding requirement for the onset of postdiapause development. Once diapause development has been initiated photoperiod no longer affects morphogenesis. This is true at least for the large majority of species which have only one dormant stage (Tauber and Tauber 1976). Insects with repeated diapauses (see p. 71) may reenter photosensitive stages again.

Withstanding unfavorable environmental conditions in both hibernal and aestival diapause requires a reduction in water content (Heinrich 1981; see also Chap. 1). Therefore, postdiapause development is closely related to water uptake, especially in embryonic diapausing species. Unfortunately, this process has sometimes been misinterpreted as a diapause-terminating stimulus. It is now well-established that water uptake is developmentally stimulating only after diapause development has been completed (for review see Beck 1980).

5 Metabolic Adaptations to Diapause

During diapause metabolism functions at a reduced level. Because morphogenesis does not occur in this developmental stage, only those functions have to be maintained which keep the insect alive. However, diapause is a multifarious phenomenon with lots of species specific strategies. For example, many insects diapausing as a pupa reduce their oxygen consumption to 2 - 4% of the normal value. However, during the reproductive diapause of the rove beetle, *Atheta fungi*, from West Europe no differences in oxygen uptake could be found compared to nondiapause individuals (Grigo and Topp 1980).

As already seen on p. 70, diapause may last a few weeks in some species, but up to several years in others. This time of dormancy, normally without any food uptake under extreme and unfavorable conditions, requires special adaptations in metabolism. A review of the tremendous amount of literature on diapause metabolism reveals many conflicting findings.

Although some general tendencies can be found, a great variety of species specific adaptations seems to exist, analogous to the differing diapause stages and numerous modes of diapause induction. Some aspects of water balance, freeze resistence and freeze tolerance, stored reserve products, oxygen consumption, and their regulatory mechanisms will be discussed on the following pages (for metabolic cold adaptations see also Chap. 1).

5.1 The Role of Water in Hibernal Diapause

A relative decrease in water content can be found in all insects during the period between induction and onset of diapause. This is due to extensive accumulation of reserve material in the fat body. Besides this decrease an absolute water loss during the diapause state can be found due to an evaporative dehydration. This can be partly avoided by a specific prediapause behavior, during which sheltered and moist places are occupied (see p. 70). Insect species with embryonic or pupal diapause, namely, Lepidoptera have adopted another mechanism: waxy body layers avoid dessication. In the Bertha army worm, *Mamestra configurata*, diapausing pupae have twice as much wax as nondiapausing pupae. In the former the period of wax secretion by the epidermal cells is prolonged from 4 to 8 days (Hegdekar 1979).

To survive sub-zero temperatures two general strategies for cold hardiness are apparent: freeze-tolerant species survive extracellular ice formation, while freeze-susceptible species decrease the supercooling point of the hemolymph by snythesis of antifreeze molecules (see Sect. 4, Chap. 1).

Controlled freezing of the hemolymph is induced by the action of nucleating agents in the first group (Zachariassen 1980), while any nucleating substances are strictly avoided by the latter one.

In both groups intracellular levels of low molecular weight compounds, such as carbohydrates, polyols, and amino acids are drastically increased. Thus, the intracellular osmotic values are markedly elevated and lower the supercooling point. Furthermore, these substances are able to bind water by their polar molecular structure. In this way remarkable amounts of the bulk water become fixed. Water in this state is thought to form shells around the subcellular components and, thus, cannot be frozen.

5.2 Stored Reserve Products

5.2.1 Carbohydrates and Polyols

In insects diapausing in the embryonic stage carbohydrates play the major role in diapause metabolism. The commercial silkworm, *Bombyx mori*, again is the best studied species so far. The basic metabolic events are reviewed by Yamashita et al. (1981). Embryonic diapause in *Bombyx* is induced by long day photoperiods in the parental generation (see p. 76) and transduced and regulated by a diapause hormone (see p. 88f). By the action of this hormone, glycogen reserves of the maternal fat body are mobilized and the hemolymph trehalose level is decreased, while the carbohydrate content of the ovaries is increased. In the ovaries glycogen is preferentially stored within those oocytes in which vitellogenesis is at its highest. After egg deposition the glycogen is abruptly utilized to form sorbitol along with glycerol at the initiation period of diapause.

These substances are thought to be freeze protectives (see Sect. 4, Chap. 1) and appear in diapause determined eggs only. When eggs with diapausing embryos were exposed to cold temperatures or treated with HCl, diapause development was hastened and terminated. After embryonic development is resumed sorbitol is exclusively used for the resynthesis of glycogen which is finally utilized as an energy source for post-diapause morphogenesis.

The three main steps of glycogen mobilization in maternal fat body and glycogen uptake into the ovary, transformation of glycogen to sorbitol in diapausing eggs, and retransformation of sorbitol to glycogen after diapause termination, are regulated enzymatically. In the first step, glycogen mobilization and uptake, the enzyme trehalase has a key function. The activity of this enzyme could be stimulated in vivo and in vitro by the diapause hormone. In the following step of sorbitol formation, glycogen phosphorylase A is the initiating enzyme. In diapause determined eggs this enzyme exhibits a rapid increase in activity from 20 h after egg deposition onward, while it remains at trace levels until 75 h after oviposition in nondiapause eggs. Nothing is known about the phosphorylase A regulation up to now.

The key enzyme in the last step, namely, reformation of glycogen from sorbitol, is NAD-dependent sorbitol dehydrogenase. This enzyme was shown to be almost inactive during the diapause stages, but became active after long-term chilling or HCl treatment.

To support the continuous conversion of sorbitol via fructose to glycogen, permanent reoxidation of $NADH_2$ is necessary, because the NAD content in the eggs is low. This reoxidation is assumed to be coupled with the respiratory chain and a hydrogen shuttle is postulated to be coupled with the enzyme since it is present in the cytosolic compartment and $NADH_2$ cannot penetrate through the mitochondrial membrane.

A diapause hormone as initiating substance in glycogen mobilization is known only for *Bombyx mori.* In all other species that have been investigated so far, low temperatures were found to directly stimulate the enzymatic pathways involved in cryoprotectant synthesis (for review see Storey 1983).

5.2.2 Lipids

While carbohydrate deposition in eggs is the most important adaptive strategy for embryonic diapause, those insects diapausing in postembryonic stages depend mainly on lipids as energy reserve. Tremendous amounts of lipids can be accumulated by the fat body (Lees 1955). Fat may form up to 56% of the dry weight in the cold-hardy noctuid *Scoliopterix libatrix* and up to 60% in hibernating females of *Culex pipiens.*

In accordance with fat accumulation some ultrastructural differences can be observed in fat body cells. In larvae of the corn borer *Diatraea grandiosella* Brown and

Chippendale (1977) found a characteristic decrease in the number of mitochondria and the amount of rough endoplasmatic reticulum, and an increase in the number of proteinaceous granules and lysosomes, indicating the shift of the fat body from a synthetic to a storage function.

Differences in the fatty acid composition between prediapausing and diapausing insects have been found in all hibernating species studied so far. In nondiapausing stem weevils, *Hyperode bonariensis*, only 43% of the fatty acids were unsaturated as against 63% in reproductive diapausing ones (Goldson et al. 1981). Similar observations were made in the alphalpha weevil, *Hypera postica* (Tombes 1966), and in the mosquito *Culex tarsalis* (Harwood and Takata 1965) (see also Sec. 3.2.6 in Chap. 1).

During the course of diapause the amount of stored lipids decreases depending on the intensity of diapause metabolism, which varies in a species specific manner (Lees 1955): 50% of the yolk fat content is consumed during the embryonic diapause of *Melanoplus differentialis*. Female *Culex pipiens* lose about 65% of their stored fat. In the beetle *Anatolica eremita* fat content decreases from 12 to 4% of the dry weight during its 8 month diapause, whereas only 2% of the energy stores are diminished after the prepupal diapause in *Gilpinia polytoma*.

Respiratory quotients of 0.7 to 0.8 clearly indicate that during diapause energy is preferably gained from lipid metabolism.

5.2.3 Amino Acids and Proteins

Increased levels of free amino acids in the hemolymph have been reported for several Lepidoptera, diapausing as larvae or pupae. Their role in diapause physiology, however, is unclear. Some authors suggest that amino acids are used as cryoprotectants just like carbohydrates and polyols. Others assume that a pool of amino acids is required for a rapid postdiapause protein synthesis. The role of the most abundant amino acids during diapause, serine, alanine, and proline, has been recently reviewed and discussed by Morgan and Chippendale (1983). Proline is thought to bind water very effectively not only in insect hemolymph but also in Crustacea adapted to high salinity and in plants. Accumulation of alanine during exposure to low temperature has been observed in many insect species and may be due to a stress-related increase in anaerobic energy metabolism (see Chap. 5). Large amounts of serine in overwintering larvae or pupae of *Diatraea grandiosella, Ostrinia nubilalis, Chilo suppresssalis*, and *Mamestra brassicae* may be used during postdiapause development for the synthesis of uric acid and other purines. Uric acid is synthesized at high rates during postdiapause. Cleavage of the β-carbon from serine yields glycine and a one carbon fragment which is the source of formic acid. De novo synthesis of 1 mol of uric acid requires 2 mol of formic acid and 1 mol of glycine. Additional nitrogen is also required, most probably from glutamine and aspartic acid. Although aspartic acid is not present in high titers in insect hemolymph, its immediate precursor, oxaloacetic acid, is usually present in high concentrations. Therefore, serine may play an important role in nitrogen excretion, since all its carbon can be used for uric acid synthesis.

Diapause-associated proteins have been reported from three insect species so far, but no striking evidence for their function is available up to now. In all species these proteins can be detected in small amounts in different developmental stages, but they become dominant in the prediapause and in the diapause state. De Loof and de Wilde (1970) described three diapause-associated proteins present in the adult Colorado potato beetle, *Leptinotarsa decemlineata.* These proteins have been also studied by several other authors (e.g., Dortland 1978) and were recently reexamined by Peferoen et al. (1982). The highest rates of synthesis of these proteins and their highest titers in hemolymph occurred in the second half of the prediapause period (day 6 to 12 after imaginal molt) when juvenile hormone titers were very low. After the onset of diapause (day 12) the amount of diapause-associated protein 1 and 3 in the hemolymph decreased significantly, due to accumulation in the fat body. All three proteins were shown by immunological methods to be different from each other. Diapause-associated protein 1 turned out to be a glyco-lipoprotein of 114,000 mol wt, whereas 2 and 3 are both glycoproteins of about 50,000 mol wt.

In the southwestern corn borer, *Diatraea grandiosella*, a diapause-associated protein was characterized by Turunen and Chippendale (1979, 1980a,b) and Chippendale and Turunen (1981). In this larval diapausing species high juvenile hormone titers induce its accumulation in the fat body. Even in specimens reared under long day conditions, under which the juvenile hormone titer is low, topical application of the hormone or its synthetic analogues stimulates diapause-associated protein synthesis. During the course of diapause the amount of the protein diminishes.

Less information is available on the diapause-associated protein of the codling moth, *Cydia pomonella.* It was found to be a monomeric glycoprotein (Brown 1980) and to exhibit constant levels throughout the whole time of larval diapause. Artificial induction and accumulation of the protein in males of this species requires both juvenile hormone application and subsequent rearing under short day conditions.

5.3 Respiratory Metabolism During Diapause

Insect diapause is characterized by a more or less reduced oxidative metabolism, depending on the different developmental stages in which it occurs. Oxygen consumption is the commonly used method to quantify oxidative metabolism rates. Hardly any differences can be found between diapausing insects from temperate regions which remain active during their reproductive diapause and the corresponding pre- or postdiapause stages. This is true for the rove beetle, *Atheta fungi*, for instance (Grigo and Topp 1980) (see p. 83). On the other hand, the great majority of diapausing insects exhibit species and stage specific reductions in oxygen consumption. Lowest oxidative metabolism rates have been found during embryonic and pupal diapause (reviewed by Lees 1955). Minimum oxygen consumption is reported for diapausing silkworms, in which it is 95-98% below the rate of the developing insect (Schneiderman and Williams 1953a; Harvey and Williams 1961).

Early observations by Bodine et al. (1934-1952) and Williams et al. (1950-1954) (cited by Lees 1955), which show that respiratory poisoning by cyanide and carbon monoxide is ineffective in some, but not in all diapausing insects, aroused interest in the properties of the cytochromes during the dormant state. In diapausing embryos the respiratory chain seems to be uncoupled and, therefore, to be insensitive to the toxins. Spectroscopic studies in nonembryonic diapausing insects revealed the absence of cytochrome C and general reductions in the content of the other mitochondrial cytochromes (Shappirio and Williams 1957a,b; Kurland and Schneiderman 1959; Shappirio 1974). Recent studies by Gruetzmacher and Keely (1982) with diapausing pupae of the cotton bollworm, *Heliothis zea*, using radioactive precursors of cytochrome during synthetic active phases have clearly shown that within the first 3 weeks of diapause a breakdown of 90% of the tracer took place in the fat body mitochondria. The residual level of tracer indicates that the degradation processes that remove the cytochromes from the mitochondria may cease after 3 weeks of dormancy. Following this, the level of cytochromes remains constant at 10% of its original value. Oxygen consumption rates at this time are 94-97% below the rates determined in newly formed pupae. A small resynthesis of cytochromes was found some days after diapause termination and a large increase in cytochrome content of the fat body only occurred in conjunction with adult emergence.

Further evidence for the reduction of cytochromes can be found in the southwestern corn borer, *Diatraea grandiosella.* In this insect fat body cells accumulate lysosomes during larval diapause, while at the same time substantial degradation of the mitochondria takes place (Brown and Chippendale 1977). Thus, cytochrome changes during diapause can be considered as significant biochemical adaptations in insect dormancy.

6 Hormonal Control of Diapause

Depending on the stage at which diapause occurs, different mechanisms of its endocrine control have been evolved. Besides the classical developmental hormones (juvenile hormones and ecdysteroids) a special diapause regulating hormone has been found to be responsive for embryonic diapause regulation. Detailed reviews on hormone-diapause interactions have been recently given by Yamashita (1983) on embryonic diapause, Chippendale (1983) on larval and pupal diapause, and de Wilde (1983) on adult diapause.

6.1 Embryonic Diapause Control

Although a large variety of insect species diapauses in the embryonic stage (see p. 69) and a diapause hormone secreted by the maternal subesophageal ganglion is generally suggested (for review see Beck 1980), conclusive data on its existence and metabolic control have been obtained only from a single species so far, the commercial silkworm, *Bombyx mori* (Yamashita et al. 1981; Yamashita 1983) (see also p. 84).

The memory mechanism storing the information for diapause induction in the embryo and preparing diapause eggs in the late pupae and adults is unknown. Nevertheless, the physiological events of diapause egg production have been elucidated.

A diapause hormone is secreted in mature females by a single pair of neurosecretory cells from the subesophageal ganglion after having been stimulated by the brain via the circumesophageal connectives. A biochemical analysis of the diapause hormone has been performed by Isobe and Goto (1980). These authors found two active forms of the hormone, called DH-A and DH-B. Both are peptides with mol wts of 3,300 and 2,000, respectively. DH-A contains two different amino sugars, while DH-B is a pure peptide. Using the *Bombyx*-bioassay DH-B was found to be three to four times more active than DH-A. The target organ of the diapause hormone action is the ovary. Three diapause specific functions of the hormone are reported:

1. Diapause hormone stimulates the accumulation of 3-hydroxykynurenine from the hemolymph in the developing eggs, which is then metabolized to ommochrome. This pigment causes the dark color of diapause destined eggs (see Sec. 2.1.5, Chap. 8).
2. Diapause hormone enhances trehalase activity in pupal ovaries which results in a higher glycogen content in diapause eggs than in nondiapause eggs (see p. 85).
3. Diapause hormone blocks the activity of esterase A, an enzmye which has a key function in the mobilization of yolk necessary for embryos to develop in postdiapause and nondiapause eggs.

Besides the diapause hormone two further groups of hormones seem to be engaged in the diapause physiology of *Bombyx*: ecdysteroids and juvenile hormones (Gharib et al. 1981, 1983). Both hormone classes show rapid turnover rates during early embryogenesis, but break down immediately after the embryo enters diapause in the germ band stage (see p. 69). After termination of diapause drastic and rapid changes in ecdysteroid and juvenile hormone levels are again observed. Induction of postdiapause development was obtained artificially when diapausing eggs were soaked in 10^{-4} M solution of 20-hydroxyecdysone for 24 h. Finally, under in vitro conditions nondiapausing embryos are able to stimulate the growth of co-incubated diapause embryos. From these results Takami (1959) concluded that diapausing embryos were under influence of a growth inhibition of maternal origin that prevented the embryos from producing their own hormones required for their continuous development. However, the effects and the interactions of the different hormones as well as the mechanisms by which they influence diapause and embryogenesis of the silkworm are still very poorly understood.

6.2 Larval Diapause Control

In larval diapause regulation juvenile hormone seems to play a predominant role during the early stages, while ecdysteroid hormones may be engaged in the onset of postdiapause development.

Most investigators have chosen Lepidoptera as test insects. The few studies performed with species from other groups are not in conflict with the findings in butterfly larvae. Therefore, the discussion of the endocrine basis of larval diapause will focus on this group.

Most larval diapausing Lepidoptera studied so far are dormant as final larval instar larvae. However, differences have been found between those which become dormant in the early final instar and may undergo stationary molts and others which diapause at the end of the last instar (prepupal) and will pupate immediately after diapause termination.

The southwestern corn borer, *Diatraea grandiosella*, and the rice stem borer, *Chilo suppressalis*, represent the former type. The endocrine balance during diapause in these species has been recently reviewed by Chippendale (1977, 1983) and Yagi (1981).

In nondiapausing individuals the juvenile hormone titer decreases rapidly at the beginning of the last larval instar, whereas in diapausing individuals the juvenile hormone content of the larvae remains at an intermediate level throughout the state of dormancy. These results indicate that juvenile hormone is the diapause regulating factor. Two further series of experiments confirmed this assumption:

1. A single topical application of juvenile hormone or a juvenile hormone analogue onto nondiapause determined larvae causes a diapause reaction with a stationary molt even under long day conditions, and
2. deprivation of the corpora allata or neck ligation in young diapause determined last larval instars causes nondiapause development, e.g., pupation. This allatectomy effect is reversible by juvenile hormone application.

Repeated applications of juvenile hormone or synthetic mimics onto diapausing individuals induce an elevation of the dormant period combined with additional stationary molts. This observation favored the idea that molting hormone production or release might also be under juvenile hormone control. Because no additional stationary molts were observed when juvenile hormone applications were performed with neck ligated larvae, it was suggested that the target organ for juvenile hormone must be located in the head. It is now generally assumed that a change in the juvenile hormone titer causes a release of the prothoracicotropic hormone (PTTH) in the brain and thereby initiates a release of molting hormone from the prothoracic gland. Last but not least, the corpora allata themselves seem to be under brain control. This is concluded from the observation that the juvenile hormone-mediated diapause control only functions with an intact nervous connection between the brain and corpora allata.

The termination of diapause development requires a breakdown of the juvenile hormone level by an unknown mechanism. Only then are molting hormones able to induce a pupal molt.

While in diapausing Lepidoptera with stationary molts the juvenile hormone level is intermediate, those insects which diapause as mature larvae (prepupae) without stationary molts exhibit a steady decline of the juvenile hormone titer to low levels during the time of arrest, combined with molting hormone contents at the trace level throughout the diapause state. This type of hormonal diapause control is represented by the codling moth, *Laspeyresia pomonella* (Sieber and Benz 1977, 1980a,b), and by the European corn borer, *Ostrinia nubilalis* (Bean and Beck 1980, 1983; Gelman and Woods 1983). In these species, juvenile hormone application to newly molted last instar larvae which are kept under nondiapause inducing conditions (e.g., long day photoperiod) caused a delay in pupation. In addition, the treatment led to the diapause state when the larvae were transferred to short day photoperiods at the same time. These findings can be interpreted in two ways: either high juvenile hormone titers induce diapause onset and short day photoperiods are required to maintain it, or the high juvenile hormone titers induce a prolongation of the last larval instar during which the experienced short day photoperiods lead to diapause onset. None of the theories can be proven at the moment.

Further studies with *Laspeyresia pomonella* diapausing larvae using the antijuvenile hormone precocene II have revealed that neither the onset of diapause nor the production of ecdysteroid hormones and pupation are controlled by juvenile hormone alone. A second unknown factor seems to be necessary, which is controlled by the photoperiod. The maintenance of diapause in this species is characterized by a general state of hormonal inactivity. Experiments to terminate diapause were carried out by ecdysteroid injections of different concentrations at different times into diapausing *Laspeyresia* larvae. Only in late diapause as little as 23% of the specimens responded with pupation. This very small number suggests that diapause termination and onset of postdiapause development cannot merely be a problem of molting hormone elevation. Nevertheless, ecdysteroids seem to be of great importance. A transfer of diapausing corn borer larvae *(Ostrinia nubilalis)* from short day diapause conditions to nondiapause long day conditions caused an increase in the ecdysteroid level after delay of 10 days, all test insects pupating successfully at day 16.

These experiments clearly show that distinct hormonal levels alone are not sufficient to regulate insect development, but that a sophisticated balance between the hormones, as well as a graduated response to environmental factors are required for the successful diapause-nondiapause regulation.

6.3 Pupal Diapause Control

In contrast to the endocrine regulation in larvae, juvenile hormones seem to be less important in pupal diapause regulation. This can be concluded from experiments performed by Shimada (1983) with the swallowtail, *Papilio machaon*. The caterpillars enter pupal diapause when a critical photoperiod of $\leqslant$14.26 h is applied to the penultimate and last larval instar. When specimens in the early penultimate instar were allatectomized they responded with a pupal molt and formed so-called precocious pupae. These pupae were diapause or nondiapause determined depending on the photoperiod experienced post allatectomy. This indicates that diapause induction is independent of juvenile hormones, at least in this species.

Molting hormones are found only at a trace level in all pupal diapausing insects studied so far. However, ecdysteroids are necessary for adult development and differentiation. Therefore, very early attention was drawn to the fate of these hormones and their regulation in the onset of postdiapause development. In a classical study with *Hyalophora cecropia* pupae, Williams (1946, 1947, 1952) demonstrated that the brain has a key function in diapause regulation. Active brains from nondiapausing pupae transplanted into diapausing ones caused the onset of postdiapause development by prothoracic gland activation. Furthermore, brains from chilled donors were able to function even in unchilled diapausing recipients. These experiments led to the now well-established opinion that pupal diapause is based on the brain's inability to synthesize the prothoracicotropic hormone (PTTH). The potency for synthesis and/or release of PTTH is regained spontaneously after a long diapause, but chilling and subsequent warming enhances this capacity markedly. Consequently, exogenous molting hormones should be able to initiate postdiapause development as well. This was proven by many studies with different pupal diapausing species. Modern methods for ecdysteroid titer determination, like radio-immunoassay, have presented further evidence for this idea. When, under natural environmental conditions the temperatures begin to rise and photoperiods become longer, the brain starts PTTH production and secretion, causing elevated ecdysteroid levels. Under laboratory conditions the ecdysteroid titer increased about 40-fold within 3 days when diapausing *Hyalophora cecropia* pupae were transferred from 6°-8°C (chilling temperature) to 27°C (McDaniel 1979). Similar observations have been made by Loeb (1982) in the tobacco budworm, *Heliothis virescens.* In this species ecdysteroid titers increased 500-fold within 11 days.

In conclusion, the pupal diapause state is characterized by the brain's inability to synthesize PTTH, which is regained immediately by favorable environmental stimuli like photoperiod and temperature.

6.4 Adult Diapause Control

Adult diapause is a delay of reproduction which is most obvious in female insects (see p. 71). Reproduction, however, is under control of the juvenile hormones (Engelmann 1983). Therefore, a linkage of adult diapause and juvenile hormones seems to be feasible. Indeed, reproductive diapause is under control of these hormones in all insect species studied so far and no fundamental differences seem to exist among the species. The results have been recently reviewed by de Wilde (1983). Most of the data obtained on this topic were gained with the Colorado potato beetle, *Leptinotarsa decemlineata.* The following outline will, therefore, concentrate on that species.

Female *Leptinotarsa* respond to day length immediately after adult emergence. Long day photoperiods (> 15 h) lead to continuous development. These animals feed actively. Vitellogenins are synthesized by the fat body and oogenesis starts immediately. Oviposition occurs at day 5 after emergence. The flight muscles are fully developed 3 days later.

Short photoperiods (< 15 h) after adult emergence, however, cause diapause induction. Feeding during the prediapause period elicits considerable fat reserves and diapause proteins (see p. 86), but neither ovary maturation nor flight muscle differentiation can be observed. The dormant state will be entered about 12 days after hatching. By this time the beetles invariably burrow into the soil and become dormant.

A comparison between the long day (nondiapausing) and short day (diapausing) individuals concerning the endogenous juvenile hormone titer as well as the in vitro synthetic activity of the corpora allata reveals significant differences in the course of adult life. Endogenous juvenile hormone titers are equal in both groups at the time of adult ecdysis. In nondiapausing animals the titer increases continuously and remains at a more or less constant level throughout the reproductive period. Under diapause inducing photoperiods only a short rise can be seen and the titer declines rapidly to subdetection limits (day 6–8). During the period of diapause, juvenile hormone is undetectable in the hemolymph. Substantiating evidence has been obtained by corpora allata in vitro incubation assays. However, glands from short day animals exhibited a reduced activity at the time of emergence and differed significantly from those dissected from long day specimens. Total inactivity is observed during the time of diapause.

Strong evidence for the role of the juvenile hormone in diapause induction was received by corpora allata extirpation experiments. Beetles from long day photoperiods (nondiapause) were deprived of their glands. These animals gained a diapause-like feature immediately. Oosorption and flight muscle atrophy have been observed, and the animals showed the typical digging diapause behavior. The effects were shown to be reversible by reimplantation of two pairs of active corpora allata.

On the other hand, diapause induced beetles cannot be activated by a single corpora allata implantation. As many as 12 active pairs of the glands are necessary to prevent the diapause state. These results drew attention to another postulated regulating system, most probably located in the brain, which was recently investigated by Khan et al. (1982a,b, 1983) and Khan (1983). These studies included disruption of the nervi corporis allati as well as short day-long day corpora allata transplantations and vice versa with specimens from different physiological conditons. The data gave clear evidence for a nervous corpora allata inhibition and a negative feedback regulation for juvenile hormone via the brain.

Furthermore, a humoral corpora allata stimulating factor from the brain was found. However, the existence of a humoral allatohibin, which is probably weaker in its action than the nervous inhibition, cannot be completely ruled out.

Besides juvenile hormone ecdysteroid hormones may also be engaged in diapause physiology of *Leptinotarsa*. Early prediapausing adults exhibit higher ecdysteroid titers than nondiapausing ones (Briers and de Loof 1981), and a steady increase was observed during diapause. The role of these hormones may be linked to the appearance of the diapause proteins (see p. 87) (Briers et al. 1982; Peferoen et al. 1982), but they are not involved in the regulation of the corpora allata (Khan et al. 1982a). They may, therefore, be regarded as less important factors in diapause regulation.

7 Conclusions

Diapause is a widespread phenomenon in the life history of insects. Its adaptive value is to survive unfavorably environmental conditions. Diapause is thought to be a modern phenomenon in phylogeny in which evolution proceeds rapidly.

Numerous differing diapause types have been described, but no general classification can ge given. Analogous to the various unsuitable environmental situations which require a dormant state, several adaptive mechanisms in diapause regulation and maintenance have been evolved. The most important factor for diapause induction is the photoperiod, while changing temperature conditions are usually signals for the onset of postdiapause development. Accumulation of reserve products (carbohydrates, lipids) and water loss are interpreted as metabolic preparations for the time of dormancy. Diapause-life is characterized by reduced respiratory rates and inactivity.

Insect diapause is under hormonal control. Ecdysteroids and juvenile hormones as well as a diapause hormone are engaged in its regulation. These hormones are thought to be under neuro-endocrine control of the brain. However, the main question of how the environmental stimuli influence the brain is still an enigma.

Acknowledgements. I am indebted to Prof. Dr. K.H. Hoffmann for his invitation to write this chapter, to P.B. Koch for helpful discussions, and to Dr. J.C. Kirkpatrick for critical reading of the manuscript.

Chapter 4

Metabolic Energy Expenditure and Its Hormoral Regulation

Rolf Ziegler[1]

Contents

1 Introduction

All living beings need energy to maintain a low level of entropy, that means to stay alive, grow, and behave as living organisms. Therefore, they continually have to ex-

1 Institut für Tierphysiologie, Freie Universität Berlin, Grunewaldstr. 34, 1000 Berlin 41, FRG

Environmental Physiology and Biochemistry of Insects, ed. by. K.H. Hoffmann

pend energy gained from the environment. Green plants gain this energy directly from the sun, while animals have to feed and use the energy of other living organisms. When animals have the possibility to feed, they often consume more than they need at the moment; they use the excess to accumulate reserves. These reserves of chemical energy can be used when the environment is less favorable, when the supply of food is inadequate, or when the animals have to be physically very active.

Animals are constantly expending energy and it is far beyond this review to cover the whole scope of energy expenditure. I will not deal with basic energy expenditure, but I will try to deal with two extreme and defined situations: first, the metabolic energy expenditure when animals do not feed, when they starve; a starving animal, therefore, depends on reserves. The utilization of these reserves has to be well-regulated, to ensure that enough are mobilized to meet immediate needs, but that they are not used up too fast. Second, the expenditure of energy during flight involves the greatest increase in metabolic rates, and the highest metabolic rate that we know of in any animal. This again poses interesting questions concering the regulation of energy expenditure.

This review will be concerned, therefore, with three main topics: the way in which insects gain the necessary energy-rich material from their diet; and the way in which they cope with "metabolic stress" when starved, or when flying.

2 Nutrition, Digestion, and Absorption

2.1 Requirement of Food

The fundamental requirement of food is that it should contain fuels which can be used for the liberation of chemical energy. Therefore, the major components of food are energy yielding compounds: carbohydrates, lipids, and proteins. These substances are also needed to construct the body of an animal, by mechanisms which themselves also need energy. Animals also need relatively smaller amounts of vitamins and minerals.

Most nutritional requirements are similar for all animals; the main differences are in what is required for a balanced diet (Dadd 1973). The diet of insects can be supplemented by symbionts. Insects need carbohydrate as a major dietary constituent, but, as a rule, they show no marked specificity for any particular carbohydrate, so most carbohydrates can fulfill this need (House 1974a).

Sufficient lipids are found in most diets, especially as phospholipids can be used too by insects (Turunen 1979). Most lipids can also be synthesized from other sources, but insects cannot synthesize the unsaturated fatty acids, linoleic and linolenic acid; these they require in their diet. No insect is able to synthesize sterols, these, too, are an essential part of the diet (Robbins et al. 1971).

Proteins are used to gain energy and to build animal tissue. For this, the 20 amino acids forming proteins are needed. Some can be synthesized by insects, but ten are essential dietary amino acids; interestingly, the same ten which are essential to the rat (Dadd 1973). The vitamin requirements of insects are more or less the same as those of mammals. Of course, insects do not need vitamin D for the formation of bones. As in mammals, vitamin C is for some insects a vitamin, for others not. Some insects, whose natural diet includes fresh plant material need exogenous ascorbic acid, while no insect whose diet excludes fresh plants has been shown to require it (Dadd 1973).

A role for vitamin A was questioned for some time. In some insects it is essential for normal growth rate, while in others development is normal when the vitamin is lacking (House 1974a). Probably in all insects, without vitamin A, visual sensitivity is very low (Dadd 1973; Shimizu et al. 1981). Most insects take up the precursor ß-carotene readily. The tobacco hornworm, *Manduca sexta*, cannot absorb it from a semisynthetic diet (Dahlman 1969), therefore, adult tobacco hornworms raised on this diet have a strongly reduced visual sensitivity compared to adults which were raised on tobacco leaves (Carlson et al. 1967).

Of the special requirements of insects, the need for exogenous sterols was already mentioned. With only few exceptions cholesterol fulfills this requirement. The exceptions are monophagous insects whose host plants have other sterols, like *Drosophila pachea* which requires Δ^7-stigmasten-3ß-ol, the sterol found in its host plant, the cactus *Lophocerus schotti* (Dadd 1973). There are other special requirements like carnitine, which in Tenebrionidae must be supplied from an exogenous source (Gilmour 1961).

Insects living on a very restricted diet usually have normal nutritional requirements. The cloth moth, *Tineola bisselliella*, which feeds and digests wool, has food requirements which resemble those of other insects (Fraenkel and Blewett 1946); it does not grow on clean wool, it needs contaminated wool. This is somewhat similar to the situation in the wax moth, *Galleria mellonella*, which can live only on contaminated beeswax (Gilmour 1961).

In *Rhodnius*, which feeds on vertebrate blood containing low levels of water soluble vitamins, we see another strategy which insects adopt; they supplement the diet with vitamins obtained from symbionts.

Holometabolous insects accumulate reserves during larval development and live off them as pupae, or even as adults too, as in some Lepidoptera, for example the cecropia moth, *Hyalophora cecropia.* Some insects, which live as larva and adult on completely different food, are able to exploit different resources, and may supplement a deficient diet of one stage by nutrients gained from an earlier stage. *Agria housei*, for example, must acquire α-tocopherol during larval feeding in order to reproduce; otherwise egg development in the female is arrested in the early blastular stage (House 1974a).

2.2 Digestion

The intake of food is controlled by many chemical and physical factors of the food and by internal factors of the insect (Gelperin 1971; Bernays and Simpson 1982). The food itself may vary greatly in consistency; from liquids, like plant saps or blood, to solid food ranging from ripe fruits to hard wood. In adaptation to these differences in nutrition we find corresponding differences in mouthparts and in the form of the digestive tract. But all insects have a foregut, midgut, and hindgut. The foregut comprises the esophagus, which passes on the food, and the crop, which mainly stores ingested food until it can be processed to the midgut. The midgut with its ceca is the main place of digestion and resorption (House 1974b). There is also extraintestinal digestion in some insects, either by injection of digestive enzymes into the prey or by excreting digestive enzymes with the feces. In the larvae of *Calliphora* the proteolytic enzymes persist in the excreta and are responsible for some liquefaction of the meat, in which the larvae live (Wigglesworth 1953b). The leaf cutting ant, *Atta texana*, grows fungi on leaves and then feeds on the fungi. Digestive enzymes in the feces of the ant help to culture successfully the fungus and so aid nutrition (Martin et al. 1975).

2.2.1 Digestion of Carbohydrates

Insects normally have digestive enzymes which can hydrolyze carbohydrates, but few of the enzymes are well-characterized (see also Chap. 1). Normally, insects have amylases and carbohydrases. *Lucusta migratoria* has in its intestinal tract at least seven different carbohydrases, with which it can digest all naturally occurring carbohydrates, except cellulose and inulin (Chippendale 1978b).

Many insects of different orders ingest a diet rich in cellulose, such as wood, foliage, and detritus, and many are able to use cellulose as a fuel. Examples of such insects are found among the Thysanura, Orthoptera, Isoptera, Coleoptera, and Hymenoptera. Cellulose digestion is accomplished by a collection of enzymes believed to be similar to those responsible for cellulolytic abilities of fungi. One or other of these enzymes may be made by the insect, or the insect can use enzymes of symbiontic bacteria and/or protozoa of the mid- or hindgut. Some insects can use the cellulase of fungi which either grow on the wood on which they feed, or of fungi which they cultivate (termites) (Martin 1983). It is difficult to determine conclusively whether an insect synthesizes the enzymes itself, or whether they are made by symbionts.

2.2.2 Digestion of Lipids

Lipases which hydrolyze triacylglycerols are very common in insects (Downer 1978). In mammals lipid digestion is supported by emulsifying lipids by the help of bile salts. As bile salts are derivatives of sterols which insects cannot synthesize, bile salts cannot be used to emulsify lipids. Some insects use fatty acylamino acid complexes to facilitate lipid digestion (Downer 1978).

2.2.3 Digestion of Proteins

The proteases found in insects appear to be similar to mammalian trypsin and chymotrypsin; they work at neutral pH. As the pH in insect gut is usually close to neutral or slightly basic, pepsin-like proteases are not common, but in blowflies the pH of midgut may be as low as 3 (House 1974b). Carboxypeptidases and aminopeptidases were shown to occur in insects, too.

2.3 Absorption of Food

In insects food is mainly absorbed by the midgut and midgut ceca. The crop appears to be impermeable to most substances except some lipids (Treherne 1967; House 1974b; Downer 1978). Absorption of nutrients starts immediately after feeding. Water and ions are absorbed in the hindgut (see Chap. 7).

The uptake of nutrients from the gut appears to be by facilitated diffusion. Thus, diffusion is assisted by rapid removal of water from the gut lumen, resulting in high concentrations of solutes in the lumen, and by rapid removal of absorbed nutrients from the hemolymph by conversion to other solutes, or by storage in the fat body (Treherne 1967).

Water which is taken up with food is removed from the gut in the ceca and the rectum. In *Schistocerca gregaria*, fluid above the combined output capacity of these organs is excreted in the feces. Two to 4 h after the intake of food has stopped, most of the crop fluid has been absorbed and the input of fluid into the midgut from the crop is smaller than the absorptive capacity of the ceca. So fluid from the Malpighian tubules moves forward carrying solutes into the anterior midgut and the ceca, forming a countercurrent. As the anterior midgut and the ceca reabsorb water, nutrients are taken up more completely. So the advantage of a countercurrent is used when food is scarce, but a countercurrent will also accumulate noxious substances. This cannot be tolerated for a long time, so if there is enough food no countercurrent will be formed (Dow 1981).

2.3.1 Absorption of Carbohydrates

Carbohydrates are broken down to monosaccharides. For glucose, the most common monosaccharide, there is no evidence of an active transport system; glucose diffuses rapidly into the hemolymph and it is quickly taken up by the fat body and converted to trehalose. The rate of absorption appears to be limited by the rate of conversion of glucose to trehalose and by the rate at which fluid containing glucose is released from the crop into the midgut (Treherne 1967; Chippendale 1978b).

2.3.2 Absorption of Lipids

In vertebrates dietary lipids are degraded to free fatty acids and 2-acylglycerol, which form in combination with bile salts negatively charged aggregates (the so-called micelles) and these penetrate the mucosal wall. Inside the cell triacylglycerols and phosphoglycerols are reconstituted. Together with cholesterol, cholesterol esters, and proteins they are incorporated into the chylomicrons, which are discharged into the lymphatic system (Johnston 1970). In insects, there are few studies on lipid resorption, but they indicate that insects use different absorptive mechanisms. The cabbage butterfly, *Pieris brassicae*, converts the phospholipid lecithin to lysolecithin which is surface-active and so probably could not only enter the mucosa, but could also function in the emulsification of less polar lipids (Turunen 1979). Free fatty acids can be taken up by the crop (Downer 1978).

2.3.3 Absorption of Proteins

Protein is usually digested to amino acids which are absorbed. There seem to be exceptions; in *Pediculus* and *Rhodnius prolixus*, katahemoglobin is found in hemolymph. Amino acids again are taken up most readily by the midgut ceca. There does not seem to be an active transport; the uptake appears to work by diffusion (House 1974b).

3 Starvation

No animal is constantly feeding; all animals have interruptions of feeding activity when they live on accumulated reserves. Even in insect larvae which sometimes appear as "feeding machines" as they seem to spend all their time feeding, feeding periods are interrupted by postprandial rests. The nonfeeding periods may be very short like in *Manduca sexta*, where these interruptions last only for minutes (Reinecke et al. 1980), or long as in *Rhodnius* which feeds only once per instar. For some time after feeding has stopped, an animal will be able to resorb as much nutrients from the gut as it utilizes, but, as nonfeeding is continued, it can no longer gain sufficient nutrients from the gut and it begins to starve; a situation involving a change of metabolism towards utilization of reserves. In the development of all insects, one finds periods when they fast. Typically, during molting insects cannot feed, and have to rely on reserves.

If an animal continues to starve it might be able to reduce its energy expenditure by reducing those parts of metabolism which are at that moment unnecessary, or postponing them, or by reducing total metabolism. Many insects reproduce soon after they have reached the adult stage. Some insects do not need to feed to reproduce, they are said to be autogenous, like some Lepidoptera, which cannot feed as adult, while most insects are anautogenous and have to feed before they can reproduce. If an anautogenous adult insect is starved, it will stop maturing eggs and will resorb the yolk of partially mature eggs (Engelmann 1970). Starvation of adult insects results in a decline of juvenile hormone synthesis by the corpora allata which interrupts egg maturation. If feeding is resumed, juvenile hormone synthesis restarts, and egg maturation can proceed (Tobe and Chapmann 1979).

While starving, man can reduce his resting metabolic rate by about 15% (Owen et al. 1979). Insects can reduce their metabolism, measured as total respiration, to a much greater extent during starvation. In adult *Manduca sexta* repiration decreases during starvation by more than 50% within 3 days (Siegert and Ziegler 1982). In starving larvae, the decrease in respiration is even faster and greater: if larvae of the third day of the fifth instar of *Manduca sexta* are starved, respiration decreases within about 20 h to 25–30% (Fig.4.1a) (Siegert and Ziegler 1982). During the molt from the fourth to fifth instar, when the larva cannot feed, a similar decrease is seen (Fig. 4.1b). At the end of the last instar, when the animals stop feeding there is again a very pronounced decrease in respiratory rate (Siegert 1983).

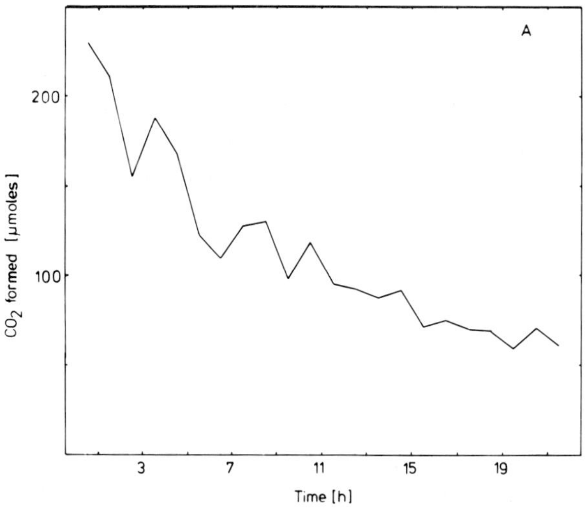

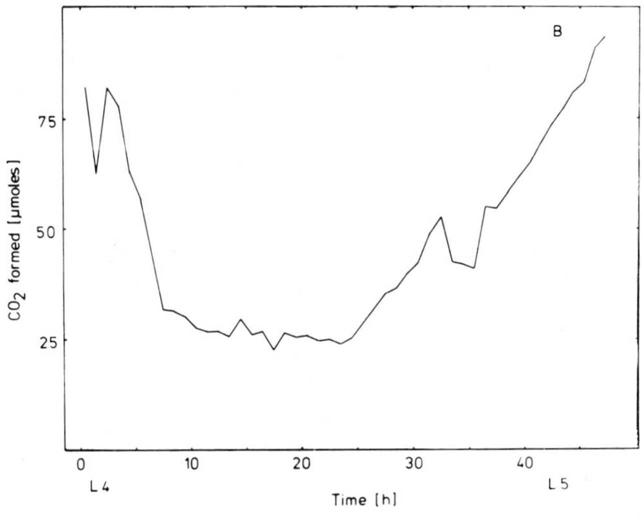

Fig. 4.1. A Respiration of a starving larvae of *Manduca sexta* (3rd day of the 5th instar); **B** respiration of a molting larvae of *Manduca sexta* L4 is the 4th instar and L5 the 5th instar

The decrease of respiration in larvae of *Bombyx mori* starts about 1.5 h after a meal. Within 24 h of starvation, respiration drops in the fourth and fifth instar larvae to about 30% of the value of fed animals. The steepness and the extent of the drop depends somewhat on the age of the larva (Bosquet 1971).

3.1 Influence of Starvation on Larvel Development in *Manduca sexta*

Most insects have a more or less fixed number of larval instars during which the insect has to be able to survive and grow. All insect larvae need to accumulate reserves for use during the molting process, but holometabolous insects, in addition, have to be able to survive the (sometimes long) pupal period without food, and at least in some species where the adult does not feed, these reserves must be sufficient to sustain adult activities, such as flight and reproduction. If larvae of Lepidoptera are starved from the beginning of the last instar, they die without pupating or molting; if they are starved during the penultimate instar or at the begin of the last instar and then refed, they form supernumerary larvae. Larvae starved during the last instar from small pupae (Nijhout 1975; Bhaskaran 1981). The molt to the next instar (including pupation) is controlled by hormones: JH (juvenile hormone), ecdysone, and prothoracicotropic hormone (Riddiford and Truman 1978). In *M. sexta* the size of the head capsule indicates whether or not at the end of the fifth instar metamorphosis will occur. If the head capsule exceeds a threshold width, the titer of JH declines during the fifth instar and the tissues are reprogramed for pupation (Nijhout 1975). Prolonged starvation of *Manduca* larvae at the begin of the last instar prevents the decrease in JH titer; such animals do not metamorphose, but molt to a supernumerary larva. It is thought that the JH titer does not decrease in starved larvae because the normal increase of JH esterase, which inactivates JH, does not occur (Cymborowski et al. 1982), but it could also be that the corporate allata are not inactivated (as they are in fed animals) and so continue to produce JH (Bhaskaran 1981). These mechanisms ensure that animals which are too small, and so do not have sufficient reserves, get another chance to accumulate reserves, while animals which have sufficient reserves, proceed in their metamorphosis.

3.2 Hormonal Activation of Glycogen Phosphorylase in Larval Fat Body of *Manduca sexta* During Starvation and Molt

Larvae of *M. sexta* molt to the fifth and final instar 10 days after hatching from the egg. The inactivity during the transition from fourth to fifth instar lasts more than 24 h. In the fifth instar, larvae resume feeding for another 4 to 5 days before they stop feeding again, empty their gut, and become wandering larvae searching for a place to pupate. Pupation occurs on day 10 after the larva has reached the last instar (Reinecke et al. 1980). About 20 days after pupation, the adult will eclose.

Except for the periods around ecdysis, larvae of *M. sexta* are more or less constantly feeding, but when they become wandering larva they will not feed for at least 25 days until they emerge as adults.

In *M. sexta*, hemolymph sugar and fat body glycogen decrease during the molt from fourth to fifth instar and when the larva stops feeding on the 5th day of the fifth instar, but during the active feeding period of the fifth and last instar a marked increase in both carbohydrate stores occurs. Hemolymph sugar not only increases in total amount (taking into account the increase of hemolymph volume), but also the concentration increases from ca. 10 to about 18 mg/ml. Fat body glycogen increases during the feeding period of the last instar from 1 to 100 mg (Ziegler et al. 1978; Siegert 1983).

Glycogen phosphorylase (GP) is the enzyme, which breaks down glycogen to glucose-1-phosphate, which in turn can be used to form hemolymph sugar or to enter glycolysis and yield energy. If glycogen is synthesized during the feeding period it might be expected that GP would be inactive, while if glycogen decreases during nonfeeding it would be likely that GP is activated. GP exists in an active and an inactive form; these can be rapidly interconverted according to the need of the animal and is under the control of hormones (Steele 1982). In larvae of *M. sexta* during feeding, only about 10% of GP is in the active form, while when the animals do not feed during normal development 30-50% is in the active form (Siegert 1983; Ziegler 1984). The described changes during molt look as if caused by fasting, but because the changes during the molt cycle might be developmental changes independent of starvation, we tested the hypothesis directly: animals were injected with JH and ecdysone, and animals were starved during their feeding phase. The hormone injections never activated GP (Ziegler 1979), but starvation of larvae on the 3rd day of the last instar (these are large enough for the experiments and still 2 days from the end of the feeding phase) causes a decrease in glycogen of the fat body, and GP is activated.

The activation of GP is rapid: within 1 h of starvation the activity of GP is increased and after 3 h is maximal with 30-50% of the enzyme in the active form. After 36 to 48 h, GP is inactivated again even though starvation continues (Siegert and Ziegler 1983a). The reason for the inactivation is unknown, but it might explain why at the end of the larval feeding phase, when the animals will not feed for at least 25 days, GP is activated only for 24 - 48 h.

In mammals the activation of liver GP during starvation is controlled by hormones. In the corpora cardiaca (CC) of insects there is a peptide known which is a good candidate hormone for controlling fat body GP during starvation. It was called hyperglycemic factor because it increases hemolymph sugar in *Periplaneta americana*, and similar peptides occur in many insects (Goldsworthy and Gäde 1983), including *M. sexta* (Ziegler 1979). Removal of CC on the 2nd day of the fifth instar of *M. sexta* prevents activation of GP when these animals are starved on the 3rd day. Injection of CC extracts into these animals activates GP. This suggests that the activation of fat body GP during starvation is controlled by a hormone from the CC, and this hormone was called glycogen phosphorylase activating hormone (GPAH) and not hyperglycemic, because changes in hemolymph sugar in *Manduca* are very small (Ziegler 1979; Siegert and Ziegler 1983a).

What is the specific signal that induces this rapid activation of GP? The most obvious change at the onset of starvation is the cessation of feeding. To dissociate feeding activity from the metabolic effects of the uptake of nutrients, animals were fed with a diet which contained only agar and water. *Manduca* feeds well on agar, but cannot gain nutrients from it. In animals fed with agar, GP is activated just as in starved ones. Termination of feeding is, therefore, not the signal for activation of GP. Of course, a decrease in hemolymph sugar could be the specific signal as it is in mammals, but total hemolymph sugar does not decrease during the first 36-48 h of starvation; the level is even somewhat increased. Hemolymph sugar in these insects is mainly trehalose (more than 90%) and it is the level of this sugar which increases during starvation. So a decrease of the trehalose level cannot be the signal for the activation of fat body GP; on the contrary, trehalose levels are kept high by the activated GP. Larvae of *Manduca* have glucose as a minor hemolymph sugar, whose level in fed animals varies depending on age between 250 and 1000 μg/ml. The glucose concentration decreases rapidly during starvation; a decrease is seen within 1 h and after 3 h the level of glucose is down to less than 100 μg/ml (Fig. 4.2) (Siegert

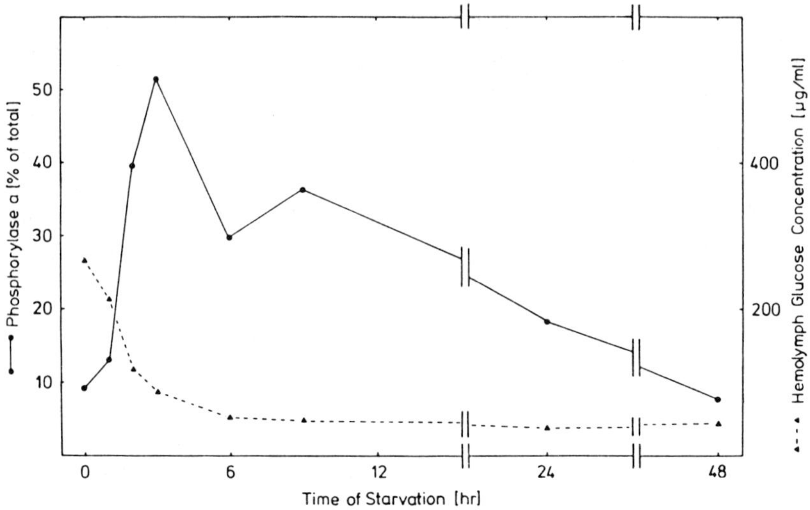

Fig. 4.2. Changes in the activity of glycogen phosphorylase and of hemolymph glucose concentration during starvation of larvae of *Manduca sexta* (3rd day of 5th instar). (Modified from Siegert 1983)

1983). This decrease in hemolymph glucose parallels the increase in GP activity (Fig. 4.2). Thus, the decrease in glucose levels in the hemolymph could be the specific signal for the secretion of GPAH, and thus for the activation of fat body GP. Injection of glucose into starving larvae prevents activation of GP, while injection of trehalose has only a minor influence (Siegert 1983; Siegert and Ziegler 1983b). The amount of glucose that needs to be injected to prevent GP activation is very large compared with the amount of glucose in the animal, but the rate of conversion is high (Sec. 2.3), and if larvae of this age are feeding, they will ingest within 3 h more sugar, than the amount injected.

If glucose is the specific signal which triggers secretion of GPAH, then it is to be expected that when fat body GP is activated during development, hemolymph glucose would be low. That is indeed so; hemolymph glucose decreases from ca. 1100 µg/ml at the end of the fourth instar to 200 µg/ml when the animal has slipped the head capsule in preparation for ecdysis. The following day (the first of the last instar), when the animal is feeding again, hemolymph glucose increases to about 1100 µg/ml and GP is inactivated (about 10%). Also, at the end of the larval feeding phase fat body GP is activated again during normal development, and the hemolymph glucose level has fallen by this time to below 200 µg/ml (Siegert 1983). High levels of glucose prevent the secretion of GPAH rather than reduce the responsiveness of the fat body, although the responsiveness changes during development (Ziegler 1984), but injection of CC extracts into glucose-injected starved larvae activates GP (Siegert and Ziegler 1983b).

If the decrease in glucose levels is the signal for the secretion of GPAH, and so for the activation of GP, then it would be comparable with the situation in mammals. In mammals, a decrease in blood glucose inhibits the secretion of insulin and stimulates the secretion of glucagon; the latter hormone activates GP in liver and so increases blood glucose. In insects, it seems to be a decrease of glucose too, which stimulates the secretion of a hormone responsible for the activation of GP. In *M. sexta* there is no direct evidence at the moment of a hormone like insulin being involved. GPAH has physiological similarities with glucagon, both activate GP and so influence sugar levels in blood. There are reports that a peptide from the nervous system of *M. sexta* is immunological glucagon-like and induces glycogenolysis in fat body (Tager et al. 1976). It appears unlikely that GPAH is immunological glucagon-like because: (a) there is a glucagon-like peptide in *M. sexta*, but it is found in several tissues (El-Salhy et al. 1983; Maier and Ziegler, unpublished), while GPAH-activity if found only in CC (Ziegler 1979); (b) GPAH appears to be closer related to locust AKH than to glucagon, as AKH activates GP in *Manduca* in a very low dose (Ziegler, unpublished), while high levels of glucagon are needed (Ziegler 1979). Glucagon-like peptides, even in mammals, have functions other than the regulation of energy metabolism, they are thought to modulate neural activity (Stell et al. 1980). In *M. sexta* it could have a similar function.

If GP is activated in *M. sexta* by GPAH, this however does not increase the levels of glucose like in mammals, but those of trehalose. So, in this insect, glucose appears to be the measured hemolymph sugar, but it is trehalose which is regulated. This allows the maintenance of high levels of trehalose; it does not have to decrease in concentration to activate its resupply system. This may be additionally important in insects because of their open circulatory system (Sec. 4.3).

Interestingly, that whenever GP is activated in starved larvae, it has returned to the inactivated state within 24 h to 48 h. This could be caused by a decreased responsiveness of the tissue or by the termination of GPAH secretion. Although there are strong changes in responsiveness of the target tissue fat body (Ziegler 1984), the inactivation of GP appears to be caused by the termination of GPAH secretion, because GP can be activated again by the injection of extracts from CC (Siegert et al. 1982).

In mammals, glucagon not only activates GP, but also stimulates gluconeogenesis; in *M. sexta* gluconeogenesis has not been examined. In *Locusta migratoria*, gluconeogenesis is increased during starvation, but no influence of a hormone was found (Davies and Goldsworthy 1982).

3.3 Energy Metabolism in Starved Adult *Manduca sexta*

3.3.1 Changes in Carbohydrate Metabolism During Starvation

Adults of *M. sexta* can be starved without encountering problems with changes in developmental state as in larvae. Depending on the amount of reserves carried over from larval life and physical activity, starved adult females live for about 6 days (range 3-12 days), while fed females live about 11 days (range 6-20 days); at the time of death, practically all reserves have been used up and the fat body has more or less disappeared, and the whole animal appears "empty". For example, the amount of glycogen in the fat body has decreased in males from about 7 mg at imaginal molt to 0.8 mg by the 3rd day of starvation, while in fed animals it has increased to 12 mg.

GP, which is necessary for the degradation of glycogen, is more and more activated as starvation is prolonged, but the rate of change in activity is slow compared with that in larvae; in larvae, the activity increases within hours, in adults it takes days. There is also no indication of an inactivation of GP after long periods of starvation comparable to that seen in larvae (Ziegler, unpublished).

3.3.1.1 Role of Hemolymph Sugars in the Activation of Adult Fat Body GP

Hemolymph trehalose decreases during starvation; at the imaginal molt, females have about 18 mg sugar/ml hemolymph and this decreases to 5 mg/ml by the 5th day of starvation. In fed females on the 5th day, concentrations of 27 mg/ml are found (Ziegler, unpublished). Hemolymph volume decreases during starvation too, and this means that the decrease in the total amount of hemolymph sugar is proportionally greater than the decrease in concentration.

In larvae, a decrease in the concentration of hemolymph glucose stimulates the secretion of GPAH (which activates fat body GP), but in adults, hemolymph glucose is extremely low, usually <50 µg/ml, and variations do not correlate with the nutritional state of the animal or GP activity. Thus, in adults, a decrease in glucose concentration cannot be the specific signal for GPAH release. If hemolymph trehalose and GP are measured in the same animals, a correlation can be drawn between the decrease in hemolymph carbohydrate and the increase in GP activity (Fig. 4.3). Animals fed with sugar water on the 3rd and 4th day of adult life had on the 5th day a concentration of 23 mg sugar/ml hemolymph and only 28% of phosphorylase in the active form, while animals starved for 5 days had only 3 mg sugar/ml, but 47% of total phosphorylase in the active form; in adults, therefore, the decrease in hemolymph trehalose may be a signal for the activation of GP.

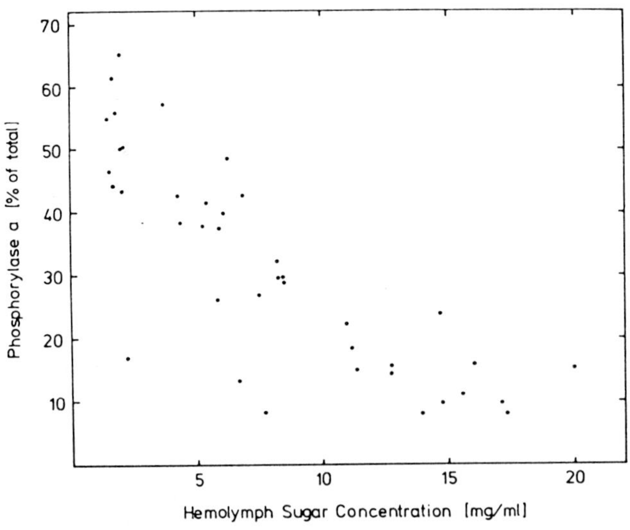

Fig. 4.3. Correlation between hemolymph sugar concentration and the percentage of fat body phosphorylase in active forms in adults of *Manduca sexta*

3.3.1.2 Role of GPAH in the Activation of GP During Starvation in Adult Manduca sexta

Injections of CC extracts into adult *M. sexta* activate fat body GP (Ziegler 1984), and so a role for GPAH in the control of the enzyme during starvation of adults might seem likely. However, in cardiacectomized adults, GP is activated during starvation (5 days) just as it is in sham-operated controls. If the cardiacectomized animals are fed, GP is not activated (Ziegler, unpublished). Clearly, GP in starving adults is not activated by GPAH from the corpora cardiaca.

3.3.2 Changes in Lipid Content During Starvation

During 2 days starvation in larvae, no change in the concentration of hemolymph lipids was detected (Siegert 1983). The concentration of hemolymph lipids in larvae is very low, 2 mg/ml, but at the end of larval life, when the animals stop feeding and prepare for pupation, the levels increase to 8-10 mg/ml; it is not possible to say whether this is a mere developmental effect or whether the cessation of feeding causes the increase.

After the imaginal molt, hemolymph lipid levels increase rapidly to ca. 40 mg/ml (Ziegler 1984). If adults are starved the level increases further, often to values > 100 mg/ml, and the lipid content of the fat body decreases. As we have seen (Sec. 3.3.1.1), hemolymph sugars are very low in starved adults. If hemolymph sugar is increased by feeding adult animals, hemolymph lipids decrease in concentration.

If extracts of CC are injected into resting adults, hemolymph lipid levels increase (Beenakkers et al. 1978; Ziegler 1980, 1984). The CC of *Manduca sexta* contain an adipokinetic hormone, which is secreted during flight and mobilizes lipids from the fat body as fuels for flight (Sec. 4.6.2). The increase in hemolymph lipids during starvation is independent of the CC because in starved cardiacectomized adults, hemolymph lipids are increased by the 5th day to 92 mg/ml compared with 22 mg/ml in fed cardiacectomized adults (Ziegler, unpublished).

Thus, the increases in hemolymph lipid levels and the activation of fat body GP during starvation in adults are not controlled by hormones from the CC, even though peptides are found in the CC of adults which can influence both (Ziegler 1984; Ziegler and Gäde 1984). Both changes appear to be controlled by decreases in hemolymph sugar during starvation; the hemolymph lipid levels in adult *Locusta migratoria* appear to be similarly regulated (Goldsworthy 1984).

3.4 Considerations on the Importance of Different Controls in Carbohydrate and Lipid Metabolism in Starving Larvae and in Adult *Manduca sexta*

It is not possible to say why larvae and adults have developed different mechanisms for the control of carbohydrate and lipid metabolism to cope with starvation, but a comparison of the life styles of larvae and adults may suggest some advantages of the different controls. Larvae of *M. sexta* feed more or less constantly to build up their body and to accumulate reserves for pupal and adult life. They support their energy expenditure directly from nutrients absorbed by the gut. Their metabolism is geared to an abundance of food, and therefore, towards the synthesis of reserves. If these animals starve, because they have no food or because they molt, they must change their metabolism rapidly from anabolism to catabolism; a larva which has run out of food must search for new food, and, therefore, needs energy, as does a molting insect. A rapid response to starvation is achieved by coupling the activation of GP to the decrease of hemolymph glucose because this depends on the balance between the absorption by the gut and the conversion of glucose to trehalose by the fat body. The level of hemolymph glucose is, thus, a very sensitive indicator for starvation.

If starvation is prolonged, GP is inactivated. It makes sense for a larva to search actively for a new food source but, if after some time a new plant is not found, a protracted search will only use nutrient reserves wastefully. It may be better for the animal before too long to reduce its energy metabolism and to form a small pupa (and latter a small adult which may lay a few eggs) rather than to continue the search for food. Of course, this makes only sense if the larva has acquired sufficient size and reserves. If small larvae are starved, they continue to search for food until they die. Larvae of *M. sexta* must reach a weight of about 4 g before they can form a pupa (Nijhout 1975); normally feeding larvae can reach a weight of 10 g. So a transitory activation of fat body GP to mobilize some, but not all, of the glycogen reserves appears to be advantageous for a starving larva.

Adults of *M. sexta* do not feed as regularly as larvae (Sec. 3). Unlike larvae, adults will not usually be able to support their energy expenditure directly from food, but will rely on reserves accumulated during larval development and only weakly supplemented by feeding.

Adult *M. sexta* reared under laboratory conditions feed very irregularly even when they have free access to sugar water, and show signs of partial starvation; many will not feed at all (Ziegler, unpublished). This could be an artifact because laboratory reared adult *Manduca*, which have been raised as larvae on a semisynthetic diet, lack carotenoids and are more or less blind (Carlson et al. 1967). Because of bad vision they may have difficulties in finding the source of sugar water, only some may find it by chance, but considering the size of the cage and the number of sources, this seems unlikely as an explanation. Additionally, if the proboscis of adults is uncoiled and placed into sugar water, only a few animals feed.

It would be of interest to know more about the feeding behavior of free living *M. sexta*. Observations of laboratory reared *Manduca* suggest that they may not feed regularly in the wild. They are at least adapted to starvation as adults; *Manduca* females are autogenous; nonfeeding females lay about 100 to 200 eggs, whereas those which do feed lay about 1000.

Adult *Manduca* have no need to accumulate reserves for a latter stage. Their responsibility is to reproduce. To this end, they can mobilize all their reserves, and their metabolism is, therefore, directed towards catabolism, not anabolism, although after feeding there are periods of synthesis.

Thus, in starving adults reserves are completely mobilized. This is ensured by the coupling of the activation of the catabolic enzymes to the decrease of the available energy substrates. That is, a decrease in the concentration of hemolymph trehalose is the signal for the mobilization of glycogen and lipid reserves in fat body.

4 Flight

4.1 Metabolic Rates in Insects and in Vertebrates

Among insects there are very good fliers which can perform fast and long-lasting flights. Flying insects offer examples of some of the highest metabolic activities known in animals (see also Chaps. 1 and 5). As insect flight muscle work is fully aerobic, their metabolic activity can be monitored by measuring the rate of oxygen uptake. Measurements of this type have shown that from rest to flight there can be an increase in metabolism by a factor of 100 (Sacktor 1975). An increase of this magnitude makes it necessary to have good controls over metabolic activity to ensure sufficient supplies of energy substrates and oxygen to the muscles. These controls are well-developed in insects and have been the subject of detailed reviews (Crabtree and Newsholme 1975; Sacktor 1975).

In mammals and birds, the metabolic scope between rest and activity varies only by a factor of about 10. A 100-fold increase, such as that in insects is very impressive, but the comparison with vertebrates is somewhat biased, although some moth and some humming birds have the same size and the same energy consuming hovering flight and only on the larger insects do we have enough data for a comparison.

In birds and mammals at rest, the basal metabolic rate is defined as the rate of energy metabolism of a fasting resting homeotherm in its zone of thermal neutrality. In insects there is no equivalent basal metabolic rate; if they are not active, they will cool down (see Sec. 5 in Chap. 1). Insects are ectothermic and some of the larger species have the possibility to heat up during intense muscular activity. The resting rate of metabolism in insects is, therefore, much lower than in homeotherms; it is in the same range as in reptiles of the same size. Small birds like humming birds and some small mammals also cool down when they are not active. If in such higher vertebrates metabolic energy expenditure is measured at a temperature of about 20°–$25^{\circ}C$, then it may be as low as in insects at rest and, if for example, birds like this become fully active, their increase in metabolic rate will be in the same range as in insects. However, there is an essential difference between insects and higher vertebrates. An insect at rest in a surrounding of $20^{\circ}C$ with a body temperature of $20^{\circ}C$ is fully active, it is only unable to fly without a warm-up phase. If a humming bird is cooled down to about $20^{\circ}C$, it is torpid and completely helpless (Bartholomew 1981).

Before flight the thoracic temperature of insects can be increased by muscular activity or by futile cycles (Crabtree and Newsholm 1975). Some small insects can fly with low thoracic temperature.

4.2 Estimates of Energy Expenditure During Flight

Energy expenditure of flying insects can be assessed either from the rate of respiratory exchange, or from the depletion of the animals' depots of fuel. Measurements of respiration have been made in many flying insects (for a list see Kammer and Heinrich 1978) and show that insect flight muscles are among the most active muscles which exist. Some insects are reported to use during flight more than 100 ml

$O_2 \cdot h^{-1} \cdot g^{-1}$ body weight. Nearly all of the O_2 is consumed by flight muslces, so consumption per g muscle will be even higher. High values of oxygen consumption are found especially among Lepidoptera, Diptera, and Hymenoptera. The values reported show large variations for single species, but this may reflect differences in oxygen consumption at the begin of flight, during prolonged flight, or between free and tethered flight. Free flight and the begin of flight consume much more energy (Heinrich 1971).

From values on respiration of *Schistocerca gregaria* during fixed flight (Krogh and Weis-Fogh 1951) it has been calculated that the overall metabolism of flight muscle of *Schistocerca gregaria* is exceedingly high, 2400 cal $\cdot h^{-1} \cdot g^{-1}$ muscle. The calculation is based on the estimate of 4.8% of body weight being flight muscle (Weis-Fogh 1952). This value seems very low, for other insects 10% to 20% are reported and Beenakkers (1965) gives a value of 18% for the closely related *Locusta migratoria*. If the value of Beenakkers is used then it will be 640 cal. This value is still high and some other insects might have far higher values, as their respiratory rate is much greater (Kammer and Heinrich 1978).

In some insects changes in depots of fuels have been estimated and so allow calculations of energy expenditure. The values for locusts are reviewed by Goldsworthy (1984), there are differences between the results of different laboratories, but expressed as total energy expenditure during prolonged tethered flight one can calculate about 50 cal $\cdot h^{-1} \cdot$ animal^{-1} (results from the laboratories of Beenakkers, Candy, and Goldsworthy), or expressed as oxygen consumption about 10 ml $O_2 \cdot h^{-1} \cdot$ animal^{-1} (1 mg of lipid is assumed to be oxidized by 2.02 ml O_2 and 1 mg of carbohydrate by 0.82 ml of O_2).

Krogh and Weis-Fogh (1951) measured for *Schistocerca gregaria* on the average 16 ml $O_2 \cdot h^{-1} \cdot g^{-1}$ (range 10-30 ml). As *Locusta migratoria* is normally heavier that 1 g, the measurements of changes in fuel depots are about 50% lower than the measurements of respiration. The reason for this difference could be caused by differences between the species. The changes in fuels could be somewhat underestimated, as only those in hemolymph and fat body were measured. For the first few hours there could be some supply of fuels from the gut. Weis-Fogh (1952) measured in *Schistocerca* that the gut with contents weighs 60 mg.

In *Manduca sexta*, O_2 consumption during free and tethered flight has been measured and can be compared with estimates of changes in fuel depots. Resting *Manduca sexta* takes up about 0.7 ml $O_2 \cdot h^{-1} \cdot g^{-1}$ (Siegert and Ziegler 1982), while during free flight 50 ml O_2 are consumed (Heinrich 1971). This is a more than 70-fold increase in respiration. By tethered flight only 10 ml $O_2 \cdot h^{-1} \cdot g^{-1}$ are utilized (Heinrich 1971). During prolonged tethered flight of *Manduca sexta*, hemolymph lipid and sugar do not change. This suggests that as much fuel is mobilized as is utilized. If *Manduca* is rested after 1 h of flight, so that fuel is no longer utilized, but mobilization of fuel continues, the hemolmyph levels increase. From these increases during rest, the amount of fuels used during flight can be estimated. As the animals have not been feeding since the end of the last larval instar, gut content cannot contribute. According to these estimates, a starved adult male of the second day utilizes about 8 mg of lipids $\cdot h^{-1}$, which corresponds to about 16 ml $O_2 \cdot h^{-1} \cdot$ animal^{-1}. *Manduca sexta* appears to utilize also some carbohydrate during flight, ca.

2.4 $mg \cdot h^{-1} \cdot animal^{-1}$ equaling ca. 2 ml $O_2 \cdot h^{-1} \cdot animal^{-1}$. As these adult starved males on the second day weigh on average 1.4 g, they utilize 11-12 ml $O_2 \cdot h^{-1} \cdot g^{-1}$ (data from Schulz 1984). This corresponds fairly well with the measurements of oxygen consumption made by Heinrich (1971) for tethered flight in this moth.

4.3 Transport Systems

As the overall metabolism of insect flight muscles is so high, they use up a lot of fuels and these fuels, at least for prolonged flight, cannot be stored in the muscle, but have to be transported in the hemolymph from the main stores in the fat body to the muscle. The fat body can contain large quantities of glycogen, triacylglycerols, and proteins (Bailey 1975). On the hormonal control of mobilization of reserves see Sec. 4.6. The hemolymph, which transports the fuels, is not confined to vessels like blood in vertebrates, but the internal organs of insects are bathed in hemolymph. The hemolymph is, therefore, not circulating as in vertebrates, but moves in a more tidal manner. This is brought about by contractions of the dorsal vessel, but also to a large extent by movements of muscles of the body. Thus, strong muscular activity as is found during flight increases circulation. The fuels have to diffuse from the hemolymph into the cells where they are needed. There is no capillary system like in vertebrates to keep diffusion distances small, but the flight muscle of insects is invaded by the T-tubule system which helps to reduce diffusion distances (Sacktor 1970). Nevertheless, in an open circulatory system the supply with fuels might be more difficult than in a closed circulatory system. The high levels of fuels which are typical of insect hemolymph might be one way to overcome this difficulty of inadequate circulation (Crabtree and Newsholme 1975).

The high concentrations of fuels in insect hemolymph might make it necessary to use transport metabolites which are different from the ones used in vertebrates. The major transport carbohydrate is not glucose but trehalose, which is a nonreducing dimer of glucose. Trehalose is found in high concentrations in hemolymph; adult *Manduca sexta* after molting have nearly 20 mg/ml. Trehalose probably offers the advantage of less influencing the osmotic pressure than the equivalent amount of glucose, being nonreducing and facilitating resorption of glucose by the gut (Sec. 2.3). There is, as already mentioned, also free gluocse in hemolymph, but the level is very low.

Lipids mobilized from the fat body and transported in hemolymph are in the form of diacylglycerols (Bailey 1975) and not of free fatty acids as in mammals. They again can be in very high levels. In starved adult *Manduca sexta* more than 100 mg/ml are found. Such high concentrations of free fatty acids would be harmful; the neutral diacylglycerols are not harmful, but they have the disadvantage of being insoluble in aqueous media and special transport molecules are needed. These transport mulecules are lipoproteins (Sec. 4.6). For a full discussion of these lipid transporting proteins see Goldsworthy (1984). At the muscles, diacyglycerols are liberated from the transport molecules and hydrolyzed by a muscle lipoprotein lipase (Goldsworthy and Wheeler 1984).

Because insect flight muscles work aerobically, they neeed a good supply of oxygen, but insect hemolymph does not usually contain respiratory pigments and plays no part in the transport of oxygen. In insects, oxygen is transported directly to the cells and carbon dioxide removed from them by an elaborate system of tracheae. Tracheae can invade the muscle fibers and come into close proximity with the mitochondria, which are exceedingly well-developed in insect flight muscle (see Sec. 1, Chap. 6). The diffusion distance is kept to a few micrometers by these arrangements (Sacktor 1980). The transport of gases is aided by ventilation, by contraction of wing and abdominal muscles, and by conductance. For a full discussion of gas transport in insects see Chap. 6, this volume.

4.4 Fuels for Flight

Insects can utilize different substrates as fuels for flight. The most familiar are carbohydrates and lipids and, to a lesser extent, amino acids. Insects are sometimes classified into groups which fly using carbohydrates, others which fly using lipids, a third group which flies using carbohydrates and lipids, and a fourth using amino acids. In classifying insects in this way, one has to remember that this is according to the main and not the sole fuel. *Cecropia* silkmoths, Lepidoptera which do not feed as adults and are commonly thought of as typical lipid fliers, can utilize carbohydrates, which are readily oxidized by homogenates of flight muscles (Beenakkers et al. 1981a). Locusts start to fly using carbohydrates, but with increasing duration of flight utilize more and more lipids; they may utilize several other substrates, too (Goldsworthy 1984).

Carbohydrates yield energy easily and a strong enough increase in glycolytic flux to allow for the energy demand during flight is possible (Steele 1981; Crabtree and Newsholme 1975). Carbohydrates can easily be transported in hemolymph, as they are quite soluble in water. They have the disadvantage that they are osmotically active, and that the energy content per weight unit is small (0.18 mole ATP/g) compared to lipids.

Lipids, on the other hand, have a much higher energy content (0.65 mole ATP/g) and are osmotically inactive. It is, thus, much more economical to store lipids for flight, especially for long flights, and in insects which do not feed for long periods. Lipids have one major drawback; they are insoluble in aqueous media like hemolymph. This is especially so for neutral lipids like diacylglycerols which are the transport form of lipids in insects. Thus, insects use special transport molecules, the lipoproteins (Sec. 4.6) (Goldsworthy 1984).

Amino acids or the amino acid proline is utilized as the major fuel only by very few insects (at least to our knowledge); by the tsetse fly, *Glossina morsitans*, and by the Colorado potato beetle, *Leptinotarsa decemlineata.* Many insects of the advanced orders have high levels of amino acids in hemolymph with a general predominance of proline and/or glutamate. Proline can be introduced into the Krebs cycle by way of glutamate and metabolized to pyruvate which by transamination is converted to alanine, accepting the amino group from glutamate (Bursell 1981). This partial combustion of proline gives a high energy yield; 14 mole of ATP/mole of proline or, on a weight basis, 0.52 mole ATP/g which is nearly as much as from lipids.

In the tsetse fly about 20% of proline is completely oxidized, while the remainder is recycled. Proline is resynthesized from alanine with the help of acetyl-CoA from fat degradation. So one can say, in the tsetse fly lipids are combusted by way of proline (Bursell 1981). Thus, proline appears to be the ideal fuel: it is recycled from lipids which are ideal to store, has an energy content about as high as lipids, and is soluble; but because it is utilized by so few insects, it is very likely not so ideal. Before a final conclusion on this issue can be made we should know the major fuel for flight for more insects, and we need to know more details on metabolic energy expenditure of insects during flight. As a minor fuel, proline is utilized by many insects; it is thought to replenish intermediates of the Krebs cycle in the mitochondria, or it could serve as a shuttle for the oxidation of extramitochondrial NADH (Bursell 1981) (see Chap. 5).

4.5 Changes in Fuel Depots During Flight

At the begin of flight locusts utilize mainly carbohydrates, but its utilization decreases during the first 30 min from 270 $\mu g \cdot min^{-1}$ to 10–40 $\mu g \cdot min^{-1}$. At the same time the utilization of lipids is increased from 38 $\mu g \cdot min^{-1}$ to 80–85 $\mu g \cdot min^{-1}$, after 30 min of flight when mainly lipids are used (Goldsworthy 1984). During the first 30 min of flight hemolymph trehalose levels decrease by ca. 50% and then remain stable. GP which can contribute to hemolymph sugar by breaking down fat body glycogen, is activated from 10% to about 25% within 5 min of flight (Marrewijk van et al. 1980; Goldsworthy, Ziegler, and Gäde, unpublished). GP activity further increases slightly to 30% within 100 min of flight. Despite the rapid activation of GP, however, glycogen of fat body contributes to the energy budget only after 30 min of flight (Beenakkers et al. 1981b; Goldsworthy 1984).

Hemolymph lipid levels are low in resting locusts, ca. 5 mg/ml, but with the begin of prolonged flight lipids are mobilized from the fat body, and after 30 min when lipids are the major fuel, hemolymph levels have increased about fourfold and remain constant during prolonged flight. This increase in hemolymph lipid concentration is caused by the adipokinetic hormone (Sec. 4.6).

In other insects we do not have such detailed information on the level of fuels during flight. In the monarch butterfly, *Danaus plexippus*, hemolymph lipid levels increase during the first 2 h of flight, while those of carbohydrates decrease (Dallman and Herman 1978). The main fuel for flight in monarch butterflies is lipids, but the level of hemolymph lipids at the start of flight is already about as high as it is in locusts after 30 min of flight.

In another Lepidoptera, the sphingid *Manduca sexta*, hemolymph lipid levels depend on the nutritional state (Sec. 3.3.2). Starved males on the second day of adult life have ca. 50 mg lipids/ml hemolymph. During the first 30 min of flight, hemolymph lipids decrease by about 25 mg/ml (Fig. 4.4) (Ziegler and Schulz 1982; Schulz 1984) to a level which is still equivalent to the increased levels in locusts after 30 min of flight. After 30 min of flight in *Manduca sexta*, hemolymph lipid levels stabilize for at least another hour of constant flight. This does not indicate that after 30 min lipids are no longer utilized as fuels for flight, but that increased amounts of lipids are mobilized from fat body; if *Manduca* is rested after 1 h of flight, hemolymph lipids increase rapidly and, within 30 min, they reach the original resting level, but increase within another 90 min to 25 mg/ml above the initial

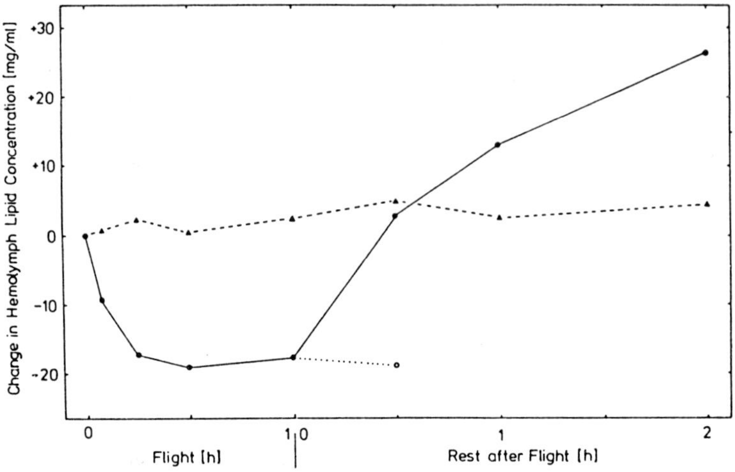

Fig. 4.4. Changes in hemolymph lipid concentration during flight and during rest after flight in *Manduca sexta*. ●——● Adult males flown for 1 h and then rested; ●········○ adult males flown for 90 min; ▲– – –▲ unflown controls. (Modified from Schulz 1984)

levels (Fig. 4.4). The lipid content of the fat body decreases at the same time (Schulz 1984).

Hemolymph carbohydrates decrease in concentration rapidly during the first 5 min of flight bei 8 mg/ml; if flight continues, their levels remain more or less unchanged. During rest after 1 h flight, hemolymph sugar levels increase only slighty (6 mg/ml in 1 h) and do not regain their initial preflight values. GP, which mobilizes the glycogen reserves of fat body, is activated within 30 to 60 min from 10%–35% in the active form. By this activation, about 3 mg$\cdot$h^{-1} of glucose could be liberated from the fat body. The speed at which carbohydrates are used during the first 5 min of flight is much greater; if utilization continued at the same rate as during the first 5 min, *Manduca* would need 19 mg of glucose per hour. This suggests that after 5 min of flight the use of carbohydrates is greatly reduced (Schulz and Ziegler 1983; Schulz 1984). The beginning of flight is more energy consuming than prolonged flight (Heinrich 1971). It could be that the heating-up of the thorax at the beginning of flight demands extra energy in the form of carbohydrate oxidation.

In the blowfly *Phormia regina*, which utilizes carbohydrates as substrate for flight, glycogen of flight muscles is utilized first, thoracic muscle GP is activated at the initiation of flight, but after 10 to 15 min the muscle glycogen is used up and other reserves have to be mobilized. Fat body glycogen is largely unchanged during the first 5 min, but after 15 min the depletion is pronounced (Sacktor 1975). Thus, the animal utilizes glycogen from the flight muscle initially, but fat body glycogen is utilized subsequently. The glycogen from fat body is transported to flight muscles as hemolymph trehalose. In another blowfly, *Calliphora erythrocephala*, hemolymph trehalose levels are unchanged during a 45 min flight, but only as long as the hormonal control by the corpora cardiaca (CC) is intact, if the CC are removed, hemolymph levels decrease dramatically (Vejbjerg and Normann 1974).

In the two insects which we know utilize proline as a major fuel for flight, regulation of proline supply appears to be different. In the tsetse fly, proline reserves are rapidly and continuously depleted until flight ceases after a few minutes; proline is resynthesized during rest (Bursell 1981). In the Colorado beetle, proline content of the thorax decreases rapidly, but it stabilizes after 5 min of flight and after 10 min is back to preflight levels. This indicates an increased synthesis of proline during flight, and this synthesis continues after flight (Weeda et al. 1979).

4.6 Hormonal Control of Flight

There are indications of a hormonal regulation of flight in different insect species, but only in locusts do we have anything approaching a comprehensive picture. In locusts hormones influence the mobilization of fuels from reserves – mainly from fat body (Goldsworthy 1984), the transport of fuels by hemolymph (Goldsworthy 1984), and the combustion of fuels by flight muscles (Candy 1981; Goldsworthy 1984).

4.6.1 Hormonal Regulation in Locusts

In locusts, energy metabolism of flight is mainly regulated by the adipokinetic hormone (AKH) from the glandular lobes of the CC (Goldsworthy and Wheeler 1984). Locust AKH I is a blocked decapeptide with the amino acid sequence: Glu-Leu-Asn-Phe-Thr-Pro-Asn-Trp-Gly-Thr-NH_2 (Stone et al. 1976). There is also an AKH II, which is an octapeptide whose amount in the CC is much smaller (10–20% of AKH I) (Carlsen et al. 1979).

AKH is secreted during flight. In locusts flown for only 2 min and then rested, hemolymph lipid levels increase. During the first 15 min of flight AKH concentration in hemolymph is very low and some authors assume that AKH is not released that early. From existing measurements of hemolymph AKH levels, it is not possible to answer this question; Goldsworthy and Wheeler (1984) point out that comparatively small volumes of hemolymph from flown locusts are usually extracted, and low levels of AKH would, therefore, have passed undetected. Results from electron microscopy indicate strongly that AKH is secreted early during flight, because increased secretory activity in the glandular lobes of the CC is seen after only 5 min of flight (Beenakkers et al. 1981a). There is the possibility that early mobilization of lipids is caused by octopamine and not by AKH. Octopamine is reported to increase hemolymph lipid levels. It is found in insect nervous systems, and its concentration in hemolymph is increased during the first 5 min of flight from 0.3 to 1.7 nM (Candy 1981). During longer flight, it is definitely AKH which influences hemolymph lipids. After 15 min of flight there is a marked increase in AKH titer (for a fuller discussion see Goldsworthy 1984; Goldsworthy and Wheeler 1984).

AKH is assumed to activate an adenylate cyclase and thus increase cAMP levels to activate a protein kinase. This protein kinase is assumed to phosphorylate and activate lipase. The effect, however, of AKH on this lipase has not been shown, but AKH induces the release of stereospecific diacylglycerols from triacylglycerols stored in fat body (Beenakkers et al. 1981b; Goldsworthy and Wheeler 1984).

Diacylglycerols released from fat body into hemolymph are taken up by lipoproteins. In resting locusts, the hemolymph lipid levels are low (about 4–10 mg/ml) and lipids are transported by a lipoprotein called A_{yellow}. At times of increased lipid mobilization, that is during flight, after the injection of AKH, or during starvation, a new lipoprotein complex appears in the hemolymph. It is called A^+ as it elutes in gel filtration ahead of A_{yellow}. A^+ is most likely a combination of A_{yellow} with a smaller protein called C protein. A^+ can carry much more lipid than A_{yellow} (Goldsworthy 1984).

In flying locusts the percentage of fat body GP in the active form is increased from 10% or 15% to 30%. As GP can be activated by AKH it is assumed that GP is activated during flight by AKH (Marrewijk van et al. 1980), but this was never proven directly. The time course of GP activation (Marrewijk van et al. 1980) and the time course of AKH secretion during flight (Cheeseman and Goldsworthy 1979; Orchard and Lange 1983) do not speak in favor of this role of AKH.

AKH also influences flight muscle metabolism. Locust flight muscle at the begin of flight preferentially metabolizes carbohydrates, but under the influence of AKH utilizes mainly lipids. Octopamine, which might be important at the beginning of flight, favors the oxidation of carbohydrates (Candy 1981; Goldsworthy 1984).

4.6.2 Hormonal Regulations in Insects Other than Locusts

In the blowfly *Calliphora erythrocephala* the level of hemolymph trehalose is controlled by a hormone from the CC (Vejbjerg and Normann 1974). The cockroach *Periplaneta americana*, which is not a good flier in the laboratory, utilizes mainly carbohydrates, and no adipokinetic effect is found if extracts of CC are injected, but a strong hypertrehalosaemic response is present (Steele 1981). In the Colorado beetle, which utilizes proline as the major fuel for flight, proline synthesis is stimulated during flight by a hormone from the CC (Weeda 1981).

In two Lepidoptera which utilize lipids as fuels for flight, adipokinetic activity has also been demonstrated; in *Danaus plexippus* (Dallman and Herman 1978) and in *Manduca sexta* (Beenakkers et al. 1978; Ziegler 1980). As already described (Sec. 4.5), in early flight hemolymph lipids are decreased in *Manduca sexta*, but during continued flight larger amounts of lipids are mobilized (which means that during rest after flight an increase of hemolymph lipid levels occurs). Adipokinetic activity is only found in the CC of *Manduca* and if these glands are removed, *Manduca sexta* does not fly well, but it does fly and shows a decrease of hemolymph lipids during flight, but not an increase after flight (Ziegler and Schulz 1982; Schulz 1984). So the adipokinetic activity in the CC of *Manduca sexta* is due to an adipokinetic hormone.

Recently, the lipid transporting proteins in the hemolymph of adult *Manduca sexta* were characterized by density centrifugation (Shapiro and Law 1983). The density of this lipoprotein is decreased by the injection of a very high dose of locust AKH-I (200 pmol; in locusts 1 or 2 pmol give a full response). This change appears to be caused by an increase of lipid-loading and the binding of a smaller protein. The dose of AKH used is exceedingly high, but locust AKH is chemically distinct from that of *Manduca* and a high dose of one-tenth of a pair of CC from *Locusta migratoria* is needed to get a full adipokinetic response in *Manduca sexta* (Ziegler and Gäde 1984). As one pair of CC from locusts contains about 100 to 200 pmol, one would expect a full response in *Manduca* with as little as 20 pmol, but 100 pmol failed to give a maximal response (Shapiro and Law 1983). Nevertheless, these results indicate that although hemolymph lipids decrease during flight in *Manduca sexta* and do not increase like in *Locusta migratoria*, the system of lipoproteins for the transport of lipids may be similar in both insects. Preliminary experiments with preparations of negatively stained hemolymph lipoproteins from *Manduca* (Ziegler and Goldsworthy, unpublished) support such a speculation.

In *Manduca sexta*, GP in the fat body is also activated during flight and, as in *Locusta migratoria*, GP can be activated by the injection of CC extracts. If CC are removed in *Manduca*, GP is activated during flight just as it is in intact animals (Schulz and Ziegler 1983; Schulz 1984). So the mobilization of fat body glycogen in adults of *Manduca sexta* during flight is not controlled by a hormone from the CC.

Because there is an AKH in locusts and an AKH in *Manduca sexta,* the question arises, as to whether they are identical molecules. The amino acid sequence is known for locust AKH, but not for *Manduca* AKH, so a direct comparison is not yet possible. Cross-injections show that locust material is active in *Manduca sexta*, and material from *Manduca sexta* is active in *Locusta migratoria*, but very different amounts are needed. Extracts of CC from *Manduca sexta* are active in *Locusta migratoria* only in very high concentrations, while CC extracts from *Locusta migratoria* are more active in *Manduca sexta* (Ziegler and Gäde 1984). On the basis of CC equivalents, extracts of *Manduca sexta* injected into *Manduca sexta* are about as active as CC extracts from *Locusta migratoria* injected into *Locusta migratoria.* So the differences in the results of crossinjections can only be explained by the existence of different molecules of AKH.

5 Conclusions

Insects principally need the same nutrients as other animals, but some insects acquire their nutrients and so the energy which they need for living from unusual sources. To survive they have to expend the acquired energy in a controlled way. This is especially obvious during starvation and during intense activity like flight when animals live on reserves.

All animals may face the possibility of various degrees of starvation at some stage during their life, but in insects there are developmentally programed times of fasting: during molting, the pupal stage, and, in some insects, during the whole adult life.

Many insects live completely separate and different lives as larvae and adults. Larvae accumulate reserves, while adults, like in *Manduca sexta*, often live mainly on the reserves accumulated by larvae. If they have to starve, larvae try to conserve reserves; this is regulated in *Manduca sexta* by a hormonal mechanism which regulates carbohydrate mobilization. In adults of *Manduca sexta*, on the other hand, all reserves are used up during starvation and they even reproduce at such times. In adults of *Manduca sexta* mobilization of reserves during starvation does not seem to be controlled by hormones.

For flight, however, fuels have also to be mobilized; this is controlled by peptide hormones from the corpora cardiaca. In different insects, even if the hormones have to mobilize the same fuel, they appear to be chemically distinct.

Acknowledgements. I am grateful to Dr. G.J. Goldsworthy who critically read the manuscript. Original research described in this review is supported by the Deutsche Forschungsgemeinschaft.

Chapter 5

Anaerobic Energy Metabolism

Gerd Gäde[1]

Contents

1 Introduction

When considering anaerobic energy metabolism, many people are unaware that it occurs in insects at all. This may be easily explained by the general knowledge that the energy metabolism of insect flight muscles is highly aerobic. It is well-known that insect flight muscles are metabolically the most active tissues known in nature. Their performance can be quite incredible; in some species more than one million successive wingbeats, with rates over 1000 contractions per second, have been found. These repeated contractions of muscles involve oxidative processes for energy provision and, therefore, a very good provision of oxygen. It has been shown, for example, that some blowflies on initiation of flight consume oxygen at a rate of about 3000 $\mu l \cdot min^{-1} g^{-1}$; thus, elevating their resting oxygen consumption approximately 100-fold. In comparison, hummingbirds in flight show a respiratory rate five times that at

1 Institut für Zoologie IV, Universität Düsseldorf, Universitätsstraße 1, 4000 Düsseldorf, FRG

Environmental Physiology and Biochemistry of Insects, ed. by K.H. Hoffmann

rest. It therefore is not an exaggeration to say that the respiratory rate of certain insects is the most intensly by known in biology. From these few remarks it is evident that insect flight muscles are the tissue of choice to study insect metabolism and its control in detail; concerning this subject the reader is referred to previous reviews (for example, Sacktor 1965, 1970, 1975) (see also Sec. 4, Chap. 4).

It is also evident from the above remarks that the high performance of insect flight muscles can be achieved only when oxygen is available in sufficient amount and when the transport of oxygen to the sites of oxidation, the mitochondria, is not impaired. For certain insects (e.g., fruit flies and locusts), it has been shown that despite maintaining flight for hours, no, or only a small, oxygen debt occurs, indicating that the availability of oxygen is not limited. Oxygen is transported directly to the flight muscles without the use of respiratory pigments through a sophisitcated tracheal system which invades the muscle fibers and is in close contact with each mitochondrion. The transport of the respiratory gases is discussed in detail in Chap. 6.

However, since insects have successfully colonized more or less every habitat on this planet, we have to consider whether the availability of oxygen is unlimited for all species, or, if some species have undergone special adaptations in their energy metabolism to cope with the various conditions at different habitats. Furthermore, we should examine the proposition that all tissues of insects are well organized for an oxidative metabolism as the flight muscles: perhaps the flight muscles are more the exception than the rule in having their exceptionally high aerobic metabolism.

At a superficial level, we may already find answers to these questions. We know that some insect larvae live in aquatic environments (midges, for example), or in the soil or decaying matter of terrestrial habitats (flies and beetles, for example), where oxygen tensions may be low. On the other hand, some adults of the Odonata dive beneath water and stay for over 30 min submerged while laying their eggs on (or into) the stems of water plants. Or, to give another example, the migratory locust, when prevented from flying, can jump only 10–15 times successively, indicating that the provision of oxygen to the leg muscles is probably not highly developed. Finally, I want to draw the reader's attention to experiments where highly active terrestrial insects, such as bees or blowflies, for example, can be subjected experimentally to a nitrogen atmosphere for some hours, but restoration of all body functions occurs during aerobic recovery. In these cases, the total animal is deprived of oxygen and even the highly developed nervous tissues have to rely on anaerobic energy metabolism. How is nervous function maintained? In general, it is believed that brain function, for example, depends on aerobic metabolism. This chapter will discuss many of these examples, but will not review in detail the abundance of available data, rather it will attempt to review the anaerobic energy metabolism of those insects for which a comprehensive account can be presented.

2 Environmental and Functional Anaerobiosis

First, we should consider some general features of anaerobic metabolism. When we speak of anaerobic metabolism, this mainly refers to experiments in the laboratory where the partial pressure of oxygen (P_{O_2}) can be closely controlled and, by using nitrogen gas, a true anoxic atmosphere can be created. Physiologically, this may be a rather unusual condition in the habitat itself. Most habitats, or microhabitats, will have residual oxygen content and are therefore hypoxic, but not anoxic. It is also clear that the adaptation of an organism to its habitat is not determined by a single factor in the environment, but rather by interactive combinations of many factors. In relation to oxygen tension, for example, temperature and salinity play an important interactive role in the aquatic environment; whereas 210 ml $\cdot$ l^{-1} of oxygen is present in air, seawater saturated with oxygen contains only about 5.5 ml $\cdot$ l^{-1} at 20°C. These aspects cannot be covered in this short chapter, but some information is given in Chap. 1.

Concerning adaptation to reduced oxygen tensions, we must differentiate between functional and environmental anaerobiosis. Functional anaerobiosis is caused by the animal itself and occurs characteristically in muscle tissues, when animals capable of high locomotory activity use their muscle fibers for "burst" work lasting only for a limited time. Under these conditions, oxygen consumption may exceed oxygen delivery so that the muscle tissue becomes anoxic and relies on anaerobic energy metabolism for provision of energy for muscle contraction. As other tissues are not affected, anaerobic end products, like lactic acid in the working vertebrate skeletal muscles, can be transported via the blood to other sites of greater oxygen availability for (in the case of vertrebrates) oxidation in the heart or for resynthesis of glucose in the liver.

Environmental anaerobiosis is defined as exposure of the whole organism to hypoxic or anoxic conditions in the natural environment and is, therefore, also called organism-level anaerobiosis. In the microhabitat it is caused by external physical factors. In general, air breathing animals have few problems in obtaining adequate amounts of oxygen because the P_{O_2} is high (see above), and the phenomenon of environmental anaerobiosis occurs characteristically in aquatic animals. Compared with air, the oxygen content of water is low (e.g., maximally 1/30), but temperature, salinity and the amount of decaying organic matter can further alter the amount of dissolved oxygen.

3 Environmental Anaerobiosis in Insects

3.1 Availability of Oxygen in the Habitats

In insects, environmental anaerobiosis characteristically occurs only in aquatic species; most live in freshwater, very few in brackish water, and virtually no insect is truely marine. It is well-known that some of the lowest levels of oxygen are found in limnic biocenoses: the P_{O_2} of the deeper portions of temperate lakes are low for many weeks during summer and winter stagnation periods, and this may also be true for other water layers where complex diurnal and annual variations may occur depending, among others, on light intensity, photoperiod, convection currents, and disintegration processes. The muddy bottom of shallow waters is frequently devoid of oxygen and this is particularly true for highly eutrophic waters with little or no through flow, where a high concentration of H_2S is produced by the action of microorganisms in the decomposition of the large input of plant material. Winter anoxia also occurs when ice coverage of lakes and ponds in cold temperate regions may cut the insects off of their surface oxygen supply.

Terrestrial insects in soil (adults or developing stages) may encounter low oxygen tensions naturally when the ground becomes water-logged or flooded, or covered either with decaying matter like the wood of trees, or with ice and snow, especially in high mountain regions.

3.2 General Adaptational Strategies

In general, to overcome times of hypoxia or anoxia, insects exhibit several kinds of adaptational strategies which are especially important for the survival of less active species (highly mobile species will probably try to leave the challenging and unfortunate environment). Such adaptations can be of morphological (structural), physiological and/or biochemical nature. We shall consider here a few examples.

3.2.1 Morphological Adaptations

In aquatic insects with a closed tracheal system (and thus rely entirely on oxygen dissolved in the water), such as members of the Ephemeroptera, Trichoptera, Odonata, and Plecoptera, either the permeability of the cuticle for oxygen is increased or special structure to aid in the extraction of oxygen from the water are developed; tracheal gills, or rectal gills present an enlargement of the surface area for oxygen uptake (see also Sec. 1, Chap. 6).

3.2.2 Physiological Adaptations

Aquatic insects maintain the rate of oxygen uptake by increasing the ventilation rate and/or increasing the efficiency with which oxygen is removed from the water. As a result, $\dot{V}_{O_2}$ remains fairly constant irrespective of a decline in ambient oxygen tension, down to a critical oxygen level, below which consumption falls with concentration. However, no clear-cut picture emerged from various experiments. As reviewed in detail by Keister and Buck (1974) mayfly nymphs of various species as well as larvae, pupae, and adults of various terrestrial insects, show a complete spectrum from great sensitivity to changes in environmental oxygen concentration to great insensitivity. To name but one figure: adult (*Aedes*) mosquitoes or (*Phormia*) blowflies have constant respiratory rates down to between 5% to 2% oxygen (P_{O_2} = 35 to 14 Torr; 4.7 to 1.9 kPa). Another physiological adaptation to ambient oxygen lack may be characterized by the use of respiratory pigments as an oxygen store, or the existence of respiratory pigments with a very high oxygen affinity. The latter provision is made, for example, in *Chironomus* larvae, which possess a hemoglobin species that is still saturated at 0.6 mmHg (0.8 kPa) at 17°C (for comparison: 27 mmHg (3.6 kPa) at 37°C for human hemoglobin).

3.2.3 Biochemical Adaptation

This third example concerns an adaptation which serves when all other strategies to compensate for declining oxygen tension have failed. Here, the insect relies on anaerobic metabolism for energy expenditure; the substrate hydrogen atoms are not transferred to molecular oxygen, but to other organic compounds resulting in different kinds of fermentation products. The best known of these processes is the production of lactic acid.

3.3 Anaerobic Metabolism of Aquatic Insect Larvae

As we have seen, many aquatic insect larvae encounter periods of low tensions in their natural habitat. We will now consider some examples in detail, and discuss how their energy metabolism is adapted.

3.3.1 Strategies of Chironomids

Data concerning the Chironomidae stem from the work of Wilps (1976, 1978), Wilps and Zebe (1976), and Wilps and Schöttler (1980). The modern extant family of Chironomidae contains a great number of species: whereas only 200 fossil forms were found during the Tertiary, more than 2000 species and subspecies are known today. They live in a wide range of habitats, such as hot springs, glacier lakes in high mountains, bottom sediments of polluted lakes and ponds, and in the mud basins of sewage farms. Correspondingly, the various species have modified their structural organization and physiological properties during evolution. For example, tubuli increase the preanal surface, and their size is directly correlated with the ion concentration of the surrounding medium. Furthermore, there is a direct correlation between the permeability of the integument and the tubuli with respect to the quality of the surrounding water; chironomids living in mud are reported to possess a cuticle highly permeable to water and oxygen. Also, hemoglobin occurs only in such species living in poorly oxygenated habitats; all other species have a well-developed tracheal system available.

The larvae of *Chironomus thummi thummi* live in the decaying mud of eutrophic lakes. They build U-shaped tubes, where the insect lies with its head in the entrance opening. Rhythmic movements of the body generate a constant flow of water which passes through the burrow over the body of the insect and the oxygen is extracted by diffusion. Although this species possesses hemoglobin, it has a respiratory function only when the partial pressure of oxygen of the surrounding medium is very low; in well-oxygenated water, the insect's oxygen demand is provided only by diffusion. An oxygen storage function for the hemoglobin has been suggested, but rejected because the oxygen bound to hemoglobin would only be sufficient to support the insect's aerobic metabolism for about 10 min. The hemoglobin, however, has a very high affinity for oxygen. Although estimates vary between chironomid species and different authors, 100% saturation is obtained by between 7–72 mmHg (0.9–9.6 kPa), and the P_{50} is between 0.1 and 0.6 mmHg (0.013 and 0.08 kPa).

From these data it is clear that chironomid larvae living in habitats with low oxygen tensions show good adaptations for coping with hypoxic situations. Anoxic conditions are not tolerated for a very long time (up to 60 h) as already reported by Harnisch (1953) and Neumann (1962), because the larvae take up water by up to 40% of their body weight. This is most likely caused by an inadequate provision of energy for essential osmoregulatory processes during anaerobiosis (the larvae are homoiosmotic living in a hypotonic medium; thus, water diffusing through the permeable cuticle into the body must constantly be removed at some cost in energy expenditure). However, in the laboratory this process can be counteracted by the addition of salt (NaCl, 8‰), and this also increases their resistance to anoxia. Under such conditions the mortality during anaerobiosis is highly decreased.

Biochemical investigations of anaerobic metabolism in different chironomid species were made more than 40 years ago by Harnisch (1938, 1939, 1942) who found that glycogen breakdown increases tenfold and is accompanied by the production of organic acids (not specified) and fatty acids (mostly butyric acid). However, the consumption of glycogen is much higher than the production of the different acids. When Harms (1972) found a highly active alcohol dehydrogenase in larvae of *Chironomus tentans* a new insight into the anaerobic metabolism of these animals was gained and this has been subsequently investigated in detail by Wilps (citation see above).

It was found that larvae of *Chironomus thummi thummi* utilized about 600–700 μmol of glycogen/g dry weight during 48 h of anoxia, with the maximal rate of glycogenolysis occurring during the first 12 h when about 25–30% of the stored glycogen is broken down. This indicates a high energy demand which probably is related to the already mentioned osmotic problems. During prolonged anoxia, the rate of glycogenolysis is reduced. However, in total 60% of the glycogen store is depleted during 48 h of anaerobiosis, reflecting the low production of ATP equivalents in the anaerobic pathways used by this species. In contrast to various endoparasites and free-living annelids or mollusks which achieve a high ATP yield by succinate/propionate fermentation processes (reviewed by Schöttler 1980; Gäde 1983), *Chironomus thummi thummi* produce only minor quantities of L-lactate, succinate, and alanine (in the initial 12 h of anoxia). In addition, there is some production of acetate, which mainly accumulates in the incubation medium. However, ethanol is the most important end product of anaerobic carbohydrate breakdown. Only comparatively small quantities accumulate within the body, whereas the bulk is found increasingly in the surrounding water, during prolonged anaerobiosis. After 48 h anoxia, up to 1070 μmol/g dry weight accumulate in the incubation water. This amount exceeds the combined quantities of all other accumulated metabolites by a factor of about 5, compared with about 100 μmol/g dry weight of acetate. The comparatively low energy yield of enthanol fermentation is discussed as a reason for the relatively low resistance to anaerobiosis. The advantage of this pathway lies in the fact that the uncharged ethanol can easily permeate biological membranes and the tissues do not become acidified as with classical lactate fermentation. On the other hand, there is a severe loss of nutrients which have to be replenished by increased food intake in a subsequent recovery period at normoxic conditions.

The measured production of metabolites is in keeping with enzyme data in *Chironomus thummi thummi*. The activities of gylceroaldehydphosphate dehydrogenase (GAPDH), phosphoglycerate kinase (PGK), pyruvate kinase (PK), and alcohol dehydrogenase (ADH) are quite high, whereas lactate dehydrogenase (LDH) activity is low and no phosphoenolpyruvate carboxykinase (PEPCK) activity is found. The latter result indicates that anaerobic carbohydrate breakdown proceeds to pyruvate from where the different end products can be formed: succinate by the combined action of malic enzyme (pyruvate → malate), fumarase (malate → fumarate) and fumarate reductase (fumarate → succinate); acetate, via pyruvate dehydrogenase (PDH, pyruvate → acetyl-CoA) and acetic thiokinase (ATK, acetyl-CoA → acetate); lactate, by means of LDH (pyruvate → lactate); alanine, via the transaminase reaction (pyruvate + glutamate → alanine + α-ketoglutarate); and ethanol, by the combined action of pyruvate decarboxylase (pyruvate → acetaldehyde) and alcohol dehydrogenase (acetaldehyde → ethanol).

Succinate and alanine pathways have not been studied in *Chironomus thummi thummi*, but it should be mentioned that aspartate cannot be the primary amino group donor for alanine formation because of the low concentration of the former amino acid which, in addition, does not change in concentration during anoxia. The pathways for formation of acetate and ethanol have been investigated in homogenates and isolated mitochondria in vitro by use of the PDH inhibitor arsenite. From these experiments, the following scheme evolved (Fig. 5.1): pyruvate derived from glycogen breakdown during anoxia, is decarboxylated to acetaldehyde inside the mitochondria. There is no unequivocal evidence as to whether the reaction is catalyzed by pyruvate decarboxylase or by a modified PDH complex. Because there is some ADH activity present inside the mitochondrion, ethanol production may take place there, but the bulk is produced in the cytosol in the regeneration of oxidized coenzyme NAD^+ for glycolysis. No evidence is found that pyruvate is first metabolized to acetyl-CoA and subsequently, this compound reduced via aldehyde dehydrogenase, a pathway functioning in some bacteria (for a review, see Thauer et al. 1977). Acetate, however, is always formed via acetyl-CoA; arsenite poisoning results in the cessation of this process.

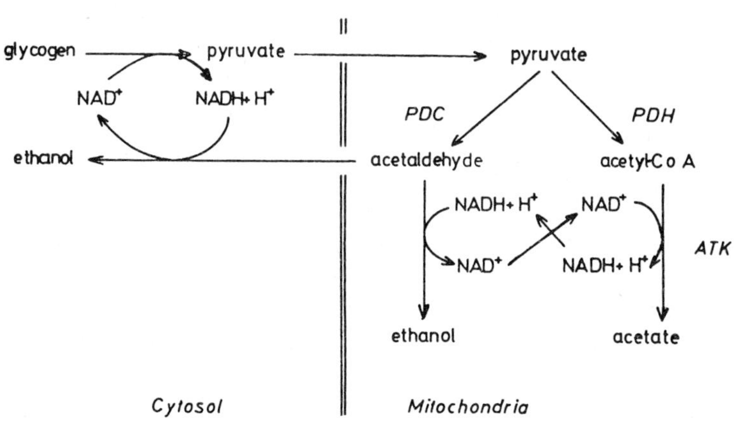

Fig. 5.1. Pathway for formation of acetate and ethanol during anoxia in *Chironomus thummi thummi*

Thus, the midge *Chironomus thummi thummi* is, during its larval life, adapted to low oxygen tensions by some morphological, and biochemical adaptations to withstand the absence of oxygen in the environment. The provision of energy by the breakdown of glycogen to ethanol (which is mainly released into the surrounding water) is an important adaptation. It is interesting to note that a vertebrate, the goldfish *Carassius auratus*, has also a metabolism characterized by anaerobic carbon dioxide and ethanol production (Thillart van den and Kesbeke 1978; Shoubridge and Hochachka 1980; Mourik et al. 1982).

3.3.2 Strategies of the Midge *Chaoborus crystallinus*

A different strategy during anoxic periods has been employed by the larvae of the midge *Chaoborus crystallinus.* The following account is based on the work of Opalka (1977), Englisch (1981), and Englisch et al. (1982). The larvae inhabit small shallow ponds or lakes where hypoxia or even anoxia prevails as described above. Their tracheal system is reduced to two small air bladders which function as a hydrostatic apparatus (see also Sec. 3.1, Chap. 8). In contrast to many other species of midges, *C. crystallinus* possesses no special organ for breathing atmospheric air. Oxygen uptake takes place only by diffusion through the cuticle supported by four rectal gills which are not well-developed. Particularly interesting, and characteristic for this species, are diurnal migrations. Thereby during the day they burrow some centimeters into the mud layer, encountering at least hypoxic conditions, but shortly after sunset the larvae move upwards until some hours later most are found in the upper water column where normoxic conditions prevail. These diurnal vertical migrations are influenced not only by light, but also by the oxygen tension. The extent of the migration depends on whether P_{O_2} is high or low; in the latter case larvae migrate to just beneath the water surface.

In the laboratory the larvae are not very tolerant of anoxia. Already up to 8% are dead after 24 h of anoxia at 14°C and the remaining animals do not survive much longer periods of anoxia (up to 56% mortality after 48 h). However, upon restoration of normoxic conditions after 24 h of anaerobiosis, most of the larvae appear to recover rapidly. Biochemical studies reveal that the carbohydrate stores are small; the glycogen content in normoxic larvae varies between 79 and 123 µmol glycosyl units/g dry weight, and the total glucan concentration (glycogen plus oligosaccharides) accounts for 118–145 µmol glycosyl units/g dry weight. In comparison with larvae of *Chironomus thummi thummi* the carboyhdrate stores in *C. crystallinus* are, thus, almost tenfold lower. During a 24 h period of anoxia ca. 80% of these stores are consumed in *C. crystallinus.* A second, but minor, substrate for energy production is the phosphagen arginine phosphate which decreases from ca. 40 to 16 µmol/g dry weight. A further energy source is malate, which occurs in extremely high concentrations in normoxic larvae (160 µmol/g dry weight) and decreases rapidly during the initial 12 h of anoxia and somewhat more slowly thereafter to ca. 70 µmol/g dry weight. Concomitantly, succinate levels rise, in an apparently inverse relationship, from 7 to ca. 100 µmol/g dry weight. If malate is converted to succinate via fumarate, by the action of fumarate reductase in the mitochondria, this reaction could yield ATP. This has been shown, for example, for isolated mitochondria from the annelid *Tubifex* ssp. (Schöttler 1980). But how is the NADH + H$^+$ provided for the reduction of fumarate? Two possibilities seem likely: first, extramitochondrial NADH + H$^+$ derives via an appropriate shuttle mechanism; second, there is a dismutation of malate, as occurs in isolated mitochondria of *Arenicola marina* (Schroff and Schöttler 1977). Here, part of the malate is oxidized to oxaloacetate and pyruvate, which is further metabolized to acetyl-CoA by the action of pyruvate dehydrogenase. Acetyl-CoA condenses with oxaloacetate to form citrate, and this is subsequently converted to succinate in the citric acid cycle. The remaining malate is converted to succinate via fumarate (see Fig. 5.2).

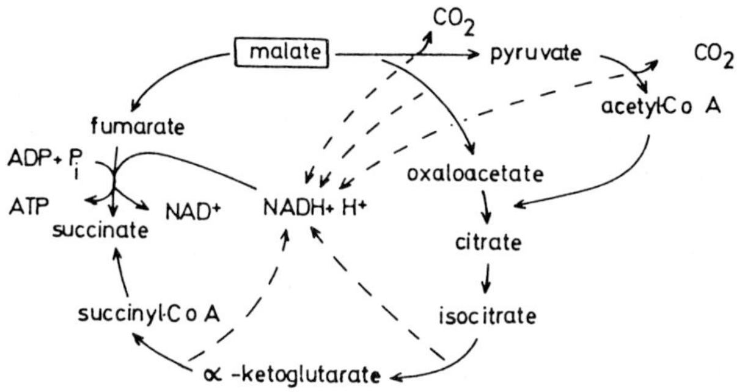

Fig. 5.2. Fate of malate in anoxic mitochondria. (Adapted from Schöttler 1980)

In other invertebrates which accumulte succinate during anaerobiosis, succinate production is thought to proceed largely by glycogen breakdown via the phosphoenolpyruvate carboxykinase reaction (see for example, Gäde 1983). However, in *C. crystallinus* this enzyme is reportedly absent. Furthermore, in other euryoxic invertebrates succinate accumulates only as an intermediate during anoxic conditions, and is subsequently metabolized to propionate which is often excreted into the surrounding medium. In *C. crystallinus* larvae, no production of propionate (or other volatile fatty acids) occurs to resemble the situation in larvae of *Chironomus thummi thummi*, and no other metabolites are excreted into the water when *C. crystallinus* are held anoxic. Thus, no valuable substrate molecules which can be utilized during a subsequent aerobic recovery period are lost.

Because the phosphoenolpyruvate carboxykinase pathway is apparently inoperative in larvae of *C. crystallinus*, and the activities of pyruvate kinase are comparatively high, it is believed that carbohydrates are broken down during anoxia via pyruvate to form lactate and alanine. Lactate accumulation takes place continuously, rising from *c.* 18 to 92 μmol/g dry weight during a 24 h anoxic period. However, the concentration of alanine increases three times faster than that of lactate from *c.* 18 in normoxia to 200 μmol/g dry weight in anoxic larvae. It seems that the enzymes LDH (lactate dehydrogenase) and GPT (glutamate-pyruvate transaminase) compete for the same substrate (pyruvate) with the conditions favoring the latter. However, this route of metabolic events presents some problems: first, the quantities of lactate and alanine found after anoxia exceed the measured quantities of consumed carbohydrates (260 vs 60 μmol/g dry weight); it is possible that trehalose, which occurs in high concentrations in the blood of many insects, may contribute to this balance, but

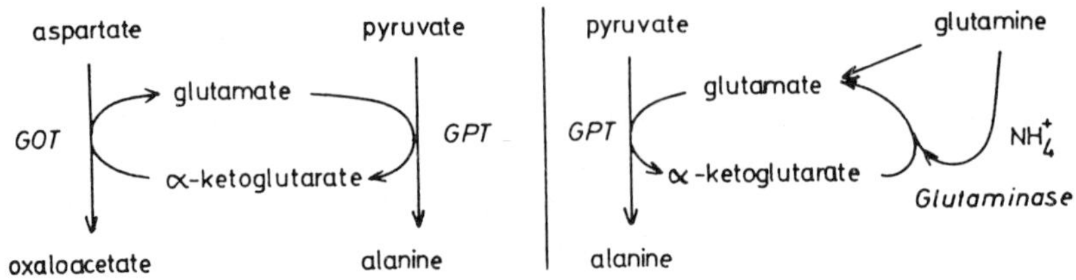

Fig. 5.3. Two possible pathways for the production of alanine

no data are available for *C. crystallinus*; second, the formation of alanine requires a substrate to donate the amino group or equivalent amounts of ammonia; because the breakdown of stored proteins does not occur during anoxia in *C. crystallinus,* the question arises as to where this amino source derives. In different euryoxic invertebrate species the combined action of two transaminases, GOT (glutamate-oxaloacetate transaminase) and GPT, with MDH (malate dehydrogenase) seem to be responsible for alanine formation (see for example, Schöttler 1980; Gäde 1983). The free amino acid aspartate functions as the amino donor (see Fig. 5.3). However, there is no 1:1 stoichiometry between aspartate fementation and alanine formation; alanine production is almost always greater than aspartate breakdown (see discussion in Gäde and Ellington 1983). The operation of this pathway in *C. crystallinus* is unlikely, because the aspartate concentration is too low and does not change appreciably during anaerobiosis. However, there is a decrease in the level of glutamine during anoxia; if its breakdown is achieved by glutaminase, glutamate and NH_4^+ will be formed. By a transamination reaction catalyzed by GPT, alanine could be formed in stoichiometry with α-ketoglutarate, which in turn could be re-aminated to glutamate via a reductive step by glutamate dehydrogenase (see Fig. 5.3). This enzyme is reported to have about the same activity as LDH and GPT in *C. crystallinus.* As yet no kinetic parameters are known for GDH in this species and further experimentation is awaited to test the above possibility of anaerobic alanine formation.

The above hypothesis that a "malate cycle" is indeed working during diurnal migrations has been tested in field experiments. The quotient in the levels of malate/succinate is low in larvae sampled from the bottom mud layer just before sunset. In all larvae caught in the free water column thereafter this quotient rises continuously until stabilizing for 3–4 h at a maximal value around midnight. After sunrise, all larvae analyzed from the bottom again show the same low quotient as animals at sunset. This result is surprising because such low values are only reached in the laboratory after 24 h of anoxia. However, it might be explained by the increased energy expenditure during burrowing in the anoxic substrate; this muscular activity may lead to anaerobic metabolism fueled by malate. Nevertheless, these experiments in the natural habitat proved the existence and functioning of the malate cycle, and show that conclusions drawn from investigations in the laboratory allow some cautious interpretation of the situation encountered by the animal in its natural habitat.

3.4 Anaerobic Metabolism of Terrestrial Insect Larvae

This section will focus attention on the anaerobic carbohydrate and amino acid metabolism of the dipteran species *Callitroga macellaria* which has been studied extensively by Meyer (1971, 1974, 1977, 1978, 1980) and Meyer and Müller (1977). The nonfeeding larvae of this species live in soil, where they may encounter reduced partial pressures of oxygen. They are quite resistant to experimental anaerobiosis (nitrogen atmosphere), surviving 24 h of anoxia, but thereafter only 50% pupate, and only a few undergo metamorphosis. Pupation is not affected until after 12 h of anoxia, whereas emergence is considerably reduced after only 4 h of anaerobiosis.

Biochemical studies of the anaerobic metabolism in this insect show that the classical pathway from glucose to lactate operates, but may not be the only pathway for anaerobic energy gain. Obviously, these larvae survive during oxygen deprivation due to mobilization of carbohydrates and of stored protein. The larvae feed on meat and store both glycogen and protein in the fat body.

It has been shown using isotopic, chromatographic, enzymatic, and other methods that after incubations in vivo and in vitro, there are indirect (= ATP saving) and direct (= ATP producing) mechanisms for energy yield during anoxia.

Energy saving reactions: Due to the (at least transient) lack of ATP (or decreased concentration of ATP), the activity of phosphofructokinase is lowered (partial inhibition of the allosteric enzyme; see Newsholme and Start 1973) and the non-ATP consuming reaction sequence of the hexose monophosphate shunt (pentose-phosphate cycle) supplying triosephosphate for the glycolytic pathway may be preferred; thus, by-passing the phosphorylation of hexosephosphate to hexosediphosphate (see Fig. 5.4). A wide range of experimental evidence has been found in *Callitroga macellaria* larvae: during semi-anaerobic conditions the RQ showed a tendency to increase; among other metabolites, fructose-6-phosphate accumulated during anaerobic conditions and after cyanide poisoning; and incubation with $^{14}C_1$-glucose or $^{14}C_6$-glucose in the presence of cyanide leads to a $^{14}C_6/^{14}C_1$-dioxide ratio of 0.43, and a $^{14}C_6/^{14}C_1$-lactate ratio of 1.4.

Energy producing reactions. Not only carbohydrates, but also amino acids, are substrates for direct energy production during anoxia. Utilization of carbohydrates results in the production of large quantities of lactate, and, the fomation of alanine, acetate (but not the volatile fatty acids propionate, butyrate, or valerate), pyruvate, polyols, and polyol phosphates (in particular a preliminary identified heptitol, glycerol, and glycerol phosphate). Glycerol phosphate is found during anoxia also in cockroaches and blowflies (Kubista 1957; Winteringham 1960; Heslop et al. 1963) and polyol production occurs in different insect species in response to anoxia, cold-hardiness and diapause (Chino 1958; Wilhelm et al. 1961, Sømme 1964, 1967, see also Sec. 4 in Chap. 1 and Sec. 5 in Chap. 3). It is known that these conditions are accompanied with at least partial blocking of the respiratory chain.

During the initial phase of anoxia the concentrations of some amino acids increase in larvae of *Callitroga macellaria*, suggesting the degradation of protein. During prolonged anoxia, amino acids such as glutamine, glutamate, and aspartate decrease considerably. This also occurs during cyanide poisoning. These amino acids must, therefore, be the substrates for anaerobic energy yield. After application

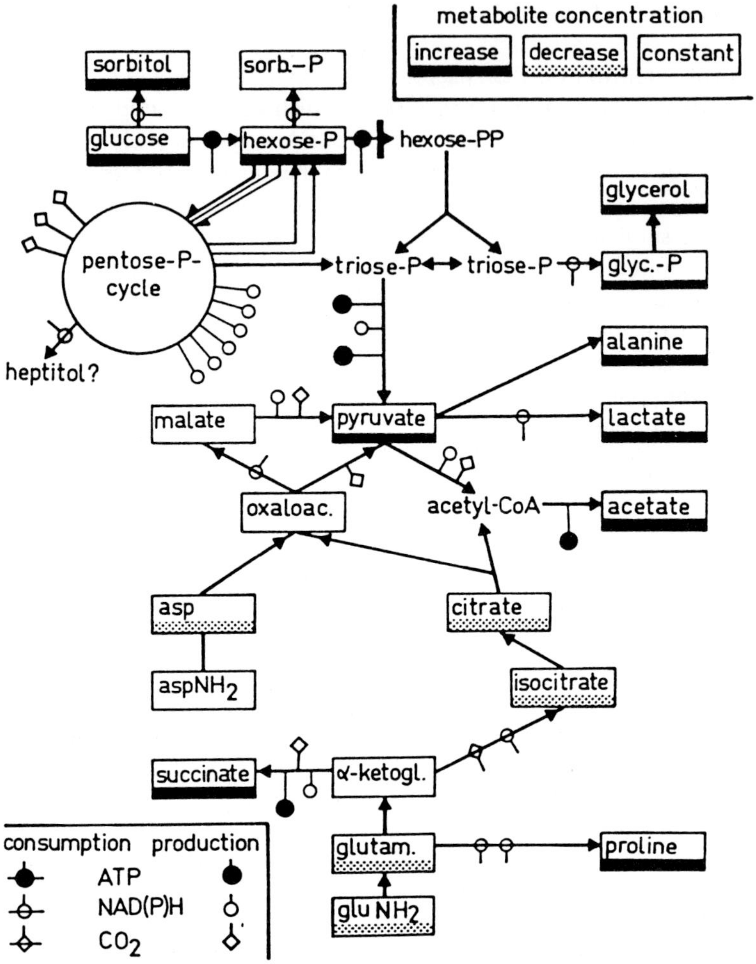

Fig. 5.4. Anaerobic metabolism in the larvae of *Callitroga macellaria*

of ^{14}C-glutamate, the radiolabel is found in acetate, pyruvate, α-ketoglutarate, oxaloacetate, succinate, and proline, but not in fumarate. The concentrations of succinate, proline, and acetate are also increased during cyanide poisoning. As seen in Fig. 5.4., in addition to ATP production during glycolysis, the reactions of succinate- and acetate thiokinases could supply one molecule of ATP each. The breakdown of glutamate is thought to proceed via α-ketoglutarate by GPT (as discussed above). The amino group acceptor, pyruvate, is partly derived from the breakdown of carbohydrates, but also from amino acids, because it is also radiolabelled after incubation with ^{14}C-glutamate. It is proposed that the subsequent fate of α-ketoglutarate involves a dismutation reaction: one part is converted to succinate via the Krebs cycle, and the other part in the reversed Krebs cycle to isocitrate and citrate. This may not be thermodynamically impossible considering the increased ratio of $NADH + H^+/NAD^+$ during anoxia. Carbon dioxide would not be detected, because one molecule would be released and one would be bound. A decrease in the levels of citrate and isocitrate has been measured during cyanide poisoning, although this causes a 99% block in the Krebs cycle. The degradation of citrate may be brought about by a reversible citrate synthase reaction (see discussion by Crabtree and Newsholme 1975), which could supply (per mole of citrate) two moles of ATP via acetyl CoA → acetate and oxaloacetate → pyruvate → acetyl-CoA → acetate, giving the acetate thiokinase reaction significant importance in this scheme. Previous data on anaerobic cockroach muscle (Kubista 1958) are in agreement with this proposal, and Hodkova and Kubista (1971) observed the disappearance of glutamate and production of acetate. However, there are still unresolved problems, and further work is needed on this system which seems rather unusual and, therefore, deserves attention.

4 Features of Locust Flight and Leg Muscle Metabolism During Muscular Activity and Anaerobiosis

It has long been known that vertebrate skeletal muscles show a great variability in function and performance, and can be classified into different metabolic types according to the quantitative contribution of the various metabolic pathways (Zebe 1961; Pette 1965). Thus, muscles which perform work more or less continuously are termed red muscles (according to their appearance because of a high myoglobin content); they characteristically contain small red fibers with a high density of large mitochondria showing a dense inner-mitochondrial structure. They have a high level of oxidative enzymes and a higher fat content compared with the white muscles. The latter are characteristically present where rapid strenuous work is performed; they contain large diameter light fibers with small mitochondria with little inner-mitochondrial structure. The preferred stored fuel is glycogen, and the level of oxidative enzymes is much lower than in the red fibers.

A similar specialization of different muscle types occurs in insects. The best known examples are the flight and leg muscles of the migratory locust (see Zebe 1961), where the differences in work-output are analogous to those of vertebrates, and similar morphological and metabolic specializations are found. Because the flight muscles have to sustain a high energy output for long periods, they are characterized histologically by fibers with short sarcomeres, and a high proportion of mitochondria (about 50% of the total muscle mass) directly in contact with ramifications of the tracheal system. Thus, access to the infinite source of oxygen in the atmosphere is guaranteed. In accordance with these morphological features are the characteristic metabolic potencies of the flight muscles (see reviews by Zebe 1961; Sacktor 1965, 1975; Kammer and Heinrich 1978); little or no activity of lactate dehydrogenase, and a remarkably high activity of α-glycerophosphate dehydrogenase (α-GPDH) as well as the mitochondrially located enzymes. Furthermore, even after exhaustive flight, no lactate accumulates, although glycogen, trehalose, and glucose levels are diminished. However, if locusts are exposed to an artifically produced N_2-atmosphere for only 90 s, levels of α-glycerophosphate and pyruvate increase about 15–20-fold compared with controls. Thus, as illustrated in Fig. 5.5 pyruvate and α-GP are the end products of anaerobic carbohydrate metabolism in

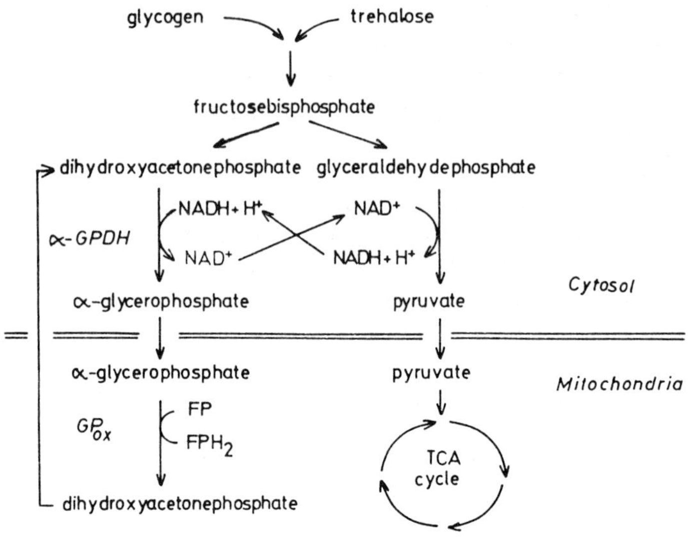

Fig. 5.5. The α-glycerophosphate shuttle of insect flight muscles

locust flight muscles, and α-GPDH instead of LDH performs the function of re-oxidation of NADH generated in glycolysis. This α-GP-pyruvate-dismutation also occurs in cockroach and fly flight muscles. The continuous operation of glycolysis, which depends on the continuous supply of NAD^+, is achieved by the α-GP cycle: NAD^+ is regenerated in the cytosol by reducing glycolytically formed dihydroxyacetonphosphate (DAP) to α-GP. This functions as a shuttle for electrons and protons into the mitochondria, where it is oxidized by the action of α-GP oxidase to yield DAP, which in turn leaves the mitochondria and is available for further oxidation of extramitochondrial $NADH + H^+$. In this way "reducing equivalents" from the extramitochondrial pool of $NADH + H^+$ pass the cytosolic-mitochondrial barrier (which is impermeable to the pyridine nucleotides) and are oxidized by the mitochondrial respiratory chain. Because the α-GP cycle is self-generating, and only a catalytic quantity of DAP is needed for the oxidation of $NADH + H^+$ that is continously formed, and because no α-GP accumulates during prolonged flight, Sacktor (1970) has suggested that glycolysis during insect flight is completely aerobic, with essentially two moles of pyruvate being formed from each mole of G-6-P.

In contrast to the flight muscles, the muscles of the locust metathoracic leg, the jumping leg, have to perform "burst" muscular activity (see Bishai and Zebe 1959; Bishai 1960; Zebe 1961; Hitzemann 1979). Morphologically this muscle has long thick fibers with a low proportion of mitochondria (about 10% of the muscle mass) and intracellular tracheolization. The enzymatic data correlate with these anatomical observations. These muscles have characteristically high LDH activity and low activities of α-GPDH and oxidative enzymes localized in the mitochondria, such as citrate synthase or thiolase, for example. Although the immediate source of energy for contraction, ATP, is stored at high levels (5 μmol/g w. wt) as in white vertebrate muscles, this is insufficient for more than a few seconds of work performance. ATP equivalents are yielded by two additional reactions: the breakdown of the phosphagen arginine phosphate, which is present at about 20μmol/g.w. wt, and the utilization of anaerobic glycolysis yielding lactate. After only two jumps, the lactate concentration is increased ca. sixfold (from 0.6 to 3.7 μmol/g w. wt) and arginine phosphate decreases, but there is no change in the level of ATP. After jumping to exhaustion (10–15 jumps), lactate increases to ca. 6 μmol/g w. wt. Hitzemann (1979) showed that locusts can jump between 50 to 100 times until exhausted. However, the intensity of these leaps differed with time. To quantify changes in jumping performance during a series of jumps, he measured the length of the jump, the force generated during take-off, and the interval between the individual jumps to calculate a parameter which he called "Sprungvermögen" (jumping power). During a series of 20 leaps, for example, the jumping power continuously decreased; after 5 jumps the decrease was about 15%, after 10 jumps an additional 12% and after 20 jumps the locust showed only 50% of the jumping power of resting locusts. All subsequent leaps were performed with about the same power until exhaustion. For the first five jumps energy equivalents are mainly regenerated by transphosphorylation, because ca. 90% of stored arginine phosphate is diminished. The second source for ATP yield is derived by glycolytical production of lactate (approx. 4.5 μmol/g w. wt). Despite these high production rates of ATP, it is insufficient to maintain the energy charge value at the high resting level (0.94), and a new value as low as 0.46 pertains. Although the jumping power decreases after the initial five jumps (which is mainly characterized by an extension of the interval between jumps), and despite the low concentration of arginine phosphate, the ATP level and the value for the energy charge does not decrease, but increases to values characteristic for resting animals. This is surprising because the animals still need energy for the jumps, although about 25% less because of the reduced jumping power. It was anticipated then that anaerobic glycolysis would be enhanced at least and, hence, enhanced lactate production would occur. However, a theoretically calculated lactate level of 14–19 μmol/g w. wt was never measured. In contrast, the lactate concentration after 20 jumps, or even in exhausted animals, never exceeded 6 μmol/g w. wt. At present there is no experimental evidence which could offer an explanation to these results. It could be argued that lactate diffuses out of the muscle into the hemolymph for oxidation in other tissues, as is the case in vertebrate skeletal muscle, but there appear to be no observations on hemolymph lactate levels after work. In this context, it is interesting to note that the activity pattern of representative enzymes of glycolysis, citric acid cycle, and β-oxidation, are almost identical in the muscles of the first, second, and third thoracic legs of locusts; thus, the pattern is not a specialization to the extreme function of the jumping muscle (Gäde 1970, 1975). The same author has also shown that the intersegmental muscles of different lepidopteran and coleopteran larvae have a high potential for anaerobic metabolism. He concludes that these muscles, including locust leg muscles, represent rather the archetype of metabolic organization in insect muscles, and that the highly aerobic flight muscles are the product of a special development in terms of functional adaptation.

5 Anoxia and Brain Metabolism

Many adult insects withstand anaerobic conditions for several hours. Total anoxia of the insect implies that the brain also becomes anoxic. It is known that the mammalian brain, a continuously active tissue with a rather high metabolic activity burning almost exclusively glucose, is absolutely oxygen-dependent. Deprivation of oxygen for only a few seconds can result in impairment of function, and after some minutes of anoxia permanent damage will occur (for a review see Wegener 1981). In contrast to this situation in mammals, several insect species appear to recover completely from prolonged anoxia, but, it may be in others that some of the complex functions of the brain may be impaired by anoxia exceeding 1 h.

Studies of Wegener and co-workers on the influence of anoxia on the central nervous system of honey bees, blowflies, and locusts shed some light on the metabolic reactions involved. When locusts are subjected to anoxia, the initial behavioral response is hyperactivity followed rapidly by paralysis.

Metabolically (Michel and Wegener 1982) the main features are a lowering of the brain's energy charge from 0.90 to 0.63 (after 5 min of anoxia) and even 0.21 (after 120 min). After 4 h of anaerobiosis the total adenylate pool is diminished by 50%. Glycogen, which is stored in the locust brain at a level of ca. 30 μmol/g w. wt, is broken down to lactate and, temporarily, to α-GP. The levels of α-GP reach a maximum after 1 h of anoxia and thereafter it is metabolized to glycerol. Recovery after 1 h of anoxia is rapid: after only 10 min in normal atmosphere, the energy charge of the brain is back to control levels. However, despite this impressive capacity to cope with anoxia, neurological damage does apparently occur. Locusts subjected to 2 h of anoxia refuse feeding in a subsequent recovery period and die within 3 to 5 days.

Experimental anoxia of up to 1 h in the honey bee and the blowfly results in an essentially similar pattern of brain metabolism (Kieffer and Wegener 1983):

1. The concentration of both high-energy phosphates, ATP and arginine phosphate, are almost completely diminished, and hence the value for the energy charge approaches zero.
2. Carbohydrates (glycogen, glucose, and trehalose in variable rates) are consumed.
3. Glycerol, alanine, and some lactate (honey bee brain), or alanine, α-GP, and glycerol (blowfly brain) are identified as anaerobic end products.
4. Because there is a loss of substrates unaccounted for by the above mentioned products, the formation of polyols is anticipated, but this has not yet been demonstrated.
5. It is also suggested that the metabolic rate during anoxia is reduced, as it is in a variety of other invertebrates during anaerobiosis (see for example, Gäde 1983).

The metabolic pathways are not unravelled in detail, but from the end products formed we may suggest that the same pathways are involved as have already been discussed for whole animals or muscle tissue in earlier sections of this chapter.

Unlike vertebrate brains, there are species within the class Insecta that possess the capacity in brain tissue to derive energy by oxidation of fatty acids (Wegener 1983). This conclusion is based on findings of a high activity of 3-hydroxyacyl-CoA-dehydrogenase in silkmoth brain, but a rather low activity of LDH. Furthermore, respiratory studies on silkmoth brain tissue reveal a respiratory quotient of 0.74 and, additionally, addition of ^{14}C-labelled fatty acids to isolated silkmoth brains results in a marked production of $^{14}CO_2$. Unfortunately, it is not yet known what metabolic changes occur in silkmoth brain tissue during anoxia.

6 Involvement of Anaerobic Processes During Flight Muscle Development and During Diapause

Flight muscle development is characterized by a progressive change and differentiation. Structurally, one of the best defined processes is the invagination of tracheoblasts into the muscle fibers. A high lactate dehydrogenase activity level is found throughout the development of tracheoblasts (Beenakkers et al. 1975). From the LDH data it is anticipated that anaerobic glycolysis may be important for the provision of energy. After tracheolization, a period of differentiation follows in which the activity of mitochondrial enzymes is strongly increased and the aerobic character of the flight muscles is expressed (see Beenakkers et al. 1981a).

Diapause is a delay in development which enables the insect to survive unfavorable conditions (for details see Chap. 3). It is beyond the scope of this chapter to go deeper into the different problems of diapause, but, briefly, one example is given where a link appears between anoxia and diapause (see also Steele 1981). Eggs of the silkworm *Bombyx mori* overwinter in a state of diapause which is induced by the diapause hormone produced by the subesophageal ganglion of the mother. Metabolically, diapause eggs are characterized by a high concentration of sorbitol and glycerol (Chino 1958) and a significant rise in the concentration of lactate (Kageyama 1976). Although the reason for the accumulation of polyols is still not exactly known, there is some evidence that anaerobic metabolism is involved. There is a sharp reduction of the activity of the electron transport system during diapause (Chino 1963). Hence, the glycolytic pathway is enhanced and becomes the main source of ATP production and there is an accumulation of lactate. NAD^+ regeneration is achieved by the reduction of dihydroxyacetone phosphate to α-glycerophosphate which, in turn, is dephosphorylated by an acid phosphatase resulting in the production of free glycerol. Support for this interpretation comes from experiments with *nondiapause* eggs. When these eggs are held anoxic, levels of lactate and polyols rise as shown in diapause eggs when oxygen is available. The mechanism of sorbitol production in diapausing eggs may be influenced mainly by two factors: reduced electron transport which stimulates glycolytic activity and the reduced activity of phosphofructokinase (Kageyama and Ohnishi 1971). Thus, glucose-6-phosphate will be redirected through the pentosephosphate pathway resulting in the synthesis of $NADPH + H^+$ required for sorbitol synthesis (the reader is referred to Fig. 5.4, since the anaerobic pathways in *Callitroga* larvae for sorbitol synthesis are thought to be identical).

7 Conclusions

Many insect species are not challenged by a lack of oxygen in their environment. There are species, however, that live in habitats where insufficient oxygen may be available. An attempt was made in this chapter to demonstrate how some of these insects have adapted their metabolism to these conditions. It is obvious that only a few examples have been studied in detail and that, at present, no general conclusions can be made. It is evident, however, that at least some features of the anaerobic metabolism of aquatic insect larvae are similar to those found in euryoxic annelid and mollusk species. It is also of interest to note that similar mechanisms occur during anoxia in the terrestrial blowfly larvae as in the diapausing silkworm eggs. Unfortunately, the data on anaerobiosis in terrestrial insects are too limited to prove whether polyol formation, for example, is a general feature. It is hoped, however, that further work is stimulated by this review and, in due time, some of the unclarified areas will be more transparent.

Acknowledgements. The author wishes to thank Dr. G.J. Goldsworthy (University of Hull, England) for critically reviewing the manuscript; Dr. S.G.E. Meyer for providing Fig. 5.4; and Mrs. K. Weidemann for her secretarial help. This chapter was written while the author was supported by a Heisenberg Fellowship awarded from the Deutsche Forschungsgemeinschaft (Ga 241/5–1).

Chapter 6

Respiration and Respiratory Water Loss

Paul Kestler[1]

Contents

1 Respiratory Mechanisms in Aquatic and Terrestrial Insects: Gills, Plastron, and Tracheae as Adaptations to Gas Capacitance and Physical Properties of the Medium

During the hundreds of million years available for genotypic, periodic, and short-term adaptations in insects the basic need for oxygen uptake and carbon dioxide release in highly changing and highly differing environments has led to a variety of structural and functional adaptations.

1 Zoologisches Institut, Universität Göttingen, Berliner Straße 28, 3400 Göttingen, FRG

Environmental Physiology and Biochemistry of Insects, ed. by K.H. Hoffmann

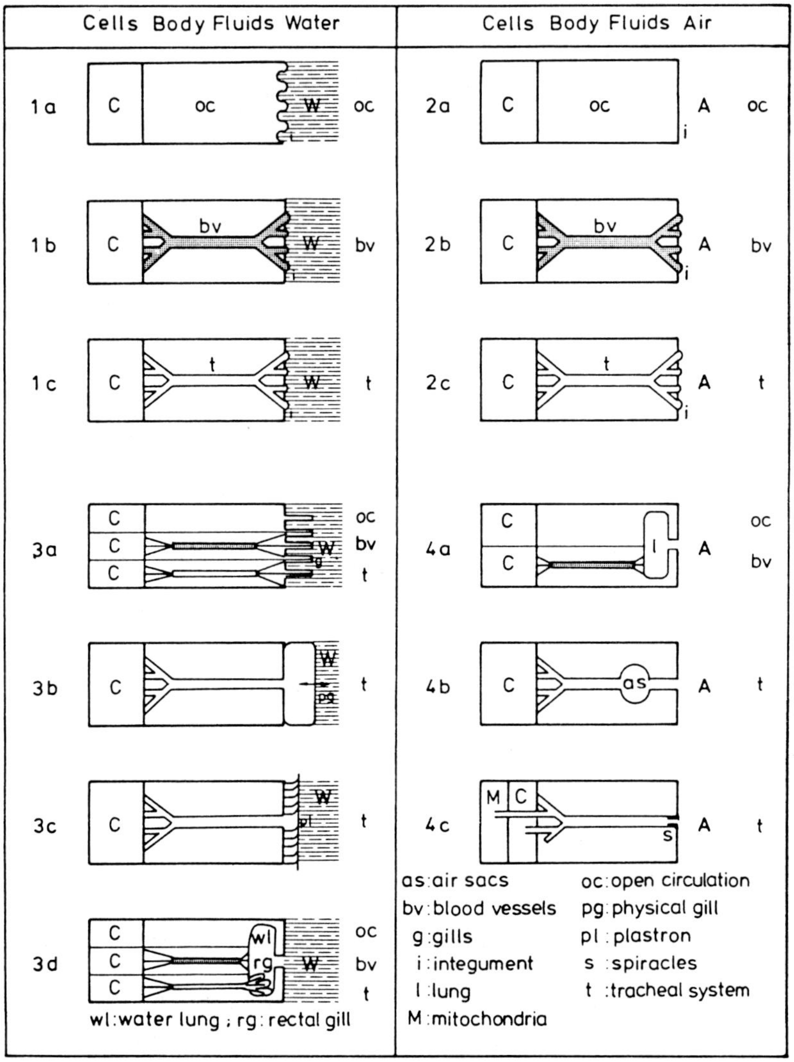

Fig. 6.1. Simple models of gas transfer between cells (*C*) and the ambient medium water (*W; left side*) or air (*A; right side*). Models 6.1.1 and 6.1.2 show the basic types (*a–c*) of body fluids involved in transfer between integument (*i*) and cells. Models 6.1.3 and 6.1.4 show these basic types in combination with different types of exchangers

The variety and success of genotypic respiratory adaptations in insects become very apparent in enumerations of general respiration principles by comparative physiologists (Dejours 1981). Most of the different types of respiration summarized in Fig. 6.1 are more or less specific for insects. A detailed description of the occurrence and physiology of the variety of types in terrestrial and aquatic insects may be found in reviews of Krogh (1941), Miller (1974), and Mill (1974). We can, therefore, focus our interest on some less well recognized but widespread principles of these respiratory adaptations.

It is a principle of oxygen and carbon gas exchange adaptations common to animals and plants that gas exchange organs have an increased evaginated surface area in the external medium water (Fig. 6.1.3a–d) and an increased invaginated surface area in the external medium air (Fig. 6.1.4a–c).

It seems apparent that the genotypic adaptation of invaginated internal surfaces in air may be due to the high drop in water potential between cells and air (Ziegler 1977). The adaptation should, therefore, reduce respiratory water loss. As pointed out by Wells (1980) invaginated surfaces could offer protection in a terrestrial environment, too. As movement of the body and the medium increase the ventilation flow rate of the water, evaginated external surfaces save energy costs of ventilation of the ambient medium water.

The question remains as to which property of the external medium the increased surface area typical of an evagination is adapted. In contrast to an invaginated surface area, an evaginated surface area has a higher volume of external medium adjacent to it. This reduces the fall in oxygen content or the rise in carbon dioxide content of the external medium in the same way as gill ventilation. If there is a physical difference in the capacitance of water and air for oxygen and carbon dioxide, invaginated exchangers indicate the better capacitance of air and evaginated exchangers the lesser capacitance of water.

The better capacitance in air may be understood together with the most impressive genotypic respiratory adaptation of insects, the tracheal system (Fig. 6.1.4b). It enabled them to use the better capacitance of the medium air successfully in ages, when the partial pressure of oxygen was below the lowest vertebrates can tolerate. The tracheal system brought the external medium air in close contact to the respirating cells. The most extreme condition has been shown by Afzelius and Gonnert (1972) to occur in the flight muscle of *Vespa crabro*. Here the external medium air may extend sometimes even up to the mitochondria (sarcosomes). It allows for their direct access to air circumventing any circulation.

Using air as an internal transport medium may have been of some special advantage as it has been conserved in the external medium water, where the need for prevention of respiratory water loss no longer supports the invagination. Gas exchange with the external medium occurs through a variety of genotypic adaptations, such as the tracheal gills (Fig. 6.1.3a), physical gas gills (Fig. 6.1.3b), plastron gas gills (Fig. 6.1.3c), and rectal gills (Fig. 6.1.3d).

To understand the genotypic adaptations of invaginated and evaginated respiratory surfaces as well as the different use of air or water as an internal transport medium necessitates a more detailed physical understanding of the differences between water and air's capacitance for oxygen and/or carbon dioxide. We have concluded that the advantage of using air instead of water as an internal transport medium to cells as well as a medium for invaginated respiratory surfaces is its better capacitance for air even at low partial pressures. Evaginated surfaces are the genotypic adaptation to a low gas capacitance of the medium water.

What does capacitance for oxygen and carbon dioxide mean physically and what are the differences in parameters which have been subject to evolution? At the same partial pressure (expressed in kPa or Torr) the maximal quantity (mmol) of oxygen in a certain volume (ml) of water is much lower than in a volume of air (cf. Sec. 2,

Table 6.1. Values of capacitance coefficient β (nmol · cm^{-3} · kPa^{-1}), effective diffusion coefficients D′ (cm^2 • sec^{-1}), and Krogh's constant K (nmol • cm^{-1} • sec^{-1} • kPa^{-1}) for oxygen and carbon dioxide in distilled water at 20°C and 1 atm[a]

	βO_2	$D'O_2$	KO_2	βCO_2	$D'CO_2$	KCO_2
Water, W	13.65	$25 \cdot 10^{-6}$	$341.3 \cdot 10^{-6}$	389.2	$18 \cdot 10^{-6}$	$7.006 \cdot 10^{-3}$
Air, A	410.5	0.198	82.18	410.5	0.155	63.63
A/W	30.07	$7.92 \cdot 10^{3}$	$2.38 \cdot 10^{5}$	1.055	$8.61 \cdot 10^{3}$	$9.08 \cdot 10^{3}$

[a] Wells 1980; Dejours 1981

Chap. 5). The capacitance coefficient is the capacitance or quantity of gas x per unit volume of the medium m and unit increase in partial pressure of gas x. In contrast to oxygen carbon dioxide has a comparable capacitance coefficient in air and water (Table 6.1, see Dejours 1981 and Wells 1980 for more details and for dependence on temperature and presence of ions). Vertebrate respiration physiologists (cf. Piiper et al. 1971 for nomenclature units and equations) include in the capacitance coefficient β not only the physical solubility, but also the chemical binding or even biological compartmentalization. The evolution of different types of hemocytes and respiratory pigments in the internal medium water are, thus, a "strategy" to increase the low capacitance coefficient of the internal transport medium water. The capacitance coefficient (mmol · ml^{-1} · kPa^{-1} or $Torr^{-1}$) of a certain gas x in a certain medium m is expressed usually as the change in concentration $\triangle c_{x,m}$ per unit change of partial pressure difference $\triangle P_{x,m}$.

$$\beta_{x,m} = \triangle c_{x,m} / \triangle P_{x,m} \tag{1}$$

The concentration c_x(mmol · ml^{-1}) is expressed as the quantity of substance M_x (mmol) in a certain volume V_m (ml) of the medium.

$$c_{x,m} = M_x / V_m \tag{2}$$

Total capacitance M_x (mmol) is then the product of volume, capacitance coefficient, and partial pressure.

$$M_x = V_m \cdot \beta_{x,m} \cdot P_{x,m} \tag{3}$$

Equation (3) helps us to understand the advantage of an increased volume of adjacent water available to everse exchangers in water. It compensates for the low capacitance coefficient of oxygen and increases the capacitance of oxygen in the adjacent medium at the prevailing ambient partial pressure and capacitance coefficient.

The capacitance coefficient of air β_g is the same for oxygen and carbon dioxide as they are usually treated as ideal gases in respiration physiology. The coefficient allows calculation of gas exchange or transport between different media which are dependent upon partial pressure differences $\triangle P_x$ (kPa or Torr) and not concentration differences. Concentration differences can only be used in the same medium and are then the product of capacitance coefficient $\beta_{x,m}$ and partial pressure difference $\triangle P_{x,m}$.

$$\triangle c_{x,m} = \beta_{x,m} \cdot \triangle P_{x,m} \tag{4}$$

The higher capacitance coefficient of oxygen in air is only one advantage of using air as an internal transport medium in animals with tracheae. The other is the higher diffusion coefficient of oxygen and carbon dioxide in air as compared with water (Table 6.1). It is responsible for the reversibility of the anaesthetization of insects with carbon dioxide. The product of the effective diffusion coefficient $D'_{x,m}$ ($cm^2 \cdot s^{-1}$) and the capacitance coefficient $\beta_{x,m}$ ($mmol \cdot ml^{-1} \cdot Torr^{-1}$ or kPa^{-1}) is the well-known Krogh's constant $K_{x,m}$ ($mmol \cdot cm^{-1} \cdot Torr^{-1}$ or kPa^{-1}).

$$K_{x,m} = \beta_{x,m} \cdot D'_{x,m} \tag{5}$$

Table 6.1 shows the different parameters of Eq. (5) for oxygen and carbon dioxide in air and water. Table 6.2 shows the ratios of the coefficients of carbon dioxide to oxygen. The data also show that internal gas transport in air is especially amplified if diffusion is involved and that the low capacitance coefficient of oxygen in water, if compared to carbon dioxide, cannot override the better diffusion coefficient as shown by the R_K value. This can be even better understood if we compare in the most simple case the molar rates of transport $\dot{M}_x$ ($mmol \cdot s^{-1}$) of gas x in a tube T in the medium m by diffusion DI or convection CO (Piper et al. 1971).

$$\dot{M}_x^{T\text{-}DI} = (A^T/L^T) \cdot D'_{x,m} \cdot \beta_{x,m} \cdot \triangle P_{x,m} \tag{6}$$

$$\dot{M}_x^{T\text{-}CO} = \dot{V}_m \cdot \beta_{x,m} \cdot \triangle P_{x,m} \tag{7}$$

Both equations contain the capacitance coefficient. Therefore, not only capacitance M_x, but also any gas transport rate $\dot{M}_x$ depends on respiratory adaptations connected to the capacitance coefficient. Thus, the use of the capacitance coefficients in the modern nomenclature of comparative respiration physiology reveals an important parameter in genotypic adaptation. Equation (6) for diffusion together with Eq. (5) and Table 6.1 show that the Krogh's constant or permeability coefficient for oxygen is 238,000 times as large in air than in water. Even carbon dioxide transport by diffusion is 9,080 times as large in air than in water. But it becomes at the same time apparent that this physical advantage of internal diffusive gas transport can only be used over relatively short distances L^T.

Equation (6) shows also that the increase in area A (cm^2) and the decrease in length L (cm) of the diffusion path increase the molar rate of diffusive DI transport of gas x in medium m. This shows the formal physical basis for the genotypic adaptation of gas exchange membranes for diffusion. Both types of increases in exchange area are concomitant with a reduction of the thickness of the exchange membranes. This has led to a further principle (Fig. 6.2.a,b) that diffusive gas exchange occurs through membranes and that convective transport with the internal or external li-

Table 6.2. Ratios R_X of the capacitance coefficients β, effective diffusion coefficients D', and Krogh's constants of carbon dioxide to oxygen in different combinations of the media air A and water W [a]

	R_β	$R_{D'}$	R_K
$CO_2(A)/O_2(A)$	1	0.783	0.783
$CO_2(W)/O_2(W)$	28.51	0.720	20.52
$CO_2(A)/O_2(W)$	30.07	$6.2 \cdot 10^3$	$1.86 \cdot 10^5$

[a] cf. Table 6.1

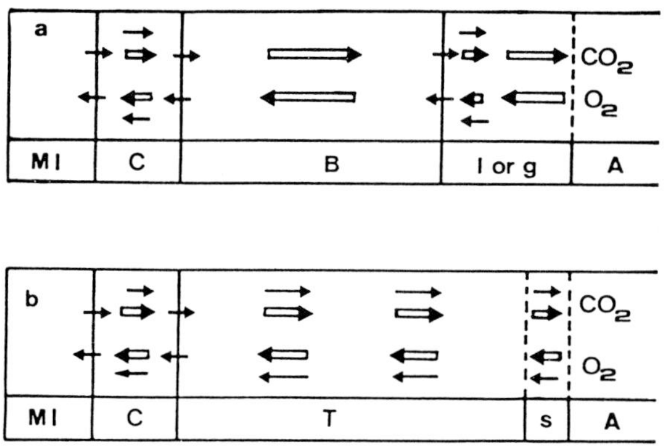

Fig. 6.2. Mechanisms of transfer of oxygen and carbon dioxide via blood (*B;* **model a**) or tracheal system (*T;* **model b**). *Simple arrow* diffusion; *double arrow* convection; *C* cytoplasm; *MI* mitochondria; other notations as in Fig. 6.1

quid or gaseous medium is realized in compartments. In small compartments like cytoplasm, alveoli or tracheae (Fig. 6.2b), diffusion becomes important, in addition.

Tracheae cannot have limited the size of tracheate animals as convective transport could supplement or even replace diffusion to the same extent it does in respiration of vertebrates. Convective transport would be even better because β is greater in air than in water. The advantage of convective transport in the medium water is the co-transport of water soluble substrates for respiration and circulation together with the respiratory gases. Giant insects would have to support energy for two different convective transport systems. This favored the indirect pumping of tracheal gases by autoventilation due to muscle contractions not specific for gas exchange. Another limitation of insect size must have the open circulation (Schmidt-Nielsen 1979).

Convective transport parameters have also been subject to genotypic evolution. $\dot{V}_m$ in Eq. (7) is the rate of volume flow (ml $\cdot$ s^{-1}) of the medium m. It is the product of the volume change $\triangle V_m$ (ml) of lung or tracheal system and frequency f of volume change (s^{-1}).

$$\dot{V}_m = \triangle V_m \cdot f \tag{8}$$

In the tracheal system and in gills both parameters can be increased to improve the ventilation rate $\dot{V}$ and convective transport rate $\dot{M}_X$.

An understanding of the evolution of two gas gill systems (Fig. 6.1.3b,c) in insects has to include, in addition to knowledge of the better capacitance and diffusion coefficients of oxygen in air, the recognition of the different physical situation for gas exchange at the interface between ambient water and air in volume constant (plastron) and volume variable (physical gills) air spaces. In a very elegant model analysis Rahn and Paganelli (1968) have shown this difference. Physical gills can only be understood, if we consider the gas exchange of the inert gas nitrogen. For plastron respiration, on the other hand, the small difference in total pressure at the interface is important. The volume decrease of the physical gills depends highly upon the diffusion of nitrogen out of the physical gills along a partial pressure gradient. The partial pressure of nitrogen is higher in the gill than in ambient water as (1) total pressure is constant in a volume variable system and (2) steady state partial pressure difference for oxygen is higher than the negligiable partial pressure difference for carbon dioxide, compensating for the differing capacitance coefficients in water. As the sum of the partial pressure differences of oxygen, carbon dioxide, and nitrogen must be zero, nitrogen has to fill up the resulting deficit in the air space in the steady state situation. This is possible without nitrogen inflow as its relative amount increases during volume decrease. Equation (9) shows this physical basis. In Sec. 3.3, Eq. (9) will be derived from Dalton's law (Piiper 1972) applied to both sides of the interface.

$$\triangle P_{N_2} = \triangle P_{O_2} - \triangle P_{CO_2} \tag{9}$$

The Fenn-Diagram (Rahn and Paganelli 1968) shows just the same in a isosceles rectangular triangle with the small sides $\triangle P_{O_2}$ and $(\triangle P_{N_2} + \triangle P_{CO_2})$.

Partial pressure differences are usually defined as positive in respiration physiology. The partial pressure difference for nitrogen leads to nitrogen loss by diffusion and limits the life of the air space of the physical gill.

In the volume constant gas gill the total pressure is not equal. This is the genotypic adaptation to the drawback of nitrogen loss. The plastron prevents volume decrease and concomitant rise of nitrogen partial pressure in the air space. In this case the total pressure difference compensates for the difference between oxygen and carbon dioxide partial pressure differences as did the nitrogen partial pressure difference before. Equation (10) shows this in the same way as the triangular Fenn-Diagram.

$$\triangle P_B = \triangle P_{O_2} - \triangle P_{CO_2} \tag{10}$$

The spaces between the hairs of the cuticular modifications of the plastron have to withstand this pressure difference, showing the physical function of these structures in this genotypic adaptation, too.

Further details of the different types of gas exchange organs in insects have been fully reviewed for aquatic respiration by Mill (1974), for terrestrial respiration by Miller (1974), and for general respiration of invertebrates by Mill (1972). They need not be repeated here. As a basic text, introducing the nomenclature, the booklet of Wells (1980) can be recommended.

This short section shall show that even the simplest equations of respiration physiology, the physical parameters of genotypic respiratory adaptations in aerial and aquatic gas exchange in insects become apparent. A better understanding of the basic physical principles of adaptation conforms with one of the main goals of comparative physiology. Adaptations other than the genotypic ones will, in addition to them, be discussed in the next sections. Tracheal respiration will be taken as an example and the adaptations to the advantage of respiration in the medium air and the disadvantage of respiratory water loss will be explicated experimentally and theoretically.

2 The Preadaptive Situation for Tracheal Respiration in the External Medium Air

Insects living in the external medium water or air use internal gas transport in air in both cases (Fig. 6.1). There seems to be no doubt that the tracheal system evolved as paired tubular invaginations of the cuticle in each segment. It is generally assumed that the tracheal system had neither intra- or intersegmental fusions, air sacs, special pumping mechanisms, nor spiracles with closing mechanisms or sieve plates (Miller 1974).

To be able to understand these adaptations to life in air one has first to look at the physical basis of gas exchange in such a simple preadaptive tracheal system. One should then ask, which physical principles of gas exchange and which parameters in the equations describing these mechanisms may have been preferentially changed by the structural and functional adaptations to result in a better adaptation to the environment.

2.1 Krogh's Model and Theory of Pure Diffusive Respiration Through a Primitive Tracheal System

As regular pumping movements were absent in the preadaptive organization of the tracheal system and as not all insects showed pumping movements, the question whether pure diffusive gas exchange is sufficient in such a tubular system, arose very early (Treviranus 1814; Graham 1833). After much discussion of controversial experimental results, compiled very thoroughly by Hazelhoff (1926), the question was finally tested by adding a model analysis to experimental work. Krogh (1920a) showed in one of the most important theoretical papers on tracheal respiration that even active insects of considerable size are able to supply cells by pure diffusion with oxygen. He suggested that ventilation (Krogh 1920b), which leads to convective gas exchange through the tracheal openings and in the distal parts, supports proximal diffusion near the tissues at higher oxygen needs.

As an answer to some heavy criticisms on the morphometric results (Nunome 1944a, 1944b, 1951) and on a number of simplifications (Buck 1962), the theory of pure diffusive gas exchange was tested by Weis-Fogh (1964). In a very detailed theoretical and experimental analysis of insect flight muscle, the most active tissue known (see, e.g., Sec. 4, Chap. 4), he came to the conclusion: "There is no reason to suggest other mechanisms than a simple, normal diffusion". He showed moreover that Nunome "has, in fact, established Krogh's result on the basis of a much more detailed analysis". The simplicity of Krogh's ingenious model is not only of didactic, but also of heuristic value to an analysis of those parameters of preadaptive diffusive gas exchange, which have been changed during adaptation. As long as we don't know more about the original morphometry, Krogh's model may show us some important drawbacks of the original design. Therefore, it shall be presented here in more detail.

Krogh's model was based on detailed measurements of morphometry and of total oxygen consumption of larvae of *Cossus, Tenebrio,* and *Scutigera.* Krogh assumed a steady state between the molar diffusive flux of oxygen through the tracheal system $\dot{M}^{T\text{-}DI}_{O_2}$ and the molar rate of oxygen consumption $\dot{M}^{C}_{O_2}$ (Eq. 11) (Fig. 6.3a).

$$\dot{M}^{T\text{-}DI}_{O_2} = \dot{M}^{C}_{O_2} \tag{11}$$

Krogh showed that the summed cross-sectional area at successive levels of branching remained constant. The model (Fig. 6.3a) lumps, therefore, the cross-sectional area of primary, secondary, or higher order n tracheae of the different spiracles $S_1 \ldots S_n$ together. The model, thus, reminds one much of a bundle of fibers in a light guide. The morphometric results of Thorpe and Crisp (1947b) in adult *Aphelocheirus* and of Locke (1958) in larvae of *Rhodnius, Ephestia, Calliphora* and *Periplaneta* supported this assumption. In contrast, Nunome's studies, which are the most extensive ever likely to be made on the tracheal system, showed a smaller area at both ends. It could well be that a constant cross-sectional area is the most primitive situation and that deviations are special adaptations.

Krogh tested only the ability of the tracheae to supply oxygen by pure diffusion. To follow Krogh's calculation in the modern nomenclature of respiration physiologists we have to introduce Krogh's constant for air as defined in Eq. (5) into the general diffusion Eq. (6) and to solve for the partial pressure difference $\triangle P^{T}_{O_2}$ between the ends of the tracheal air space giving

$$\triangle P^{T}_{O_2} = \dot{M}^{T}_{O_2} \cdot L^{T} / (A^{T} \cdot K_{O_2}) \quad . \tag{12}$$

For Krogh's measured values of 13.07 nmol · s^{-1} for $\dot{M}^{C}_{O_2}$, 6.1 · 10^{-2} cm for A^T and 0.74 cm for L^T in a *Cossus* larva and the K_{O_2} of Table 6.1 a $\triangle P_{O_2}$ of 1.95 kPa can be calculated according to Eq. (12). This value of about 2% of total atmospheric pressure or about 10% of ambient partial pressure of oxygen is reported in most textbooks. It shows that fractional concentration (part of one in dry air) of ambient oxygen $F^{A}_{O_2}$ would fall from an ambient value of 0.21 (21%) at the tracheal opening to a value of 0.19 only at the interface between tracheae and tracheoles.

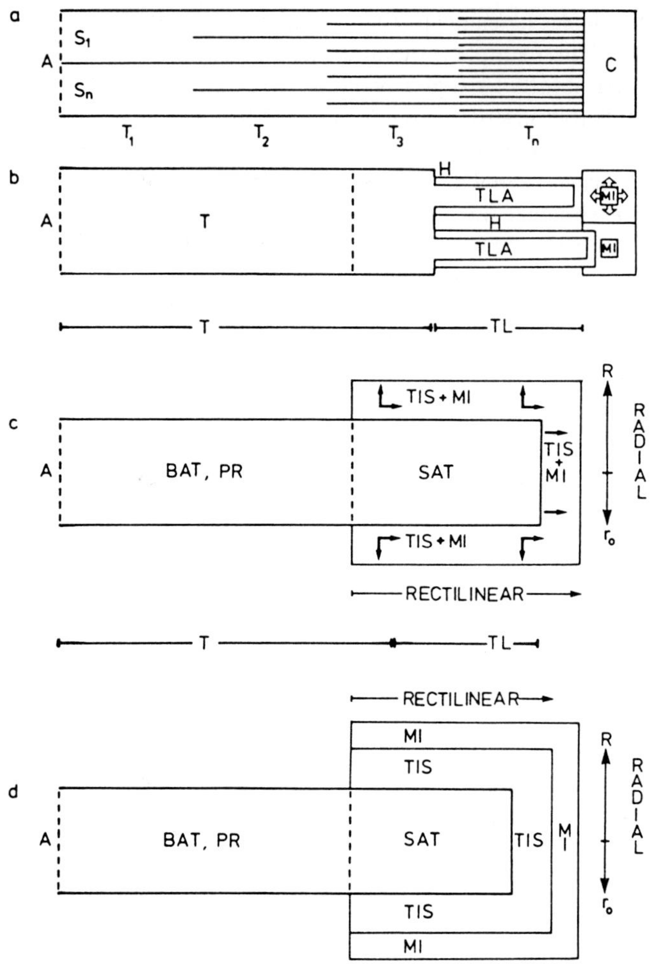

Fig. 6.3a–d. Models of gas exchange through the tracheal system. **a** Krogh's model of structure. $S_1 \ldots S_n$, different spiracular openings; $T_1 \ldots T_n$, tracheal ramifications. **b** Model of main structures involved in gas transfer. *T* tracheal system; *TL* tracheoblast; *TLA* tracheolar air; *H* hemolymph; *MI* mobile or fixed mitochondria; *double arrow* transport. **c** Weis-Fogh's mode. *BAT* big air tubes; *PR* primary system; *SAT* small air tubes; *TIS* nonmetabolizing tissues. **d** Correction of Weis-Fogh's model

2.2 A New Model of Tracheolar Respiration in Insects

It is one of the main drawbacks of Krogh's model of axial diffusion in tracheae that it neither includes diffusion in tracheoles nor in tissues. Tracheoles elongate, as intracellular tubes of the tracheoblast, the tracheal air space up to the surface of the supplied cell (Fig. 6.3b). Terminal tracheoles are attached to the cell membrane. This contact is very close, the basement membranes often fuse, and the tracheoblast may either indent the cell membrane or enter invaginations as in the tubular system of muscle (Fig. 6.3b). On the surfaces of the organs the palmate tracheal cells may, according to Wigglesworth (1972), unite and form a fenestrated membrane. This membrane, bearing the capillaries of the tracheal system, is more or less attached to the cell membrane, sometimes building a considerable blood sinus in the interspace. Wigglesworth points out: "blood must always play the part of intermediary between the lumen of the trachea and the cytoplasm of the cells". One should, therefore, suppose that convective gas transport in this part of the hemolymph should support diffusive flux of oxygen and carbon dioxide between air capillaries and mitochondria (Miller 1974).

Additional convective transport should occur due to the cytoplasmic flow in the tracheoblast and in the terminal oxygen consuming cell. If we assume diffusion through the membranes concerned, we get a picture of the suggestive flows between air space and mitochondria based on the morphology (Fig. 6.4a).

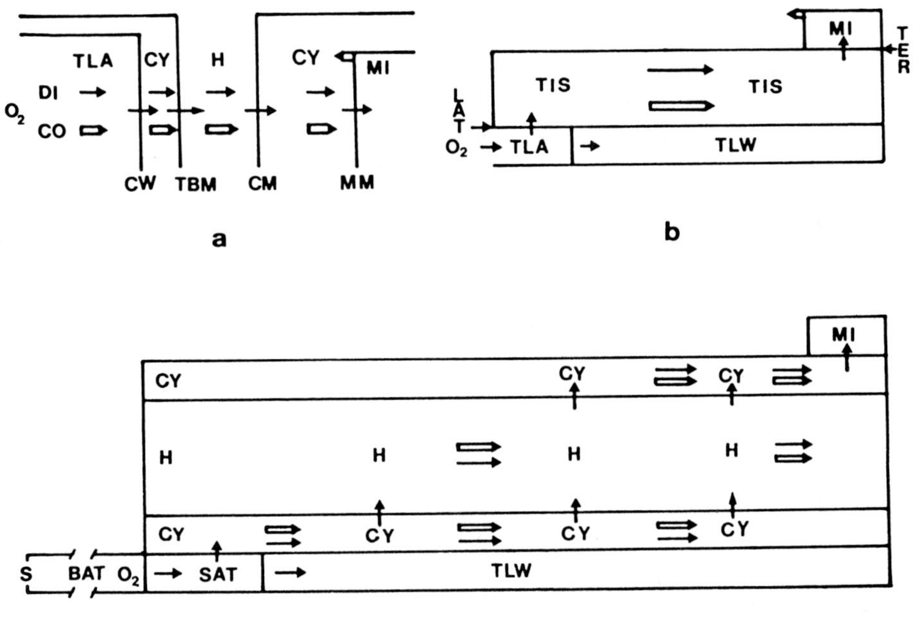

Fig. 6.4a–c. Diffusive and convective by-pass of water-filled part of tracheoles. **a** Lateral transfer of oxygen from tracheolar air (*TLA*) to mitochondria (*MI*) through cytoplasm (*CY*) and hemolymph (*H*). *CW* cuticular wall; *TBM* tracheoblast membrane; *CM* cytoplasm membrane; *MM* mitochondrial membrane; *DI* diffusion; *CO* convection. **b** Simplified by-pass model of transfer. *LAT* lateral membranes; *TER* terminal membranes; *TLW* tracheolar water. c Detailed description of by-pass transfer. *S* spiracle. Other notations as in Fig. 6.3

Mobility of the mitochondria, which can be regarded as a special type of convective transport by cytoplasmic flow may, in addition, shorten the distance for cytoplasmic transport (Fig. 6.3b; Fig. 6.4a).

If the mitochondria are more or less fixed at positions with a high consumption of ATP, tracheoles have to come close to the mitochondria. This may be achieved either by contact to the mitochondria in evaginations or invaginations of cells. The first case occurs in insect ganglia (Burrows 1980), where the distribution of tracheoles "runs parallel with the expected distribution of mitochondria" in nerve fibers. The second case occurs in insect flight muscle. In this case the extracellular liquid of the T-system is continuous with the hemolymph. According to Weis-Fogh (1964) this "must lead to muscular pumping of blood and air in and out". One may regard this as an adaptation, in which the resulting transport in the tubules replaces the transport of substances and mitochondria in the cytoplasm of the terminal oxygen consuming cell.

In contrast to morphological evidence, all theories of gas transport in tissues formulated so far in detail neglect any contribution of convective flows. They differentiate only between diffusion or ventilation in air tubes and tissue or liquid phase diffusion. Tissue diffusion was added for the first time by Thorpe and Crisp (1947a,b, and c) in their excellent study of plastron respiration, in which Crisp worked out the theory in the appendix (1947b). Later Weis-Fogh (1964) analyzed not only tissue, but also air tube diffusion in flight muscles with different patterns and geometries of tracheation and tracheolization. Like Crisp before him, he adapted different models of tissue cylinders for different types of combined rectilinear (axonal) and radial (lateral) steady state diffusion (Fig. 6.3c). This formulation provided simple equations for the relation between partial pressure difference of oxygen, length, or radius of the tissue cylinder or the air tube, and volume specific rate of oxygen consumption. Both theoretical and experimental studies confirmed Krogh's theory that pure diffusion is sufficient with a safety factor of 2–3 even in flight muscle at the highest metabolic rates of tissues. This holds even if the summed cross-sectional area is declining in some tracheae as shown by Weis-Fogh (1964).

With improved injection and freeze fracture techniques Wigglesworth and Lee (1982) showed that tracheolar endings may subdivide in insect flight muscle down to a diameter, which is within the range of the mean free path of the oxygen molecules (0.072 μm). This suggests that using the improved methods, one should find terminal tracheoles of similar size on or between other cell membranes. Wigglesworth and Lee showed that tracheoles invaginated the transversal (T-) tubules of the muscle fibers in a secondary system and ramified in a tertiary system to the low diameter of terminal tracheoles until they ended very close to each mitochondrion (Fig. 6.5). They often encircle mitochondria and build "vesicles" at the contact area. Their freeze fractures show convincingly that the membranes and compartments, which have to be passed according to Fig. 6.4a, do not change even in this most adapted system. The authors came to the result: "... the gap between the tracheole endings and the mitochondria, over which diffusion through the body fluid was postulated (Weis-Fogh 1964) has been largely eliminated". One should add: this has been achieved mainly by reduction of the size of the liquid phase compartments in Fig. 6.4a. Reduction in size of these compartments decreases the diffusive and convective transport in the liquid phase. The thin membranes between the mitochondria and the terminal tracheole may represent the shortest distance for tissue

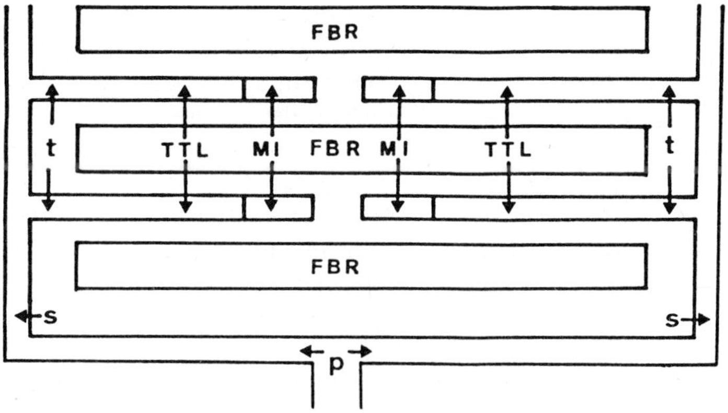

Fig. 6.5. Routes of gas transfer in flight muscle according to Wigglesworth and Lee (1982). *FBR* muscle fibrils; *MI* mitochondria; *TTL* terminal tracheoles; *p* primary tracheal system; *s* secondary system; *t* tertiary system. *Arrows* show the position

diffusion in flight muscles. As has been mentioned this is the case in intramitochondrial tracheoles in *Vespa* demonstrated by Afzelius and Gonnert (1972). It shows that even the interspace compartments may be absent.

The quote referred to above shows that Wigglesworth and Lee did not correct or criticize the assumption made in the models of Crisp and Weis-Fogh that tissue diffusion (Fig. 6.3c) takes place in an "isotropic liquid". "The rate of consumption per unit volume m and the rate of production of end products are considered constant throughout the tissue" (cit. of Weis-Fogh 1964). This assumption may be valid for air tube diffusion in the secondary system of the flight muscle, where highly ramified tracheal and tracheolar trees of different geometries are surrounded by an oxygen consuming tissue. But even there one would have to assume that mitochondria are more or less equally distributed in the tree. This assumption can surely not be made for tissue diffusion between the air-filled terminal tracheoles and mitochondria as shown in Fig. 6.4a.

This short discussion shows that we need an improved theory of tissue respiration in insects. It has to take into consideration that (1) outside the tracheoblast tissue diffusion takes place in an almost nonmetabolizing tissue situated between the small air tube system SAT and the mitochondria (Fig. 6.4b), that (2) this diffusion is accompanied by convection of respiratory gases and that (3) mobility of mitochondria shortens the distance of transport leading to the reduction in distance of tissue transport. As a fourth point (4) the theory has to take into consideration that the distance of tissue transport can be increased, if tracheoles are partly filled with water (Wigglesworth 1930, 1981) (Fig. 6.4 b,c; Fig. 6.6a,b). This occurs at low rates of oxygen consumption.

Liquid filling begins at the terminal end of the tracheoles and apparently elongates at this moment the length of the path to the mitochondria. If the lateral wall of the tracheole were impermeable to gases, the oxygen would have to diffuse through the water column. If we use Eq. (6) for the same oxygen flux through an air (A) and a water (W) column, we can equate the fluxes and solve for the unknown

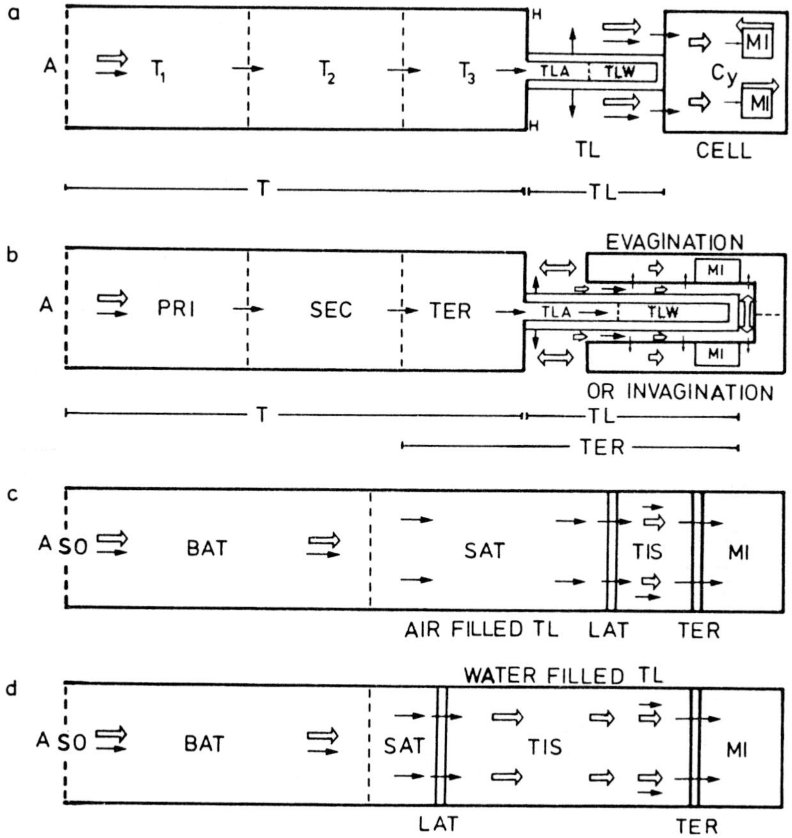

Fig. 6.6a–d. By-pass model of gas exchange along water-filled tracheoles (*TLW*). **a** Main routes of oxygen transfer to mobile mitochondria. **b** Oxygen transfer to fixed mitochondria if the target cell is either evaginated or invaginated. *PRI* primary tracheal system; *SEC* secondary; *TER* tertiary. **c** and **d** Simplified model of oxygen transfer through the tissue by-pass (*TIS*) separating small air tubes (*SAT*) and mitochondria (*MI*) (cf. Fig. 6.4b) by the lateral (*LAT*) and terminal (*TER*) membranes. **c** Air-filled tracheoles (*TL*); **d** Water-filled tracheoles. *SO* spiracular opening; other notations as in Figs. 6.3 and 6.4

maximal distance for diffusion in water.

$$L^W = (A^W / A^A) \cdot (K^W_{O_2} / K^A_{O_2}) \cdot (\triangle P^W_{O_2} / \triangle P^A_{O_2}) \cdot L^A \tag{13}$$

As both cross-sectional areas and partial pressure differences are equal we see that the distance for diffusion in the water column is negligible because of the low Krogh's constant in water (Tables 6.1; 6.2).

The large increase in air tube surface area associated with decreasing diameter would amplify the effect of even a minute lateral permeability especially at the tertiary level as has been calculated for the first time by Schmidt-Nielsen (1979).

Lateral diffusion from the air-filled tracheole into the tissue cylinder around the tracheolar liquid leads us to a description of a by-pass model of gas exchange in tracheolated tissues. In the by-pass cylinder around the waterfilled tracheole there is not only diffusion, the highly increased area compensating partly for the lower Krogh's constant but, in addition, convective transport at different levels (Fig. 6.6a,b; Fig. 6.4c). After passing the first lateral diffusion barrier around the finer air tubes oxygen molecules will, in addition to diffusing, be transported mainly by convective transport in the tracheoblast cytoplasm. During the transport along the outer cell membrane of the tracheoblast, oxygen molecules will diffuse through it. They will then be transported by convection and diffusion in the adjoining hemolymph or in the T-tubules to the cytoplasmic membrane of the terminal consuming cell. The oxygen molecules will diffuse through the outer cell membrane and will be transported by diffusion and convection to the mitochondria (Fig. 6,6a,b; Fig. 6.4c). Mitochondria will eventually (Fig. 6.6a) meet them, if they are transported by cytoplasmic flow.

To make this model more convenient for some basic considerations and calculations, one can as a first step lump some compartments together. The parts of the two cytoplasmic compartments and of the hemolymph compartment, which are involved in by-pass gas transport, can be lumped together into one effective by-pass tissue compartment (Fig. 6.4b). The tissue compartment (TIS) may be divided into the by-pass tissue and the part of the tracheole filled with water. Diffusion through this part of the tissue compartment becomes only important, if the column is very small (Fig. 6.6c). The plasma membranes of the tracheoblast and the terminal oxygen consuming cell, which are within this tissue compartment and through which lateral and terminal diffusion occur, can be in a first approximation lumped together with the mitochondrial membrane into one terminal membrane (TER) (Fig. 6.4b). The other membrane barrier, through which diffusion into the cytoplasma of tracheolar or tracheal cells occurs (Fig. 6.4b), are the lateral walls (LAT) of the tracheoles and smaller tracheae. The small air tube compartment (SAT) in Fig. 6.4c and Fig. 6.6c and d is comprised of the smaller tracheae and tracheoles with axial or rectilinear diffusion as in Fig. 6.3c and d. The primary part of the tracheal system with negligible lateral diffusion is often ventilated. It is the big air tube compartment (BAT) in Fig. 6.4c and Fig. 6.6c,d.

The by-pass model of structure includes the basic postulations for a theory of tracheolar gas transport. It reveals functional aspects of the liquid filling for control of gas transport, which seem to have been little regarded so far. The equations derived by Weis-Fogh (1964) can be applied to the air-filled part of Fig. 6.6c and d as well. The complicated geometry of the lateral by-pass in Fig. 6.4c allows at present only a discussion of some basic principles of gas exchange on the simplified model in Fig. 6.4b or Fig. 6.6c and d. As fluxes are mainly determined by the medium, it is useful in the simplified model to distinguish with the method of Weis-Fogh between tissue flux in the liquid medium and air tube flux.

In the steady state of fluxes in Eq. (14) the molar rate of oxygen consumption in the mitochondria (MI) is equal to the total flux through terminal membranes (TER). This total flux must be equal to the total flux through by-pass tissue (TIS). The total lateral tissue flux originates from and must, therefore, be equal to, the lateral oxygen flux (LAT) through the walls of the small air tube system. The total flux through the small air tube system (SAT), which shall consist of the secondary trachea and the airfilled fraction of tertiary trachea, must be equal to the flux in the bigger tracheae [primary trachea or big air tubes (BAT)], which must be supplied through the spiracular opening (SO).

$$\dot{M}^{MI}_{O_2} = \dot{M}^{TER}_{O_2} = \dot{M}^{TIS}_{O_2} = \dot{M}^{LAT}_{O_2} = \dot{M}^{SAT}_{O_2} = \dot{M}^{BAT}_{O_2} = \dot{M}^{SO}_{O_2} \qquad (14)$$

The overall partial pressure difference between outside air (A) with a $P^A_{O_2}$ of about 21.28 kPa to a minimal aerobic $P^{MI}_{O_2}$ of about 0.13 kPa at the mitochondrial level is the sum of the individual partial pressure differences (Eq. 15).

$$\triangle P^o_{O_2} = \triangle P^{TER}_{O_2} + \triangle P^{TIS}_{O_2} + \triangle P^{LAT}_{O_2} + \triangle P^{SAT}_{O_2} + \triangle P^{BAT}_{O_2} + \triangle P^{SO}_{O_2} \qquad (15)$$

Any molar flux of gas x, ($\dot{M}_x$), is usually defined in respiration physiology of animals as the product of its conductance G_x in the medium m and the partial pressure difference $\triangle P_x$.

$$\dot{M}_x = G_x \cdot \triangle P_x \qquad (16)$$

This equation can be applied to any of the molar fluxes of oxygen defined in Eq. (14). Solving Eq. (16) for the partial pressure difference of gas x and defining the resistance R_x as the reciprocal value of conductance G_x gives

$$\triangle P_x = R_x \cdot \dot{M}_x . \qquad (17)$$

The overall partial pressure difference must then be the product of overall resistance and the overall consumption or production of gas x, taken as occurring in the mitochondrial compartment (MI) only.

$$\triangle P^o_x = R^o_x \cdot \dot{M}^{MI}_x \qquad (18)$$

By analogy of Eqs. (17) and (18) to Ohm's law, the first Kirchhoff's law gives, if we insert Eq. (18) into Eq. (15) and divide by the relations given in Eq. (14)

$$R^o_x = R^{TER}_x + R^{TIS}_x + R^{LAT}_x + R^{SAT}_x + R^{BAT}_x + R^{SO}_x . \qquad (19)$$

During elongation of the water column, tissue resistance, terminal resistance, and lateral resistance increase, whereas small air tube resistance decreases. The BAT and SO resistances do not change during the preadaptive situation characterized by absence of diffusion and ventilation control (Sec. 3). During withdrawal of the column the same resistances are reversed. As all oxygen fluxes stay constant in the steady state, the related partial pressure differences change inversely to the resistances according to Eq. (17). By definition conductances change inversely to the resistances. The change in small air tube resistance in the medium air should be less than the inverse change of the other resistances. Therefore, rise of the tracheolar liquid level increases the total resistance and fall of the liquid level decreases the total resistance.

If tracheoles would show no liquid filling during reduction of metabolic rate, the overall partial pressure difference of oxygen would change proportionally to the total rate of oxygen consumption, as long as the total resistance remains constant. This means that the mitochondrial partial pressure of oxygen would rise to an eventually toxic level at low metabolic rates, as the overall partial pressure difference would go down according to Eq. (18). The rise in tissue partial pressure of oxygen should occur especially in those insects, which are adapted to a high maximal oxygen consumption and which have either no or open spiracles.

This shows that – in constrast to the theory of tracheole physiology of Wigglesworth and Lee (1982) – the liquid filling could have the primary function of controlling the lateral conductance of oxygen or carbon dioxide by changing the lateral area exposed to air. Together with the concomitant change in tissue by-pass conductance this may prevent toxic oxygen levels in mitochondria, if oxygen consumption is lower than maximal. But it may be as well that the low lateral conductance of carbon dioxide prevents a fall of the partial pressure of carbon dioxide in hemolymph or cytoplasm, which may be important either in acid-base control or in control of water loss during cyclic release of carbon dioxide (Secs. 4 and 5). In the postadaptive situation the same function may have been taken over either by spiracles (S) evolving at the spiracular opening (SO) or by control of abdominal ventilation.

The hypothesis of control of internal partial pressures of oxygen and/or carbon dioxide by adjustment of conductance is confirmed by the proportionality between length of the water column and ambient partial pressure of oxygen (Wigglesworth 1953a; 1981). Increases of the oxygen content of the gas mixture within a range of 1% oxygen to nearly 100% oxygen, cause the water column to elongate. Thus, the intracellular partial pressure of oxygen may be, even under normal conditions, the first of the events leading to a change in the water column.

If metabolites are responsible for a change in the water column, they could change not only because of an oxygen debt, as supposed by Wigglesworth (1981), but also because of an additional oxygen store or capacitance, which is, according to Eq. (3), proportional to a residual partial pressure of oxygen. Wigglesworth has to postulate maintainance of an oxygen debt even when the ambient partial pressure of oxygen rises near to its maximal level. According to Forster (1966), an oxygen partial pressure in the range of 0.13 kPa suffices for oxydation in the mitochondria. A solution of the dilemma may come from the fact that Wigglesworth's experiments demonstrate only a reaction of the water column to a change in metabolite concentrations. The metabolite(s) building the "oxygen debt" should, therefore, be of a type which (1) changes in vivo its concentration in inverse proportion to the falling or rising ambient partial pressure of oxygen and (2) influences, through its osmotic potential, the total water potential (Richter 1976), also containing the colloid (swelling) potential of the tracheoblast cytoplasm and the capillary potential of the tracheole, as has been principally described by Wigglesworth (1981). Wigglesworth's and Bult's (1939) experiments invite hypotheses. But critical biochemical experiments have to show the metabolites, which control the osmotic potential of the tracheoblast in vivo.

According to Wigglesworth and Lee's (1982) theory of tracheole physiology, liquid filling is due to an increased general permeability of the tertiary system, which is reduced in some insect tissues as an adaptation to water loss. Epidermis cells of *Rhodnius,* which need a high permeability only during secretion of a new cuticle, may normally reduce their lateral oxygen and carbon dioxide permeability and are always filled with air. This reduction may be accompanied by a reduced permeability to water and the injection fluids demonstrated by Wigglesworth and Lee (1982). But the highly increasing lateral surface area in the small air tube system (Schmidt-Nielsen 1979) leads to nearly constant water saturation, even fairly distant from the water column. In contrast to the partial pressure of oxygen and carbon dioxide, the partial pressure of water in the tracheolar cytoplasma is not influenced by the change in terminal or tissue conductance, as the partial pressure difference of water is zero there. In this case the presence or absence of liquid can neither increase nor decrease the partial pressure of water in the small air tube system and can, therefore, not have any influence on tracheal water loss. A "compromise between the need to obtain oxygen and the need to conserve water which faces all terrestrial animals" (cit. Wigglesworth and Lee 1982), can, therefore, not be achieved by changing the level of permeability of the terminal ends, where they fill with water.

In contradiction to Wigglesworth and Lee's theory, the rising liquid level would reduce the tracheolar wall area available for lateral diffusion in the small air tube system and could eventually even contribute to water conservation instead of loss, as the tracheolar liquid offers only a small surface area to air. In addition, it can be easily imagined that the epicuticular lipid layer extends to the liquid. It could cover the water surface and may reduce water loss in the same manner as in the cuticle (Beament 1964, 1965).

In any case the postulated compromise between oxygen uptake and water conservation can exist in the air-filled part of the tracheal system only if reduction of lateral permeability to water as in *Rhodnius* epidermis or reduction of lateral wall area by the liquid column together with a lipid cover of the water column lowers the partial pressure of water in the tracheal system. Measurement of the partial pressure of water in the small air tube system at different levels of the liquid column are highly desired to test these important hypotheses.

One might suppose that liquid filling had been evolved in the preadaptive situation before the evolution of occludable spiracles. In this situation liquid filling of the tracheoles increases the partial pressure differences of oxygen and carbon dioxide along the cuticle, because internal partial pressure of oxygen falls and internal partial pressure of carbon dioxide rises, whereas the partial pressure difference of water is unchanged. The improved cuticular flux of oxygen and carbon dioxide leads to a lower demand for oxygen and carbon dioxide exchange via the tracheal system. This allows the water column to elongate even further. Only if the edge of the water column would have a lowered permeability for water, may water flux into the air-filled tubes be reduced. Otherwise the reduced tracheal exchange of oxygen and carbon dioxide could only be used to reduce tracheal water loss with the help of the spirales. The functions of spiracles will be discussed in Sec. 3. As spiracles can increase the overall resistance, they are able to control the internal partial pressures of oxygen and carbon dioxide as well as the liquid column does. This suggests that an evolution of tracheolar water filling after evolution of occludable spiracles is not very likely.

The adaptational properties of the growth of the tracheal system described by Wigglesworth (1954) and Locke (1958) show another case, in which conductances can be adjusted at the tracheolar level to the oxygen demand of the mitochondria.

The results of Locke (1958) showed, in addition, that the proportions of the tracheal system are maintained by coordination of the growth of the diameter of the main tracheae and the number of new terminal branches during development. This coordinated reaction can occur unilaterally as an adaptation to the needs for oxygen at changing growth. Acclimation of *Tenebrio* larvae (Locke 1958) and of *Rhodnius* (Wigglesworth 1954) to differing ambient partial pressures of oxygen showed that this coordination occurs as an adaptation to an environmental parameter, which has surely changed during the evolutionary history of insects.

One may conclude that liquid filling and coordinated growth in the small air tube system can be interpreted, respectively, as long-term and short-term adaptations of the internal steady state needs of oxygen, carbon dioxide, and water to the availability of oxygen and water in the environment.

2.3 Tracheal Water Loss in Small Insects Without Spiracles

Krogh's model (Sec. 2.1) showed for the first time that diffusion in the air tubes of the tracheal system requires only a low partial pressure difference of oxygen and carbon dioxide. Mitochondrial partial pressure of oxygen should be low and mitochondrial partial pressure of carbon dioxide high. The by-pass model of tracheolar respiration (Sec. 2.2) shows that liquid filling can lower the mitochondrial partial pressure of oxygen and raise the mitochondrial partial pressure of carbon dioxide.

It has been discussed in Sec. 2.2 that the partial pressure of water in the small air tube system should be nearly at the saturation point, as the large lateral area should reduce the lateral resistance of the air-filled tubes to water, too. The difference in chemical activity of water needed for the flux from the tracheoblast cytoplasma to the air tubes may be neglected. Even without this lateral resistance the partial pressure of water would equilibrate to an activity of water in the range of only $a_w = 0.995$, which is equivalent to an osmolarity of 300 mosmol · liter^{-1} of the hemolymph (Edney 1977) or – to be more precise – a similar osmolarity of the tracheoblast cytoplasm.

Unfortunately, the partial pressure of water in the smaller air tubes has not been measured, but it should be in any case high, as a very small drop in temperature (less than 0.2°C) is sufficient to cause the condensation of droplets in tracheae (Beament 1964; Edney 1977).

If we assume a uniform partial pressure of water in the small air tube system (SAT), the difference to the ambient partial pressure of water will be the driving force for the tracheal water flux through the big air tube system (BAT or T) to ambient air (A).

$$\dot{M}^{T}_{H_2O} = G^{T}_{H_2O} \cdot (P^{SAT}_{H_2O} - P^{A}_{H_2O}) \quad (20)$$

In reality the water flux in the tracheal system may be composed of a diffusive and a convective component. But if we assume that the diffusive gas exchange is more primitive, we can define the tracheal conductance in the preadaptive state as a pure diffusive conductance according to Krogh's model and Eq. (6).

$$G^{T\text{-}DI}_{H_2O} = (A^T / L^T) \cdot D'^A_{H_2O} \cdot \beta^A_{H_2O} \tag{21}$$

$D'^A_{H_2O}$ is the effective diffusion coefficient of water in air or tracheal air. $\beta^A_{H_2O}$ can be defined at a certain temperature as the ratio of saturation (sat) concentration of water in air to the saturation partial pressure of water in air, which is roughly equal to β_g or $1 / R \cdot T$ (Piiper et al. 1971), if water vapor is treated as an ideal gas.

$$\beta_{H_2O} = c^{sat}_{H_2O} / P^{sat}_{H_2O} \tag{22}$$

In contrast to the steady state partial pressure differences of oxygen and carbon dioxide the partial pressure difference of water is independent of the metabolic rate, which changes either due to activity or due to size of the insect. It has been shown by an enlargement of Krogh's model (Kestler 1983) that this independence must have been of special importance in small insects for the evolution of mechanisms of respiratory water retention. With the decreasing size of an insect the steady state partial pressure difference of oxygen and carbon dioxide necessary to supply respiratory gas exchange decreases, whereas, in a constant environment, the partial pressure difference of water does not change. The isometric model assumes that the original body mass of an insect like Krogh's *Cossus* larva would change in both directions by a factor N. Its original rate of oxygen consumption $\dot{M}_{O_2}(O)$ then changes by a factor N^x. The exponent x will be 2/3, if the rate changes proportional to body area. The value of x will be 1, if the rate changes proportional to body weight. Tracheal cross-sectional area changes by a factor $N^{2/3}$ and tracheal length changes by a factor $N^{1/3}$. It follows from the general diffusion Eq. (6) that the partial pressure difference of oxygen will change by a factor $N^{x-0.33}$. As the partial pressure difference of water remains constant the diffusive conductance of water in Eq. (21) and the rate of water loss in Eq. (20) change with $N^{1/3}$, as outlined for the conductance in the next equation.

$$\begin{aligned} G^{T\text{-}DI}_{H_2O} &= (A^T(O) \cdot N^{2/3} / (L^T(O) \cdot N^{1/3})) \cdot D'^A_{H_2O} \cdot \beta^A_{H_2O} \\ &= G^{T\text{-}DI}_{H_2O}(O) \cdot N^{1/3} \end{aligned} \tag{23}$$

The ratio of diffusive fluxes of water to oxygen changes by a factor of $N^{0.33\text{-}x}$ As x has a value between 0.67 and 1, the exponent is always negative. This is the reason why the ratio increases rapidly with decreasing size when N is smaller than 1.

On the other hand, the absolute water reserves or the mass of water m_{H_2O} of an insect change in proportion to body mass m_B, if the relative water content stays constant. The rate of diffusive tracheal water loss per unit body mass may be called the rate of specific water loss with a subscript H_2O-s and it must change by a factor $N^{-2/3}$ as outlined in the next equation.

$$\dot{M}^{T\text{-}DI}_{H_2O\text{-}s} = \dot{M}^{T\text{-}DI}_{H_2O} / m^B = \dot{M}^{T\text{-}DI}_{H_2O}(O) \cdot N^{1/3} / (m^B(O) \cdot N)$$

$$= \dot{M}^{T\text{-}DI}_{H_2O}(O) / (m^B(O) \cdot N^{-2/3}) \tag{24}$$

The time of survival insofar as it depends on respiratory water loss, will decrease with a factor of $N^{-2/3}$, too.

An example will show the importance of this relation. Krogh's *Cossus* larva (Sec. 2.1) had a body mass of 3.4 g and, at a water content of 70%, a total water reserve of 2.4 g. At 20°C and 0% relative humidity it would lose 21.4 $\mu g \cdot min^{-1}$ through the tracheal system (Kestler 1983). This is 0.91% of body mass per day. In contrast, a small larva of 3.4 mg would lose 2.14 $\mu g \cdot min^{-1}$ from its water reserve of 2.4 mg through the tracheal system alone, which is 91% of its body mass per day. As many insects die after 30% loss of their original weight, the 3.4 g larva would survive 32 days, whereas the 3.4 mg larva would survive less than 8 h. The additional water loss through the cuticle should reduce the survival time even more, as it changes by the same factor of $N^{1/3}$. At decreasing body size this is much more than the usually discussed value of $N^{2/3}$, as not only cuticular area decreases by a factor of $N^{2/3}$, but also cuticular thickness by a factor of $N^{1/3}$.

This leads us to the important conclusion that insects emerging from the egg have to face the special problem of reducing diffusive tracheal water loss. This may have promoted the evolution of additional larval stages as well as the evolution of spiracular closing mechanisms and convective gas exchange.

Addition of ventilation to Krogh's model showed, in contrast to a widely held hypothesis, that the same rate of pure convective oxygen uptake along the same partial pressure difference of oxygen by ventilation would not increase, but rather decrease the respiratory water loss. To adjust both modes of oxygen fluxes to a comparable intensity, the pumping frequency must balance the diffusive oxygen uptake. In this case the net convective respiratory water loss would be 80% of the diffusive loss, which is the ratio of the diffusion coefficient of oxygen to that of water (Kestler 1983). An additional advantage to the replacement of pure diffusive by pure convective gas exchange in evolution would be the fact that the rate of specific respiratory water loss at a constant pumping frequency does not change with decreasing body size. This can be shown easily, if we calculate with the help of Eqs. (7) and (8) the rate of pure convective water loss.

$$\dot{M}^{T\text{-}CO}_{H_2O} = \triangle V \cdot f_P \cdot \beta^A_{H_2O} \cdot \triangle P^T_{H_2O} \tag{25}$$

At constant pumping frequency f_p only $\triangle V$ should change with a factor of N. If the pumping interval is long enough to allow saturation of the tracheal system, the partial pressure difference of water will be the same as for diffusion. The water loss rates changes, therefore, with the same factor N, as do the body water reserves or the body weight, thus, leaving the rate of specific convective water loss unchanged. The advantage of convective gas exchange will be the greater, the smaller the insect is. The ratio of pure convective (Eq. 25) to pure diffusive (Eq. 24) rate of water loss per unit body weight changes with a factor $N^{2/3}$.

$$R^{CO/DI}_{H_2O} = \dot{M}^{T\text{-}CO}_{H_2O\text{-}s}(O) / (\dot{M}^{T\text{-}DI}_{H_2O\text{-}s}(O) \cdot N^{-2/3}) = R^{CO/DI}_{H_2O}(O) \cdot N^{2/3} \tag{26}$$

Figure 6.7 summarizes the change in diffusive and convective rates of specific tracheal water loss and their ratio as a function of the factor N on the example of Krogh's *Cossus* larva (m^B = 3.4 g, if N = 1).

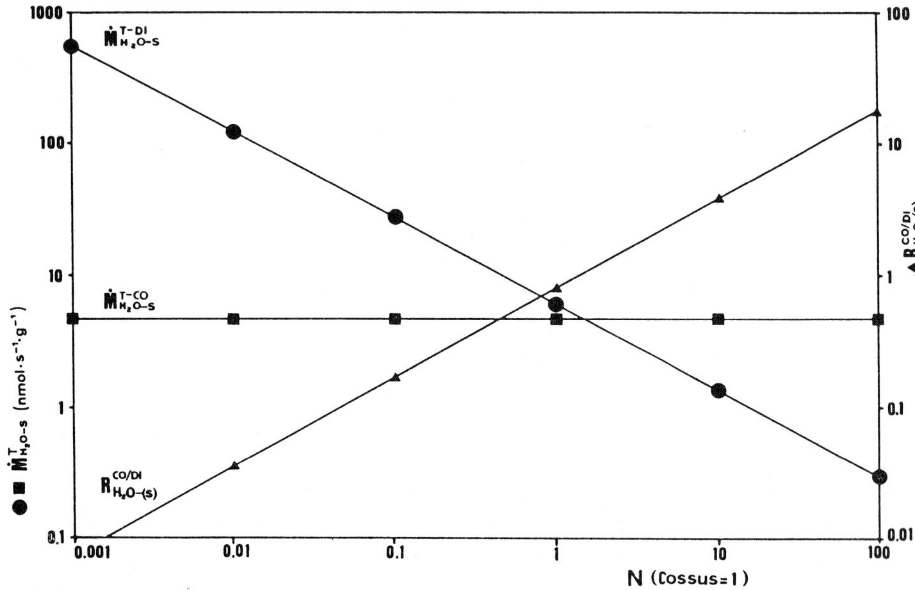

Fig. 6.7. Model of specific rate of tracheal water loss ($\dot{M}^{T}_{H_2O\text{-}s}$) by pure diffusion (*circles*) or pure convection (*squares*) if body size changes by a factor of N (*left axis*). *Triangles* show the ratio of convective to diffusive rates ($R^{CO/DI}_{H_2O\text{-}(s)}$) (*right axis*)

The larva of 3.4 g in our example would lose 80% of 0.91%, or 0.73% (17.1 μg · min^{-1}), of its body weight per day, if it could manage to exchange oxygen and carbon dioxide by convection only. The 3.4 mg larva would lose from its 2.4 mg water reserve only 17.1 ng water per minute by convection. As in 3.4 g larva this is 0.73% of the body weight per day. It could survive 29 days instead of 8 h, if it would be able to exchange oxygen and carbon dioxide by pure convection instead of diffusion and if it would keep the same pumping frequency as the 3.4 g larva whose pumping frequency is adjusted to its diffusive oxygen uptake. In the 3.4 mg larva the ratio of the rates of the convective water loss to the above mentioned diffusive water loss of 2.14 μg · min^{-1} is then 0.8%. Equation (26) calculates as 0.8 · $N^{2/3}$ giving 0.008 or 0.8%. This is the same value of 0.8%, as the relation to unit body mass can be neglected.

The unexpected result of the simple model analysis is in contrast to the suggestive argumentation of this point common to all reviews (Edney 1977). The model allows one to establish the hypothesis that replacement of diffusive gas exchange by convective gas exchange should be favored in evolution of small animals, whenever it is possible. It would prevent the rapid increase in respiratory water loss with decreasing size.

3 Gas Exchange Through Spiracles

The parallel evolution of occludable spiracles in different gas exchange systems is usually thought to have the primary function of controlling pure diffusive water loss (Hazelhoff 1926). The problem of co-diffusion during ventilation has been generally neglected. In addition, spiracles can entirely prevent diffusive gas exchange instead of just controlling it. To replace diffusive by convective gas exchange as much as possible, spiracles should prevent diffusion between the pumping strokes and reduce co-diffusion during the pumping strokes by reduction of the opening area. Experimental evidence for this hypothesis will be given later.

The next sections will analyze the preadaptive situation for gas exchange at the respiratory opening and show that even without spiracles an in-flow of air must have reduced respiratory water loss initially.

3.1 Basic Principles of Reduction of Tracheal Water Loss Through Tracheal Openings

To solve the problem of the conflicting needs of obtaining oxygen through the opening of a gas exchange system and preventing water loss through the same opening, evolution probably improved the original preadaptive situation first. From Eq. (16) we can derive the simple general Eq. (27), which is sufficient to understand which parameters must be changed to minimize tracheal water loss. In the preadaptive situation, gas exchange may be considered to occur through an open respiratory hole or spiracle situated at the same position (S) (Fig. 6.1.4c), where an occludable spiracle is later developed.

$$\dot{M}_X^S = G_X^S \cdot \triangle P_X^S \tag{27}$$

In the steady state

$$\dot{M}_{O_2}^{MI} = \dot{M}_{O_2}^S \tag{28}$$

and

$$\dot{M}_{CO_2}^{MI} = \dot{M}_{CO_2}^S . \tag{29}$$

In contrast, the total water balance of the body is normally not in a steady state. The morphological and physiological design determines, together with special physical and chemical parameters, the special conductances for oxygen, carbon dioxide, and water in Eq. (27). The conductances shall be assumed to be initially sufficient to supply several times the metabolic needs either by diffusion or convection (attributable to autoventilation by body movements). The initial diffusive or convective conductances for oxygen and carbon dioxide lead to a certain partial pressure difference of oxygen and of carbon dioxide along the virtual or real spiracle.

In contrast to the tracheal partial pressures of oxygen and of carbon dioxide, the tracheal partial pressure of water is fixed nearly at saturation at constant temperature. If ambient relative humidity is constant, the resulting partial pressure difference of water determines, together with the initial conductance, the rate of water loss. Equation (27) shows that there are only two basic possibilities of minimizing the rate of tracheal water loss; either the partial pressure difference of water or the conductance for water may become reduced.

The partial pressure difference of water can be minimized either by increasing ambient or decreasing tracheal partial pressure of water. The former may be achieved by raising ambient humidity at constant temperature, the latter by lowering ambient, and therewith, body temperature at constant water content of the ambient air. An appropriate behavior can achieve both. Spiracles sunken below the general cuticular surface and covered by hairs (Hadley 1972; Wigglesworth 1972) or spiracles covered by wings (Ahearn 1970) may support a stagnant layer of air with a higher humidity caused by the unavoidable cutaneous water loss into the unstirred layer outside the valve. Before the evolution of occludable spiracles the intersegmental position of spiracles could have had this function. In advanced state of adaptation there is an atrium external to the occludable valve and covered with a filter of hairs or teeth as in *Hyalophora* spiracles (Beckel 1958). Subelytral spiracles are another type of optimization of this mechanism of water retention.

In a first stage of adaptation the minimization of the conductance of water is also possible before the evolution of spiracles. In this stage the diffusive or convective tracheal conductances of oxygen and carbon dioxide are minimized until the partial pressure difference of carbon dioxide or oxygen reaches a maximum at the highest metabolic rates of the insect. Usually a certain tracheal threshold of the partial pressure of carbon dioxide limits this process. The resulting threshold conductance of carbon dioxide also minimizes the conductance of water, because certain parameters, such as a minimal tracheal cross-section or a minimal activity, are common to both and reduce the diffusive or convective conductance of water as well as those of oxygen and carbon dioxide. The same "strategy" should lead to pores in bird's eggshells with a minimal diameter.

This reveals a general physical principle basic to physiological water retention. The "strategy" of minimization of diffusive or convective conductance turns the adverse physical situation of a constant partial pressure difference of water to an advantage. Thus, the cause of the relative increase in diffusive water loss with decreasing size and decreasing partial pressure differences of oxygen and carbon dioxide (Sec. 2.3) is used to an advantage. This trick of avoiding water loss in a specific way does not need a differentially permeable membrane to permit oxygen uptake and carbon dioxide release at the same time (cf. Edney 1977).

If the metabolic rate is submaximal, a selection pressure for water retention should bring up the same strategy in a second stage of adaptation. In this stage the conductances of oxygen, carbon dioxide, and water are adapted to varying metabolic rates. Diffusive conductances could change with the opening area of occludable spiracles, convective conductances by a changing abdominal pumping frequency. We see that the same strategy of adaptation could lead to evolution of spiracles of insects or the pneumostome of gastropods or plant stomata, if diffusive conductance of water is minimized. But it could also lead to evolution of abdominal pumping according to the metabolic rate, if convective conductance of water is minimized.

Adaptation to the maximal and to the changing metabolic needs characterize the two steps in evolution. Hazelhoff (1926), and not Buck (1962), must have been the first one who became aware of the latter type of adaptation in the specific case of reduction of diffusive water loss, when he established the theory of diffusion control by spiracles and postulated (1) an adaptation of spiracle area to the metabolic rate and (2) selective water retention by a selective increase of the tracheal partial pressure of carbon dioxide over that of water. The specific case of reduction of convective conductance by ventilation control as a minimization of the mean pumping frequency has been analyzed experimentally by Loveridge (1968). At an initially low metabolic rate the adaptive state can be recognized by the increase in pumping frequency with increasing metabolic rate and water loss, either owing to activity or owing to an increase in temperature (Miller 1974; Edney 1977).

Revealing, with the help of Eq. (27), the two basic steps, (1) minimization of conductance of water and (2) minimization of partial pressure difference of water, show an additional important consequence. If a concomitant change in metabolic rate can be excluded, then any diffusion or ventilation control by adjustment of spiracular area or abdominal pumping frequency to ambient relative humidity would show that the adaptation to a minimal conductance of carbon dioxide is not yet perfect. It has been nicely reviewed by Edney (1977) that adaptation of spiracular area to relative humidity has been postulated by Bursell (1957; 1974b) and many other authors on circumstantial evidence only, since the state of the spiracles were deduced from water loss rates including cutaneous water loss. The reduction in pumping frequency and rate of water loss at lower humidities shown by Loveridge (1968) in *Locusta* may be the adaptation of an insect, whose water balance is not so much determined by the problem of water retention, but by the problem of getting rid of the water taken up with the food (Loveridge 1975) or retained or sorbed after having been in moist air for 24 h (Edney 1977).

This shows that a change of diffusive or convective spiracular conductance of water due to a change in ambient partial pressure of water can be better interpreted as an additional adaptation to the problem of rapid elimination of excess water, which may be not less important than water retention. This shows that we should not neglect the real ecological situation of the considered species where the optimal strategy may be a balance between water retention and elimination, which should especially be the case in those "apterygotes", which have tracheae, but no closing mechanisms (Hassan 1944).

3.2 Types of Spiracular Conductances in Insects

Besides diffusion and ventilation control flow-diffusion has been claimed to reduce respiratory water loss in insects with cyclic CO_2 release during the period of retention of carbon dioxide (Buck 1958a,b). Buck postulated a steady in-flow of air to reduce diffusive water loss. One may call the steady rate of volume-in-flow steady suction ventilation. A steady diffusive and convective gas exchange should lead to a steady state between nitrogen influx and efflux at the same time. Later, Levy and Schneiderman (1966c) have been able to prove only unsteady suction ventilation alternating with periods of pure diffusion in microcycles. They postulated what I call (cf. Miller 1982) diffusive-convective gas exchange with retention of carbon dioxide and water during a nearly closed position of the spiracles (microconstriction) and pure diffusive gas exchange during the alternating shorter nearly open position of the spiracles (microopening). During a microcycle of a microconstriction and a microopening period nitrogen would be in a steady state. This would mean that in insects diffusive conductance may be either pure diffusive or diffusive-convective. The latter may occur during steady flow-diffusion or unsteady suction ventilation.

If diffusion is unimportant, pure convective conductance may be adjusted either by bidirectional ventilation control or by unidirectional suction ventilation. This classical picture becomes even more complicated, if co-diffusion during ventilation is added. The distance for diffusion through a spiracle is in the range of 10–150 μm only. This justifies the hypothesis that during any type of ventilation, diffusion should occur either with or against the convection current. As this is also the case during steady flow-diffusion and unsteady suction ventilation the situation is less complex: we have to differentiate pure diffusive from diffusive-convective gas exchange only, the latter occurring either during ventilation or during steady or unsteady suction ventilation. The conductance of an individual gas during diffusive-convective gas exchange can then be very simply characterized as either cocurrent or anticurrent according to its diffusion with or against convection currents. Pure diffusion remains as the only alternative type of conductance.

3.3 Pure Diffusion Cannot have been the Original Mechanism of Gas Exchange

This section will show that – in contrast to the current thinking – not pure diffusive, but diffusive-convective gas exchange was probably the original in the preadaptive situation. Analyzing the physical situation further, we will see that diffusive-convective gas exchange is the normal preadaptive gas exchange during transition to land and that spiracles are not necessary to generate suction ventilation, as suggested by Buck (1958b) and Levy and Schneiderman (1958). A model will show that pure diffusive gas exchange is a unique situation of no physiological importance leaving diffusive-convective conductance as the only type in insects.

A model of gas exchange through a respiratory hole or spiracle of the tracheal system is shown in Fig. 6.8. The position where an occludable spiracle will be or is situated shall be defined as a real or virtual spiracular tube (S). The partial pressure of tracheal gases in the big air tubes (BAT or T) is assumed as constant because of mixing induced by heart movements or gut movements in a small insect. The partial pressure of water is assumed to be at saturation. The mixing would increase the partial pressure gradients along the unstirred and initially open spiracular tube (S) to the ambient air (A). This amplifies the diffusive gas exchange

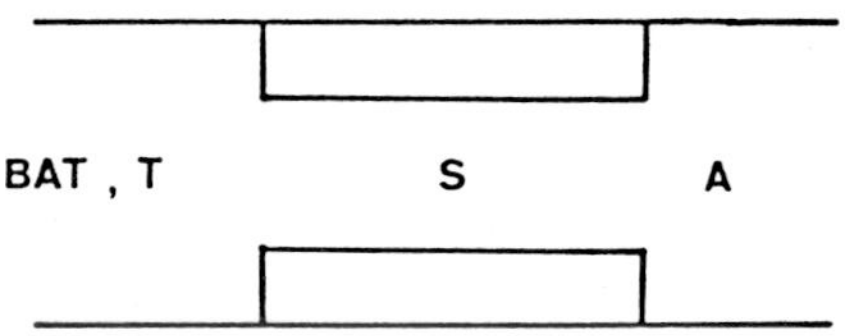

Fig. 6.8. Model of spiracular structure according to Buck (1958b). *S* spiracular tube

through the spiracular tube, but exaggerates the problem of diffusive water loss, shown in Sec. 2.3, to be attributable to a constant partial pressure difference of water, even more.

As, according to Dalton's law, the total or barometric (B) pressure P_B in the uniform compartments T and A is the sum of the individual partial pressures, we have

$$P_B^T = P_{O_2}^T + P_{CO_2}^T + P_{N_2}^T + P_{H_2O}^T \tag{30}$$

and

$$P_B^A = P_{O_2}^A + P_{CO_2}^A + P_{N_2}^A + P_{H_2O}^A \,. \tag{31}$$

The barometric pressure difference may be defined as

$$\triangle' P_B^S = P_B^A - P_B^T \,. \tag{32}$$

Inserting Eqs. (30) and (31) into Eq. (32) gives

$$\triangle' P_B^S = (P_{O_2}^A - P_{O_2}^T) + (P_{CO_2}^A - P_{CO_2}^T) + (P_{N_2}^A - P_{N_2}^T) + (P_{H_2O}^A - P_{H_2O}^T). \tag{33}$$

We can define the difference $\triangle'$ of the partial pressures of gas x similar to the barometric pressure in Eq. (32)

$$\triangle' P_x^S = P_x^A - P_x^T \tag{34}$$

and get

$$\triangle' P_B^S = \triangle' P_{O_2}^S + \triangle' P_{CO_2}^S + \triangle' P_{N_2}^S + \triangle' P_{H_2O}^S \,. \tag{35}$$

To be compatible with the conventional definition of a positive transfer rate of carbon dioxide in respiration physiology (Püper et al. 1971), used so far in this chapter, we have to insert the absolute value of $\triangle' P_x$ as the partial pressure difference of gas x into Eqs. like (27) to get always positive fluxes. If we insert $\triangle' P_x$ for the partial pressure difference of gas x we get the biophysical convention of negative effluxes and positive influxes.

Pure diffusion excludes bulk flow of air, which is induced by a barometric pressure difference. Equation (35) shows that pure diffusion can only occur if the sum of the $\triangle'$-partial pressure differences of oxygen, carbon dioxide, nitrogen, and water is zero.

In insects the $\triangle'$-partial pressure difference of oxygen will be positive and lead to an influx, characterized by a positive sign. The $\triangle'$-partial pressure differences of carbon dioxide and water will be negative and lead to an efflux, characterized by a negative sign. The $\triangle'$ -partial pressure difference of nitrogen depends upon the function of nitrogens as an inert filler in the tracheal gas resulting in a certain tracheal partial pressure of nitrogen. Any positive $\triangle'$-partial pressure difference of nitrogen in the steady state must lead to a diffusive nitrogen influx and vice versa. As the stores or capacitance (Eq. 3) of tissues cannot change in the steady state and as nitrogen is neither produced nor consumed in metabolism any influx or efflux of nitrogen would raise or lower the tracheal partial pressure of nitrogen. It can only be held constant in the steady state by a convective counterflux of nitrogen leading to no net flux of the inert gases. A diffusive nitrogen efflux would have the tendency to lower the tracheal partial pressure of nitrogen. This would lower the total tracheal pressure and induce an in-flow of air with a convective influx of nitrogen and vice versa. A steady state $\triangle'$-partial pressure difference of nitrogen can, therefore, only exist in combination with an opposed $\triangle'$-pressure difference characterized by a changed sign. This presupposes that the partial pressure differences of oxygen, carbon dioxide, and water cannot change in the steady state.

The partial pressure difference of water is assumed to be constant. Accepting a negative efflux of carbon dioxide, we can insert the $\triangle' P_{CO_2}$ according to Eq. (34) and mitochondrial transfer rate according to Eqs. (28) or (29) into Eq. (27). We solve for the tracheal partial pressure of x (oxygen or carbon dioxide) and get Eq. (36). It shows that the tracheal partial pressure of oxygen or carbon dioxide is constant at a constant metabolic rate and conductance.

$$P_x^T = P_x^A - (\dot{M}_x^{MI} / G_x^S) \quad (36)$$

We can rearrange Eq. (35), leaving only the constant partial pressure differences of oxygen, carbon dioxide, and water on the right side.

$$\triangle' P_{N_2}^S - \triangle' P_B^S = - \triangle' P_{CO_2}^S - \triangle' P_{H_2O}^S - \triangle' P_{O_2}^S \quad (37)$$

To get the same results with the conventional absolute $\triangle$-partial pressure differences of oxygen, carbon dioxide, and water, Eq. (37) must be written

$$\triangle' P_{N_2}^S - \triangle' P_B^S = \triangle P_{CO_2}^S + \triangle P_{H_2O}^S - \triangle P_{O_2}^S . \quad (38)$$

It has been shown by Buck (1958b) that the total pressure difference necessary to compensate the partial pressure difference of nitrogen in the nitrogen steady state is very small. As both $\triangle' P$-values have opposed signs, the left side of Eq. (38) will have the sign of the partial pressure difference of nitrogen if the pressure difference is always smaller. The sign of the partial pressure difference of nitrogen is determined according to Eq. (38) by the difference between the sum of the partial pressure differences of carbon dioxide and water and the partial pressure difference of oxygen. This difference is decisive for the flux type. Figure 6.9 shows this on a seesaw model, which can help to distinguish between the three basic situations:

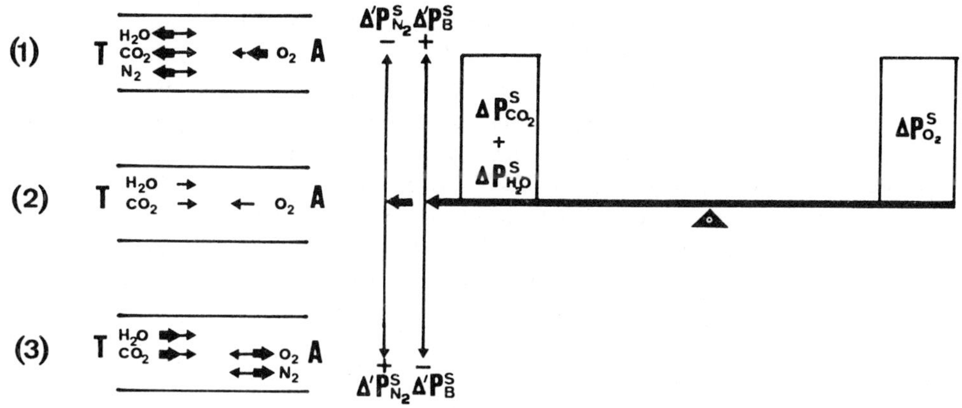

Fig. 6.9. Seesaw model showing the dependence of pure diffusion **(2)**, suction ventilation **(1)**, and pressure ventilation **(3)** from the partial pressure differences according to Eq. (38). *Thin arrows* diffusion; *thick arrows* convection

1. Diffusive efflux or influx coupled to convective influx owing to suction ventilation;
2. Pure diffusion without diffusive flux of nitrogen and without convective fluxes as a unique condition;
3. Diffusive influxe or efflux coupled to convective efflux owing to pressure ventilation.

The seesaw model makes clear that pure diffusion is a situation unlikely to occur in insects, as the metabolic rate, according to Eq. (36), would have to be adjusted to any changes in ambient partial pressure of water.

If the initial situation in evolution was pure cutaneous respiration without tracheae, the partial pressure difference of oxygen must have been higher in the steady state than the partial pressure difference of carbon dioxide, as Krogh's constant for carbon dioxide is higher than Krogh's constant for oxygen in tissues (Dejours 1981) or water (Table 6.1). At least at high humidities this situation favors suction ventilation in the first tracheoles developing from tracheoblast mutants of epidermis cells.

Without any cutaneous gas exchange the situation favors pressure ventilation through an air tube, as Krogh's consant of carbon dioxide is smaller in air than Krogh's constant of oxygen (Table 6.1). This situation has been shown to occur in diffusion through the pores of hen's eggs (Ar et al. 1980), where "cutaneous" gas exchange through the shell itself is zero.

We come to the important conclusion that the preadaptive situation for tracheal gas exchange favored water retention even before evolution of spiracles and abdominal ventilation, because cutaneous gas exchange occurs even now after the evolution of cutaneous water retention. As pure diffusive and pure convective conductance are ruled out, any diffusive fluxes in insects are either cocurrent or anticurrent to convective fluxes. Ventilation improves the convective fluxes which should be of adaptive value for water retention.

3.4 Buck's Theory of Flow-Diffusion Fails to Explain Water Retention

If we understand how the out-diffusion of water molecules has been impeded by the convective in-flow of gas molecules in the preadaptive situation, we may be able to understand which parameters of the physical mechanism of diffusion against convection currents must have been changed to improve the preadaptive situation.

In spite of concerning cyclic CO_2 retention, Buck's theory of flow-diffusion (Buck 1958a,b) has been the first attempt to describe cocurrent or anticurrent diffusive-convective gas exchange. Buck was the first to assume correctly:

1. Retention of carbon dioxide and water by a bulk in-flow of air;
2. An additional oxygen uptake by a convective influx with the in-flowing air supporting the diffusive influx of oxygen;
3. A convective influx of nitrogen with the in-flowing air raising the tracheal partial pressure of nitrogen;
4. A diffusive efflux of nitrogen being in a steady state with the convective influx.

The following section will show that Buck failed to show the physical mechanism of flow-diffusion because he used the wrong equations to describe diffusion with or against convection currents and failed as a consequence, to show convective reflux as the most important parameter of flow-retention.

In his main paper Buck (1958b) makes on p. 123 a trial and error calculation of the volume-flow rate of air-in-flow on the basis of the postulated steady state of nitrogen fluxes. He uses this calculated volume-flow rate (cit.: "R_fAIR = $1.2 \cdot 10^{-7}$ cm^3/sec/spiracle") as the rate of flow-retention of carbon dioxide (cit.: "R_fAIR = RET_fCO_2 = $1.2 \cdot 10^{-7}$ cm^3/sec/spiracle") in his further calculations. This results in a fundamental error, as the rate of back-flow of carbon dioxide or water is a molar flux (mmol $\cdot$ s^{-1}), which may be expressed in STPD-volumes, but not in volume units of the rate of air-in-flow. If Buck had calculated flow-retention of water in the same way, he would have noticed that the rate of water retention must then be the same as that of carbon dioxide, which would be inconsistent with his calculation of water balance over the cycle of CO_2 release (Buck 1958a).

Moreover, it is inconsistent to treat the efflux of nitrogen differently than the efflux of water and carbon dioxide as all three diffuse against the suction ventilation current. If Buck had calculated the flow-retention of nitrogen in the same way as that for carbon dioxide he would have noticed that his rate of flow-retention of nitrogen was higher than the out-diffusion of nitrogen, which was 4/5 of the volume-flow rate. A consistent calculation would not lead to a steady state, but a net in-flow of nitrogen.

On the other hand, if Buck had treated the balance between efflux and influx of carbon dioxide and water in the same way as that of nitrogen or oxygen, he would have noticed that convective influx of water and carbon dioxide is zero in air free of carbon dioxide or water, as the product of the volume-flow rate and zero concentration would be zero. This would leave the rate of out-diffusion of carbon dioxide or water unchanged in relation to a situation without a suction ventilation current.

This shows that Buck's theory of flow-diffusion fails to explain water retention in that ecological situation where it is most needed. The theory is unable to answer the basic question of how the air-in-flow caused by the suction pressure impedes the diffusive escape of water.

The solution of this problem is that diffusion against convection currents differs from pure diffusion in being diffusion with back-flow or anticurrent diffusion (Kestler 1982). The difference in the movement of the out-diffusing molecule in both situations is much like swimming in stagnant water or against the current. The automotive swimming (diffusion) is the same, but the rate of advance is different. Buck's theory fails to explain the anticurrent diffusion of nitrogen, carbon dioxide, and water occurring within the spiracular tube. This neglected back-flow is not the considered convective in-flow into the spiracular tube, which depends upon the ambient partial pressure of the gas. The convective reflux must depend upon the partial pressure at that position within the spiracle, where back-flow occurs.

3.5 Theory of Diffusive-Convective Gas Exchange

The last section showed that we need an improved physical description of the diffusive-convective gas exchange in one dimension. The simple tubular model of spiracle structure introduced by Buck (1958b) and shown in Fig. 6.8 may be sufficient. The physical description should supply equations to respiration physiologists containing basic parameters, which have been subject to optimization during evolution of insect gas exchange.

One would be surprised, if the problem of diffusion with or against convection currents had not been treated in natural sciences. Reviewers of at least parts of the problem are Frank and Mises (1935) or Jost and Hauffe (1972) in phyiscs, Cunningham and Williams (1980) or Geankoplis (1983) in chemical engineering, Jacobs (1935, 1967), Hutten (1971), and Laquet (1976) in general physiology. A closer inspection shows that the general equation for diffusive-convective gas transport and the solutions given by Frank and Mises (1935) or Jost and Hauffe (1972) are used even at present by these reviewers.

If one follows this treatment one must for a first approximation intentionally neglect other physical models, which have been summarized by Cunningham and Williams (1980). Respiration physiologists may learn from these physical methods and mathematical solutions, especially those used by engineering chemistry. The rather complicated, but detailed solutions they offer would be perhaps more accurate, but could be at present scarcely integrated into the normal treatment of diffusion and convection problems in the leading journals for respiration physiologists.

The classical solution of the general equations for diffusive-convective gas transfer for steady state diffusion against convection current in a tube goes back to Hertz (1922, 1923), who analyzed the separation of gas mixtures by a counter-current flow. The methods for derivation of the classical general equations and for their solution for physiological boundary conditions are explained most explicitly for respiration physiologists by Jacobs (1935) and are used in this section. Together with correction of minor errors the equations are transformed to be compatible with the accepted conventions and the nomenclature of respiration physiologists

(Piiper et al. 1971). From a molar balance of the diffusive and convective fluxes into and out of a volume element within the spiracular tube we get, according to Frank and Mises (1935), an equation for the change in concentration of gas i as a function of time t and distance x.

$$\frac{\partial c_i}{\partial t} = D_i' \cdot \frac{\partial^2 c_i}{\partial x^2} - v' \cdot \frac{\partial c_i}{\partial x} \tag{39}$$

D_i' is the effective diffusion coefficient, v' is the vector of the flow velocity of the medium in the direction of diffusion along x. If flow velocity is zero we have pure diffusion and the convection term at the right side becomes zero, leaving the well-known Fick's second equation. His second law describes the accumulation of gas i in the volume element as a function of distance x and time t. In the general equation the flow velocity vector in the convection term is multiplied by the concentration gradient along the distance x.

If gas i diffuses in the opposite direction to the absolute flow velocity v, v' becomes negative and the sign between the diffusion and convection term changes. If we replace, in addition, the concentration as in Eq. (4), by the product of the capacitance coefficient and the partial pressure of gas i, we get the general equation for diffusion against the convection current of gas i.

$$\frac{\partial P_i}{\partial t} = D_i' \cdot \frac{\partial^2 P_i}{\partial x^2} + v \cdot \frac{\partial P_i}{\partial x} \tag{40}$$

If gas i diffuses with the convection current induced by the absolute flow velocity v we get

$$\frac{\partial P_i}{\partial t} = D_i' \cdot \frac{\partial^2 P_i}{\partial x^2} - v \cdot \frac{\partial P_i}{\partial x} \ . \tag{41}$$

Both equations describe the partial pressure of gas i during diffusive-convective gas exchange in a spiracle either during ventilation, in which the individual gas changes between these equations between inspiration and expiration, or during suction ventilation. During suction ventilation oxygen is the only gas with cocurrent diffusive-convective gas exchange according to the partial differential Eq. (41). The partial pressures of the other gases with anticurrent gas exchange is described by Eq. (40).

The only solution of Eq. (39) for the steady state rate of gas transfer I could find in the literature is that used by Hertz, describing the anticurrent diffusive flux against a convection current only for the boundary condition that the concentration at the inflow end of the tube is zero. This would mean that this equation could just be applied to the situation of diffusion of carbon dioxide or water with convective back-flow within the spiracle at zero partial pressure of ambient air, which Buck's theory could not describe, as outlined in the previous section. Curiously he cited Jacobs (1935), where he could have found on p. 134 a chapter and on p. 135 the "Hertz-Jacobs equation" (161) describing flow-retention during flow-diffusion.

At non-zero partial pressures of gas i at the ambient and at the tracheal end of the spiracular tube one has to derive new solutions of Eqs. (40) and (41) for the partial pressure gradient in the tube. One may use the methods given by Jacobs (1935) and state that the change in partial pressure of gas i per unit time will be zero in the volume element, if a steady state is reached. For the resulting equilibrium between the diffusion and the convection term in Eqs. (40) and (41) exists a general solution. By insertion of the boundary conditions of the partial pressures at both ends of the tube into this solution, one can obtain equations for the stationary partial pressure at any position x within the tube either during anticurrent or cocurrent gas exchange. If one differentiates these equations with respect to x one gets equations for the partial pressure gradient at position x. If these equations for partial pressure gradient are inserted into Fick's first law,

$$\dot{M}_i^S = -A^S \cdot \beta_i \cdot dP_i^S / dx_i \quad , \tag{42}$$

one can obtain the desired new equations for the steady state cocurrent or anticurrent rate of diffusive gas transfer at any position in the spiracular tube.

Neglecting the convective influx or efflux at the opening of the spiracular tube for a while, one gets at this point the following equations for anticurrent (ANCU) diffusion (DI) of gas y with backward-flow (Eq. 43) and for cocurrent (COCU) diffusion of gas x with forward-flow (Eq. 44).

$$\dot{M}_y^{DI\text{-}ANCU} = A^S \cdot v \cdot \beta_y \cdot \triangle P_y^S \cdot \exp(-v \cdot L^S / D_y') / (1 - \exp(-v \cdot L^S / D_y')) \tag{43}$$

$$\dot{M}_x^{DI\text{-}COCU} = -A^S \cdot v \cdot \beta_x \cdot \triangle P_x^S \cdot 1/(1 - \exp(v \cdot L^S / D_x')) \tag{44}$$

In spite of describing the special types of diffusion only, both equations contain the air velocity because it influences the partial pressure gradient for diffusion. Spiracular cross-sectional area (A^S), length (L^S) of the spiracular tube, and the effective diffusion coefficient are the same as they would be in the general diffusion Eq. (6). The ratio of the diffusion coefficient to the length is replaced by the term on the right side of the partial pressure difference term. As exp(+n) in Eq. (44) will be always higher than 1, the cocurrent diffusive transfer will become positive, like the anticurrent diffusive transfer rate (Eq. 43). Equation (43) is a more general equation than the Hertz-Jacobs equation as we may insert an ambient partial pressure of zero into the equation, defining $\triangle P_y^S$ to get the Hertz-Jacobs equation in the usual nomenclature of respiration physiologists, having P_y^T instead of $\triangle P_y^S$.

The addition or subtraction of the simultaneous convective influxes into and effluxes out of the spiracular tube according to Eq. (7) results in the Eqs. (45) for anticurrent and (46) for cocurrent diffusive-convective gas exchange of gas y or x at the opening A of the tubular valve, where the partial pressure of gas x or y is P_i^A.

$$\dot{M}_y^{S\text{-ANCU}} = A^S \cdot v \cdot \beta_y \cdot (\triangle P_y^S \cdot \exp(-v \cdot L^S / D_y') /$$
$$(1 - \exp(-v \cdot L^S / D_y')) - P_y^A) \tag{45}$$

$$\dot{M}_x^{S\text{-COCU}} = - A^S \cdot v \cdot \beta_x \cdot (\triangle P_x^S \cdot 1/(1 - \exp(v \cdot L^S / D_x')) - P_x^A) \tag{46}$$

During steady suction -ventilation in insects, oxygen would be the only gas of type x with cocurrent diffusive-convective gas exchange, whereas carbon dioxide, water and nitrogen would be of type y with anticurrent diffusive-convective gas exchange.

Spiracular conductance according to Eq. (27) can be derived from Eqs. (43) and (44) only, as Eq. (27) is valid for bidirectional net convection (ventilation) only.

In the nitrogen steady state of the flow-diffusion postulated by Buck (1958a,b) the net spiracular nitrogen exchange will become zero and Eq. (45) for nitrogen can be solved for the air in-flow velocity v in the steady state diffusive-convective gas exchange.

$$v = -(D'_{N_2} / L^S) \cdot \ln(P_{N_2}^T / P_{N_2}^A) \tag{47}$$

Equation (47) shows that the air in-flow velocity during steady state suction-ventilation will be increased.

1. If the length of the spiracular tube is minimized and
2. if the tracheal partial pressure of nitrogen is maximized.

The short length of the spiracular diffusion path, usually less than half a millimeter, would be of an immense disadvantage for diffusion control according to Hazelhoff's theory even if compared with the problematic tracheal diffusion (Sec. 2.3) with an 50 to 100 times longer diffusion path (Kestler 1982). This shows that evolution seems to have optimized Eq. (47), as even a compilation of tracheal gas determinations in insects made by Levy and Schneiderman (1966a) showed that the tracheal partial pressure of nitrogen is unusually high.

It may be added that during longer or shorter closure of the spiracles any described volume decrease of the tracheal system (Herford 1938; Brockway and Schneiderman 1967) will increase the tracheal partial pressure of nitrogen, too, as the fractional or percentual concentration of nitrogen rises much more than the total pressure falls. Thus, any constriction of the spiracles prepares the partial presure of nitrogen for an optimal retention of carbon dioxide and water during the next opening, as it leads to an increased air in-flow velocity (Eq. 47).

4 Resolution of CO_2 Release, Water Loss, and Tracheal Pressure Changes of Undisturbed Insects During Spiracle Opening and Ventilation

The principles of adaptation of respiration and respiratory water loss to the environment discussed and analyzed theoretically for the steady state in Secs. 2 and 3 lead to conclusions summarized in Sec. 5. They will be compared with measurements of the result of evolution in undisturbed insects, which are described in this section.

4.1 Principle of the Physiological Method

As any influence of experimental handling or "stress" can induce activity or raise the metabolic rate (Kestler 1971; Matthews and Downer 1973; Downer 1979; cf. Miller 1982), it must have an influence on the gas exchange, which depends on the metabolic rate (cf. Secs. 3.2.1 and 3.2.2, Chap. 1). The full range of gas exchange adaptations can, therefore, only be detected in undisturbed insects living in their natural environment. This necessitates a time-consuming experimental approach in two successive steps:

1. The normal respiration behavior must be recorded automatically in an environment which is accepted by the experimental animals as natural. Ecological factors should be hold at their optimum.
2. The influences of the special experimental conditions on the respiration behavior occurring normally must at least be checked as a perturbation control.

This is even important for the presence of an experimenter with recording or observation devices as my own unpublished observations have shown that a locust changes its respiration behavior if one is entering the laboratory. The only experimental work on insect respiration in which such a strict perturbation control has implicitly been made are the classical experiments of Schneiderman and co-workers and of Buck and co-workers in diapausing Saturniid pupae of *Hyalophora cecropia* and *Agapema galbina.* They have been best summarized by Miller (1974), Edney (1977), and Schmidt-Nielsen (1979) and are now described in all relevant textbooks. In contrast, adult insects are either pinned down to avoid locomotion during recording of respiration movements or loaded with recording devices. In most cases a perturbation control is not made. In the next paragraphs of this section the mechanisms of gas exchange and water retention of undisturbed adults of *Periplaneta americana* are, therefore, compared with that of diapausing *Cecropia* pupae.

4.2 Resolution of Respiration Movements in the CO_2 Release of Undisturbed Insects During Rest and Activity

The best method for automatically recording the respiration behavior in a freely movable insect is to resolve the influence of spiracles and pumping systems on the gas fluxes of oxygen, carbon dioxide, or water vapor of an individual insect at rest. As CO_2 transfer occurs in the same direction as water transfer, and as the most sensitive methods exist for its continuous recording, it has been preferred. It is typical of most continuous recordings of CO_2 release in individual insects of different orders, made either with thermal conductivity devices (Daynes 1925 and 1933 as cited by Edwards and Patton 1967; Punt 1944, 1950, 1955, 1956; Punt et al. 1957; Kanwisher 1966; Edwards 1970) or with infrared gas analyzers (Wilkins 1960; Kanwisher 1966; Rothe 1983) that respiratory movements are deduced by circumstantial evidence only. Hamilton (1959, 1960, 1964, 1972) was the only one who observed bouts of abdominal ventilation in teneral locusts, while recording a burst of CO_2 release with an infrared gas analyzer, but he was not able to resolve single pumping volleys and assumed, in 1972, a closed state of the spiracles between the volleys. Observation of spiracle movements in unrestrained insects was possible only in pupae. Schneiderman (1960) has been the only one who has done the laborious work of observing the same spiracle over hours and recording the acitivity by hand. This enabled Levy and Schneiderman (1966c) to correlate spiracular constriction period and the following flutter period with the best records of the release pattern resolved up to that point by Punt et al. (1957). As has been discussed by Buck (1958b), this interpretation of their recordings with a thermal conductivity "diaferometer" is in contradiction with the postulated oxygen debt, because the constriction period would have to follow the spiracle opening in the flutter period. Punt's recordings of the more continuous oxygen uptake resolve the constriction period (Buck 1958b; Kestler 1971) in contrast to other methods for continuous recording of oxygen uptake (Schneiderman and Williams 1953b, 1955; Welbers 1976; Scheid et al. 1982 and pers. comm.), which measure during the constriction period the volume decrease. Unfortunately, Punt's sensitive recordings of oxygen consumption and CO_2 release lack calibration and show an appreciable zero shift.

An even better resolution of spiracle movements can be obtained by the method developed by Kestler (1971). Figure 6.10 shows the cyclic CO_2 release of a *Cecropia* pupa at the end of the diapause when cycles are much shorter than the 36 h once measured at minimal metabolic rate. In those cycles the flutter period is mainly elongated (Schneiderman and Williams 1953b, 1955; Burkett and Schneiderman 1974b). The method employs a differential infrared gas analyzer (DIRGA). If compared to the record of spiracle movements (Schneiderman 1960), it shows that the CO_2 release rate is directly proportional to the degree of opening of the spiracles. The CO_2 recording resolves the closed state of the spiracles during the constriction period (C). During the flutter period (F) the efflux during single microopenings can

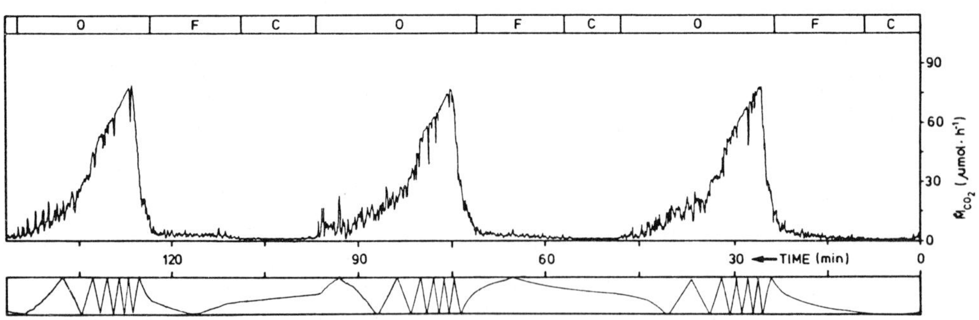

Fig. 6.10. CO_2 release of the CFO type. Record of the rate of release from a *Hyalophora cecropia* pupa at 20°C. All recordings in Figs. 6.10–6.17 are from *right to left.* The relative amount of CO_2 released during a specific behavior can be obtained from the integrator units. *C* spiracles constricted; *F* spiracles fluttering; *O* spiracles open

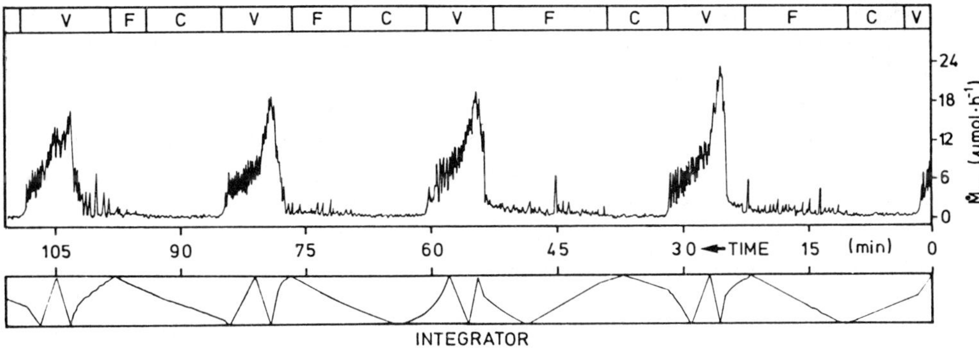

Fig. 6.11. CO_2 release of the CFV type. Record of the rate of release from a quiescent male *Periplaneta americana* at 20°C (cf. Fig. 6.10). *C* spiracles constricted; *F* spiracles fluttering; *V* bout of abdominal ventilation

be resolved. The CO_2 retention due to the reduction of spiracular opening area becomes apparent if compared with Schneiderman's observations. Proportionality between release rate and degree of opening is also given during the open period, showing even the short constrictions during the decline period in the last part of the open period (O) of this CFO type of cyclic CO_2 release. The record is also a perturbation control for Schneiderman's observations, excluding the influence of the removal of the spiracular filters for observation. This method of recording respiration movements in the unrestrained insect has been used in further experimental work on the CFO type in cooperation with Baaske (unpubl.) in my laboratory and with Scheid using a respiratory mass spectrometer (Bridges et al. 1980). The method has been developed for recording respiration of undisturbed adult insects to show the CFV type of resting respiration in *Periplaneta americana, Blaberus discoidalis, Carabus problematicus,* and *Schistocerca gregaria* (Kestler 1971, 1978b). Discontinuous ventilaton (V) had been described before in *Byrsotria fumigata* (Myers 1958; Myers and Fisk 1962; Myers and Retzlaff 1963), in *Periplaneta* at 29° C (Paulpandian 1964), in *Schistocerca* (Hamilton 1959, 1960, 1964), and in *Blaberus giganteus* and *Blaberus craniifer* (Miller 1966). CO_2 recording (Fig. 6.11) and simultaneous observation of the CFV type in adult *Periplaneta americana* (Kestler 1971) showed two unexpected results: (1) ventilation periods are connected with the same movement pattern of the spiracles, a constriction period (C) and a flutter period (F), as it is typical for the diapausing saturniid pupae; (2) the small CO_2 bursts caused by single pumping volleys are added onto a more basic and more continuous efflux of CO_2, which is proportional to the mean opening area of the thoracic spiracle 2 during expiration both showing at the same time the characteristic peaks which can be seen in Fig. 6.11. Additional controls at higher resolution suggest that cocurrent diffusion is added to convective expiration when the tracheal partial pressure of carbon dioxide is high.

The behavior of the external spiracular valve of the second thoracic spiracle of *Periplaneta* is, even in the details, similar to that in the CFO type of *Hyalophora* described later by Burkett and Schneiderman (1974b). This suggests that the control of spiracular movements in cockroaches (Case 1957; cf. Kaars 1981; Miller 1982) may be similar to that in pupae (Beckel 1958; van der Kloot 1963; Burkett and Schneiderman 1974a). As in the (C)FO type (Schneiderman 1960), the constriction period disappears in the (C)FV type at higher temperatures or metabolic rates (Kestler 1971; 1978b). The easiest way to induce a lasting increase in the metabolic rate is to attach a recording device on the insect. This may be the reason why Wasserthal (1981) could show only FV cycles in adults of *Attacus atlas*, resting at 21°C, as thermistors and other recording devices were attached to the insect. He showed, nevertheless convincingly, a coupling between discontinuous ventilation and oscillating hemolymph flow by contact thermography. The results confirm that the (C)FV cycles are the typical resting respiration (Kestler 1971; cf. Miller 1982). The cycles occur from 5°C up to 35°C if the animal is at rest. The persistence of ventilation in *Periplaneta* even at temperatures as low as 5°C and the occurrence of ventilation in small inactive phorid flies (Miller 1981) is in contrast to the widely held view (Krogh 1920b; Wigglesworth 1972; Wasserthal 1981) that (1) ventilation is only necessary for large and active insects to support their higher metabolic needs and that (2) smaller or inactive insects use diffusion because ventilation would lead to higher water loss. Kestler (1983) showed that this is theoretically wrong, as has been outlined for water loss in Sec. 2.3.

Ventilation with the high frequency of the metathoracic pacemaker of cockroach ventilation (Farley et al. 1967) can only be observed after flight, at extreme temperatures, or if *Periplaneta* is captured. If it is pinned down in a special way, struggling replaces parts of the resting ventilation period and is followed by fast bouts of stress or emergency ventilation, each one replacing one resting pumping stroke. This shows that Hazelhoff's (1926) theory of ventilation in active, and diffusion control with the spiracles in resting, cockroaches is based on the apparent lack of a sufficient perturbation control: in his experiments on pinned down cockroaches the intrinsic form of the irregular bouts of constriction and fluttering was described correctly (cf. Kestler 1971), but their regular cycling and function as subsystems of the (C)FV cycle had been overlooked in spite of mentioning the observation of free insects as a control. Likewise the important experimental contributions of Wigglesworth (1935), Herford (1938), and many others (cf. Miller 1974, 1981, 1982) on spiracle opening and ventilation in more or less immobilized insects may show itself as valuable research on subsystems, too. One may also suggest that the irregularity of the discontinuous resting ventilation if compared to the regular (C)FV cycles may be due to mild experimental stress induced either by the recording device for ventilation (Myers 1958; Myers and Fisk 1962; Paulpandian 1964; Hustert 1974; Komatsu 1977; Koch 1983) or by electrodes for recording myograms (Miller 1973, 1981; Hustert 1974; cf. Miller 1966, 1980). The insensitivity of the burrowing cockroaches used by Myers and Miller to the dorsal contact with recording devices may be due to an adaptation to similar stress acquired in the natural resting place. This allows perturbation control even in neurophysiological experiments on subsystems of respiration, as it enabled Myers and Retzlaff (1963) to isolate the abdominal pacemaker for discontinuous resting ventilation in the Cuban burrowing cockroach *Byrsotria fumigata*. The combined (C)F and the V periods of cyclic ventilation have been shown by Kestler (1971) to be in inverse relationship to the mean metabolic rate (cf. Miller 1980, 1982). This suggests that in the O and V periods spiracle opening may be triggered by the $P^T_{CO_2}$ which depends (1) on the only slightly greater than normal CO_2 capacitance of the tissues (Bridges and Scheid 1982) and the hemolymph (Buck and Friedman 1958; cf. Edwards and Patton 1967) and (2) on the spiracular loss in the F period. But beginning and end of ventilation, in contrast, may be triggered independent of spiracular release in the F-period if a certain amount of CO_2 has been produced (Kestler 1971).

Cockroaches interrupt the continuous (C)FV cycling only during the well-known diurnal activity period. The activity can be resolved into single bouts of grooming and orientation with local search (cf. Bell 1982). Each behavior can be recognized by a specific irregular pattern of CO_2 volleys due to irregular short openings of the spiracles. The mean amplitude of the CO_2 volleys is proportional to the mean metabolic rate. Especially during grooming sequences total closure seems to be maximized and the short rapid spiracle openings can be observed to be coordinated with the heterogeneous motor sequences of grooming. The irregular CO_2 release pattern specific for a certain behavior is, thus, indicative of autoventilation. Tracheal pressure changes seem to be used economically during autoventilation for convective gas exchange. Spiracles seem to open only if the partial pressure of oxygen or carbon dioxide reaches an extreme threshold. It could well be that the widespread coupling of ventilation to more regular homogeneous forms of behavior (cf. Otto and Weber 1982) could serve water conservation in a similar way.

This shows that insect respiration specific for a certain behavior has been analyzed only in a few cases, as demonstrated for resting respiration in this section or for flight in the outstanding experimental work of Weis-Fogh (1967) (cf. Miller 1974).

4.3 Tracheal Pressure Changes During Resting Respiration

The existence of two forms of cyclic CO_2 release in the resting state, (C)FO cycles in diapausing pupae and (C)FV cycles in adult insects of different orders, suggests (Kestler 1971, 1978b) that the specialized gas exchange of diapause respiration represents a basic adaptation to the resting state independent of the metabolic rate. It may be compared to sleep in vertebrates. Moreover, the cyclic changes in tracheal partial pressure of oxygen and carbon dioxide measured by Levy and Schneiderman (1957, 1958, 1966a, 1966b) and the cyclic pressure and volume changes in the tracheal system (Levy and Schneiderman 1966c; Schneiderman and Schechter 1966; Brockway and Schneiderman 1967; Burkett and Schneiderman 1974b) in diapause respiration of pupae suggests that during the (C)F periods, passive suction ventilation (Miller 1974) should occur in both types. But outside Schneiderman's group there has still been no confirmation of the outstanding measurements of subatmospheric pressure changes in the tracheal system, which showed directly that suction ventilation occurs in the CFO type.

The partial pressure measurements are confirmed in *Hyalophora* by the measurements of Kanwisher (1966) and of Scheid et al. (1982 and pers. comm.), but both papers doubt the existence of diffusive convective-gas exchange in the flutter period, postulating pore diffusion or pure diffusion, respectively. Both use, as Buck did,

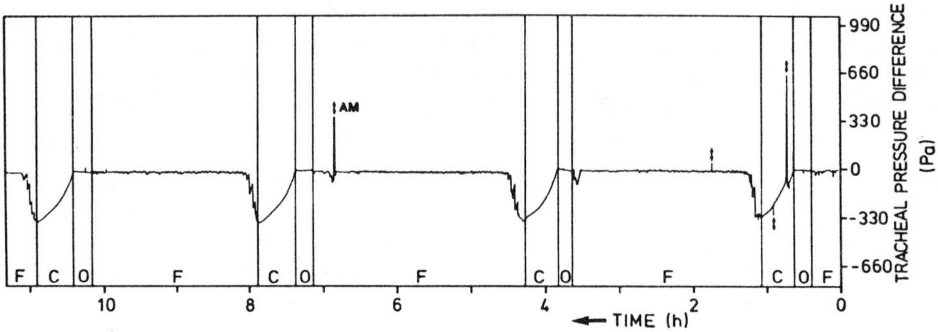

Fig. 6.12. Quantitative record of the difference between tracheal and ambient barometric pressures from a *Cecropia* pupa at 20°C showing subatmospheric pressure cycles of the CFO type. Abdominal movements *AM*; (cf. Fig. 6.10)

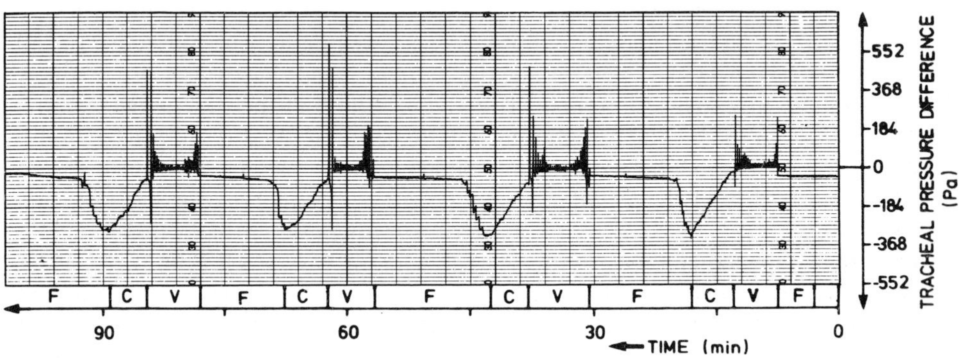

Fig. 6.13. Quantitative record of the difference between tracheal and ambient barometric pressures from a quiescent suspended male *Periplaneta americana* at 20°C showing pressure cycles of the CFV type with a subatmospheric steady state in the F period (cf. Fig. 6.11 and text)

equations based on a constant partial pressure gradient (cf. Secs. 3.4 and 3.5). In contrast, Kestler (1980) confirmed the occurrence of subatmospheric pressures in the C and F period of the (C)FV type in *Periplaneta.* The improved technique uses transducers with less volume displacement of the membrane, a negligible dead volume of air in the connecting tube and an improved temperature control. This enables the first quantitative measurements, if compared to the last technique used by Burkett and Schneiderman (1974b), which also used solenoid valves to control the pressure difference to air. Figure 6.13 is a record of the changes in tracheal barometric pressure difference to air during CFV cycles in a *Periplaneta* male. It is compared in Fig. 6.12 with the first confirmation of the classical measurements in the CFO cycles. In both curves the tracheal pressure falls during the C period by a few hundred Pascal units (1 Torr = 133 Pa). The fall is not linear because it is accompanied by a volume decrease (CFO: Brockway and Schneiderman 1967; CFV: P. Kestler unpubl.). With each of the first microopenings of the flutter period, the pressure rises by steps towards the atmospheric pressure and then fluctuates around a steady state of subat-

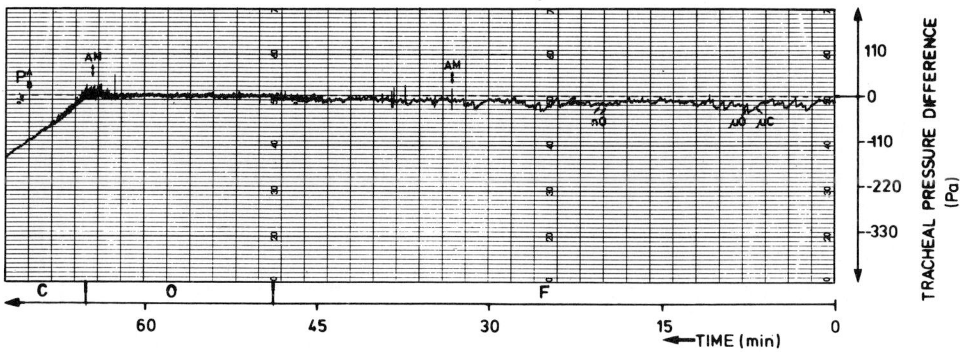

Fig. 6.14. Tracheal pressure difference before and after an open (*O*) period. The resolution is higher than in Fig. 6.12 showing nearly permanent subatmospheric pressure in the flutter period with pressure falls during microconstrictions (*μC*), pressure rise during microopenings (*μO*) and fluctuations caused by nanoopenings (*nO*). Spikes are due to abdominal movements *AM*; P^A_B ambient reference pressure at O Pa

mospheric pressure because the spiracular openings become less synchronized. In CFO cycles the pressure only sometimes reaches the atmospheric pressure as can be seen in Fig. 6.14. One gets usually similar records in well hydrated cockroaches, whereas one can find the pattern of Fig. 6.13 in good preparations in which the suspended cockroaches survive for a month and allow for dehydration experiments. Figure 6.14 shows very clearly a steady state subatmospheric pressure in the flutter period after the initial pressure rise due to short microopenings. This pressure pattern resembles the CO_2 release pattern of unhandled cockroaches after similar treatment, where the minute CO_2 bursts occurring during microopenings vanish and where the typical flutter movements, which are pulsations of the valve in the microconstriction period and do not normally break the liquid film sealing the valve, lead to very small and very brief openings (nanoopenings). Nanoopenings have also been observed by Burkett and Schneiderman (1974b), especially at low metabolic rates, and did not cause changes in intratracheal pressure. The steady subatmospheric pressure is exactly what has been postulated in Buck's theory to be the cause of steady flow-diffusion in the CF period. This is in contrast to the objections against Buck's postulate, made by Levy and Schneiderman (1966c) in a detailed discussion of the pressure changes, in which two additional assumptions were made. The authors postulated a steady inflow of air in the constriction period and pure diffusion during the microopenings. The former statement has been shown by Bridges et al. (1980) to be wrong, as the tracheal system is tight to inert gases in the constriction period. The latter statement must be questioned as (1) air flows in as long as the pressure rises (cf. Fig. 6.14) and (2) anticurrent nitrogen diffusion is directed only outward, whereas spiracular fluxes of oxygen, carbon dioxide, and water are in a quasi steady state with the sources or sinks in the tissues. Therefore, even if the spiracles stay open for a second, a slight vacuum, undetectable with the amplification used here, should persist. It would be sufficient to induce a convective influx of nitrogen, replacing the out-diffusing nitrogen as in the steady state described by Eq. (47). This hypothesis will be tested with an improved version of the present method and may show

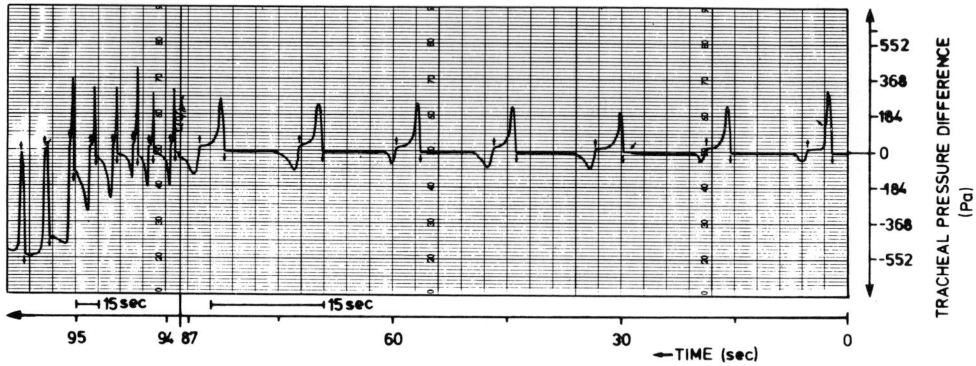

Fig. 6.15. Tracheal pressure measurements during the last 15 pumping strokes of a ventilation (*V*) period recorded at higher time resolution than in Fig. 6.13. *Down-arrow* expiration movement; *up-arrow* inspiration movement; *oblique arrow* microopening. Note the leaking of the spiracles indicated by an exponential fall or rise. It is caused by nanoopenings. Atmospheric pressure had been reached 10 strokes earlier between the expiratory and inspiratory movement only (cf. text). The last pumping stroke was ineffective

unsteady foms of diffusive-convective gas exchange also in the open period. The deviations from the steady state during microconstrictions or microopenings may be an adaptation to higher metabolic rates needing a higher opening area. The microconstriction, like the constriction period, contributes to a longer lasting pressure rise in the microopening period. The second function of both constriction periods is to raise the partial pressure of nitrogen in the tracheal system, concomitant with the tracheal volume decrease due to the compliance of the whole system. As has been outlined at the end of Sec. 3.5 this prepares a higher steady state velocity of inflow. The positive pressure pulses during ventilation (Fig. 6.13 and Fig. 6.15) can be easily recorded. A compression phase was described earlier by McCutcheon (1940), Watts (1951), and Weis-Fogh (1967). Its possible function for gas transport from bigger to smaller air sacs has been demonstrated in a model by McCutcheon (1940). Additional functions, such as the support of hemolymph transport or expiration by allowing pressure to equalize before the spiracles open, have been reviewed by Miller (1974). The positive pulses in Fig. 6.13 and Fig. 6.15 are due to a fast constriction of the valves during the "expiration" movement. If the spiracles open during the expiratory plateau phase or expiratory pause, pressure falls with the expiration of a blast of air to ambient pressure. This fall is often interrupted by the begin of the passive abdominal "inspiration" movement, which leads to a more rapid pressure fall to a subatmospheric pressure if spiracles are closed during the inspiration movement. This is typically the case at the end of the ventilation period (Fig. 6.15) and either leads to negative pulses, if spiracles open or are leaky in the "inspiratory" pause after the end of the "inspiration" movement, or supports and replaces the slow building of the subatmospheric pressure in the constriction period, occurring also in the CFO type. This pressure originates from the lower capacitance coefficient of the tissue water for oxygen if compared to carbon dioxide (Tables 6.1 and 6.2).

The replacement of the contribution to subatmospheric pressure of the slow pressure decrease by the fast pressure decrease (Fig. 6.15) allows the slow pressure potential to be used entirely for volume decrease and a concomitant rise of the partial pressure of nitrogen. This shows the two ways in which negative pressure pulses support water retention by anticurrent diffusion. If the spiracles don't open at all during one of these last "pumping" movements of the V period, the subatmospheric pressure returns to its original position (Fig. 6.15). The amplitude of the positive pulses is higher at the beginning and end of a ventilation period and shows an inverse relationship to the opening area of spiracle 2, which has been mentioned in Sec. 4.2 as opening in proportion to the CO_2 release. The adominal "inspiration" movements with closed spiracles lead to active suction ventilation (Kestler 1980, 1982) either in the inspiratory pause or in the first microopenings of the flutter period. Positive pressure pulses prevent, together with the reduced opening of the valves, the cocurrent diffusion of water with carbon dioxide during expiration by increasing the convective part of CO_2 release and simultaneously decreasing the diffusive part. This behavior makes sense only if less water is lost by convection than by diffusion, which has been shown in Sec. 2.3. Both types of pressure pulses are reduced in amplitude after drinking in proportion to the CO_2 release. The adominal "inspiratory" movements with charge to the spiracles of dragonflies (Miller 1964, cf. Miller 1982) may have in the CFV type the function of controlling the ratio of diffusion to convection in the F and V period, represented by the parameters area and velocity, respectively, in Eqs. (45) and (46).

The pressure pulses of the discontinuous ventilation in the CFV cycle of *Periplaneta* have been first recorded by Watts (1951). He attributed them to postactivity ventilation, thus, repeating Hazelhoff's mistake as has been discussed in the last section. Davey and Treherne (1964) recorded the pulses in the hemolymph, but they were not able to detect any ventilation. The supposed function in crop emptying may be an additional example of the simultaneous economic use of the same movements for different purposes, as it occurs also during autoventilation, oscillating hemolymph flow (Wasserthal 1981), ecdysis (cf. Miller 1981), or if the "compression" pulses support hemolymph transport (cf. Miller 1974). The short positive pulses interrupting the entirely subatmospheric pressure have been recorded in the hemolymph of *Tenebrio* by Slama (1976) and Slama et al. (1979). Provansal et al. (1977) suggested, in addition, that the observed reduction of the number of pulses would economize respiratory water loss after dehydration. Such patterns (cf. also Davey and Treherne 1964) can also be recorded in tracheal and hemolymph pressure of extremely dehydrated and stressed cockroaches (Kestler unpubl.). In this case the deflation of the intersegmental membranes, which can be observed during "inspiration" movements with closed spiracles, is caused in the dehydrated insect by the loss of hemolymph. The apparant lack of any knowledge about the tonic function of the abdominal muscles in resisting the subatmospheric hemolymph and tracheal pressures should encourage neurophysiological work on the control mechanisms (cf. Miller 1981).

4.4 Problems of Measuring Tracheal Water Loss

There is nearly no paper or review on cyclic CO_2 release which does not emphasize that prevention of respiratory water loss is the main purpose of this complicated type of gas exchange. The same purpose is also emphasized for the spiracles and their function. Edney (1977) has best reviewed the evidence for respiratory water retention and concluded "... the evidence is too scanty, or too indecisive ...". Mullins (1982) summarizes for water retention caused by cyclic CO_2 release, as measured by Wilkins (1960) in the cockroach: "... but evidence supporting this view is lacking ...".

The main reason are technical difficulties. Buck and Keister (1955) could show that intubation of the spiracles increases weight loss, but they could not resolve the weight loss during the open period in untreated pupae. Brockway (1967) resolved the weight loss in the open period. But neither could show how much of the weight change is due to water. The increase in water loss during the open period has been specifically resolved for the first time by Kanwisher (1966) with his thermal conductivity device. But calibration is lacking and he did not show the water loss through spiracles and cuticle in the flutter period before. Brockway measured the weight loss after sealing the spiracular openings with vaseline. But this is no control for the part of the weight difference caused by the high rate of oxygen uptake and the small rate of carbon dioxide release. Similarly, a lack of resolution and the same indirect control by sealing the spiracles are common to the classical and more recent measurements of general "tracheal" or "cutaneous" water loss in cockroaches and other insects (Gunn 1933; Koidsumi 1934; Ramsey 1935; Jacovlev and Krüger 1953; Edney and McFarlane 1974; Coenen-Staß and Kloft 1976; Gilby and Rumbo 1980). Edney (1977) summarizes that spiracle sealing "... inevitably leads to physiological injury in long experiments – and this is a general dilemma."

A solution of the dilemma is possible if one lets the insect do the sealing. As Bridges et al. (1980) showed, by replacing nitrogen with argon, that the spiracles are tightly closed to inert gas influx in the constriction period, it follows that cutaneous gas exchange can be measured in the constriction period. The first step in answering whether the suction ventilation of air in the flutter period leads to water retention is to measure the difference in water loss rate between the C and F period, assuming that the cutaneous loss does not change. Figure 6.16 shows the weight loss of a cockroach recorded with a recording ultramicrobalance. The cockroach can move freely in a gauze cage made of stainless steel. The excursions of the recorder pen are caused by "inspiration" and "expiration" movements as resolved in Fig. 6.17 at a higher recording speed. The short expiratory and the longer inspiratory pauses can be discerned. It can be seen that the weight change at the end of the V period is the same as during the beginning of the constriction period. This suggests that the negative and positive pressure pulses shown in Fig. 6.13 and Fig. 6.15 prevent the water loss. If the weight loss in the second half of the ventilation pause (CF period) is compared to that in the first half, a slight increase in weight loss may be suggested. Additional experiments have to show whether it becomes more prominent at lower humidities, thus, indicating water loss.

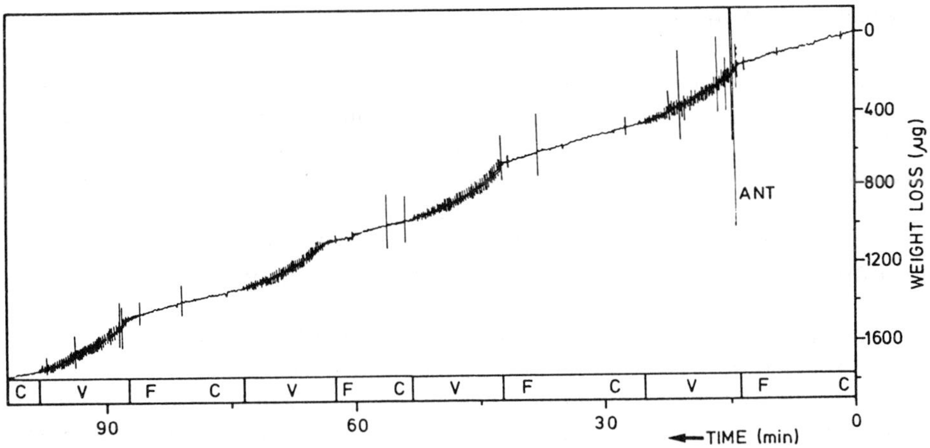

Fig. 6.16. Record of weight loss from a quiescent male *Periplaneta americana* during CFV cycles. Body mass 1.03 g; temperature 24.8°C; relative humidity 76%; *ANT* antennal movements

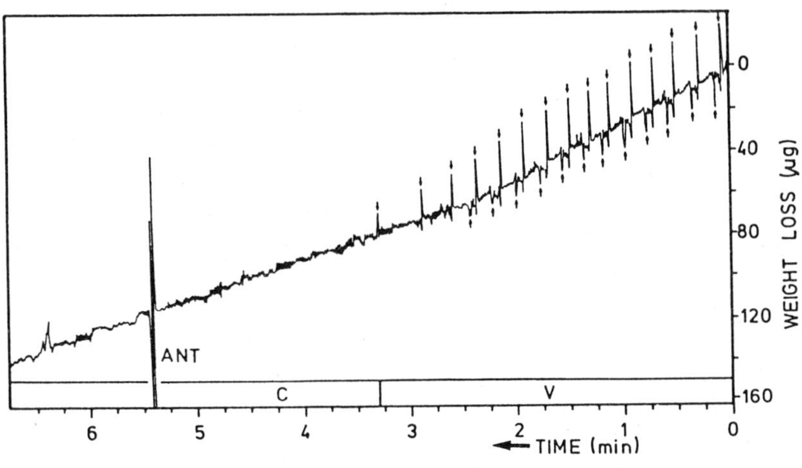

Fig. 6.17. Weight loss from a quiescent male *Periplaneta americana* at the end of a ventilation (*V*) period and the beginning of a constriction (*C*) period, recorded at two speeds. The control does not show the oscillations of the constriction period. Notations as in Figs. 6.15 and 6.16. The lack of inspiratory excursions of the pen during the last pumping strokes (cf. Fig. 6.15) indicates that the excursions are caused by air flow

The problem remaining is that we do not know the contribution of the other gases to the weight change. Even if there is no spiracular or cutaneous water loss or gain, any change in the recorded apparent body weight during a certain time interval is caused by the amount of oxygen taken up, the amount of carbon dioxide released, the amount of nitrogen taken up or released, and the apparent weight change due to volume changes of the animal which are, according to the Archimedic principle, equal to the easily resolvable weight of the displaced ambient air. It has been shown by Kestler (1978a) that (1) the CFO and the CFV cycles persist in water-saturated air if an air flow prevents the accumulation of metabolic heat or carbon dioxide and that (2) the contribution of the other gases and of the volume changes to the apparent weight balance can be recorded over the CFO and CFV cycle in water-saturated air. This shows that the continuous CO_2 release in water-saturated air demonstrated by Buck or Schneiderman and their co-workers is not an adaptative response to water-saturated air.

An improvement of the recording technique (P. Kestler unpubl.) allows are resolution of 0.5 μg and a simultaneous recording of the carbon dioxide release with the DIRGA in a flow through system, thus, showing directly the presence and duration of the constriction period in the (C)F period. The water loss can be obtained at the desired temperature by the difference between the weight change in the desired humidity and that in water-saturated air. This allows, in addition, the calculation of the rate of oxygen uptake in the F and V or O period if the weight loss of CO_2 recorded with the DIRGA is added to the weight balance in water-saturated air and corrections are made for volume change. A quantitative presentation and the necessary detailed discussion of the results would be too lengthy for this volume and will be published elsewhere. The results will show that (1) the water loss in the flutter period is only slightly increased and (2) respiratory water loss by diffusion is higher than by convection.

5 Conclusions

This chapter shows that a successfull approach to a better understanding of the physical parameters, changed during adaptation of insect respiration to the ambient and internal media water and air necessitates, as has been successfully done by vertebrate respiration physiologists (Rahn and Paganelli 1968; cf. Dejours 1981), a quantitative formulation of simple models, which can be tested experimentally and further improved.

The theory of Weis-Fogh (1964) on diffusive gas exchange through tracheae must be corrected because of the latest morphological evidence on tracheal supply of mitochondria. A simple by-pass model reveals adaptational functions of liquid filling of tracheoles (Sec. 2.2). The classical theory of Krogh (1920a) (Sec. 2.1) can be applied successfully to show the physical basis for a need for respiratory water conservation and the evolution of spiracles especially in small insects, too. Adding pure convective gas exchange to the model shows the unexpected result that ventilation is favorable for water retention especially in small insects and at low metabolic rates (Sec. 2.3).

The main parameters of tracheal and spiracular gas exchange, which must be adjusted to reduce respiratory water loss, are shown and important principles of adaptation discussed (Secs. 3–3.2). Diffusive-convective gas exchange with suction ventilation can be shown to be the initial preadaptive situation (Sec. 3.3). Buck's theory of CO_2 and water retention must be corrected (Sec. 3.4) and replaced by a theory of diffusive-convective gas exchange of which basic formulations are given (Sec. 3.5).

The principles of adaptation of respiration and respiratory water loss to the environment discussed and analyzed theoretically for the steady state in Secs. 2 and 3 are sufficient to establish the following hypothesis: if adaptation has to avoid respiratory water loss,

1. spiracles should be closed during activity and rest until the capacitance and therewith the partial pressure (Sec. 1) of all tissues is minimized for oxygen and maximized for carbon dioxide. Therefore, gas exchange should be discontinuous especially at rest. A minimal metabolic rate should, in addition, guarantee maximal duration of the resulting spiracle constriction.
2. Diffusive-convective gas exchange should be optimized by minimizing the opening area and length of spiracles and maximizing the rate of air-inflow and the tracheal partial pressure of nitrogen. To guarantee anticurrent diffusion of water, a subatmospheric barometric pressure in the tracheal system should be interrupted only during regeneration of the higher carbon dioxide capacitance. The lower oxygen capacitance should persist, as sufficient oxygen uptake can occur during suction ventilation by cocurrent diffusion and convection.
3. If maximal CO_2 capacitance is reached, this state should either persist or a regeneration of the capacitances should be achieved as fast as possible. The latter would allow for a new constriction period. During regeneration convective gas exchange should be preferred. A cocurrent diffusive gas exchange of carbon dioxide should support convection only as long as the partial pressure difference of carbon dioxide is high.

The measurement of cyclic CO_2 release, tracheal pressure cycles, and cyclic weight loss in quiescent undisturbed insects clearly confirm the hypothesis given above, showing that resting respiration is a sequence of periods of the three types listed above. During diapause of pupae metabolic rate can be minimized under hormonal control (see Sec. 5.3, Chap. 3). This allows for the dangerous predominantly diffusive CO_2 release in the (C)FO cycles. Adult insects must be prepared by a higher basal metabolic rate to become instantly active. Therefore, convective gas exchange, being advantageous for water retention, supports CO_2 release in the shorter cycles.

Acknowledgements. The author gratefully acknowledges the valuable assistance and the helpful discussions and comments of the former students H.-D. Baaske and F. Killmann. Thanks are also due to all those who assisted in the preparation of the figures and the manuscript. Supported by grants from the German DFG.

Chapter 7

Water and Salt Relations

Frank Hevert[1]

Contents

1 Introduction: A Physical Approach

When considering the relations between an insect (as well as other living organisms) and the environmental water and ions, it is advantageous to start from a very physical point of view. Thus, the insect in its natural environment is a compartment with thermodynamic open-system characteristics containing an aqueous solution of different salts and bounded by a membrane with specific properties. Relative content of water and ionic concentrations are different from the environment. A more or less

1 Institut für Allgemeine und Spezielle Zoologie, Universtität Giessen, Stephanstraße 24, 6300-Giessen, FRG

Environmental Physiology and Biochemistry of Insects, ed. by K.H. Hoffmann

defined composition is specific for a species and a more or less pregnant deviation from this "physiological" composition is not compatible with life of the individual.

Since such systems are subject to forces as diffusion and osmosis, they have the tendency to dissipate by losing their high energy level. To minimize the perilous problems of this dissipation two evolutionary possibilities exist: One strategy leads to a very close correlation between a certain species and a certain habitat (the ecological niche), i.e., to an adaptation over long periods, the other leads to the capability to tolerate various and varying environmental conditions, i.e., enables an acclimatiziation within a short time.

In principle we are speaking of concentration differences or gradients between the insect and its environment which must obviously be generated and maintained by the living animal itself, requiring input of energy (cf. Chaps. 4 and 5). The mechanisms to do this are either passive (= structural) or active (= functional). The best example for a structural mechanism is the lowering of permeabilities and the adaptation of epithelial surfaces, the best example for a functional mechanism is active ion transport in conjunction with regulatory processes.

2 The Principal Pathways for Water and Ions: Importance of Control

Figure 7.1 illustrates this idea of the insect as a compartment in (with regard to the energetics it would be better to say "beside" or even "over") its environment. It is separated from the environment both structurally (integument, epithelia) and functionally (control mechanisms, regulatory mechanisms). The environement is the pool from which the insect gains water and ions and to which it loses water and ions,

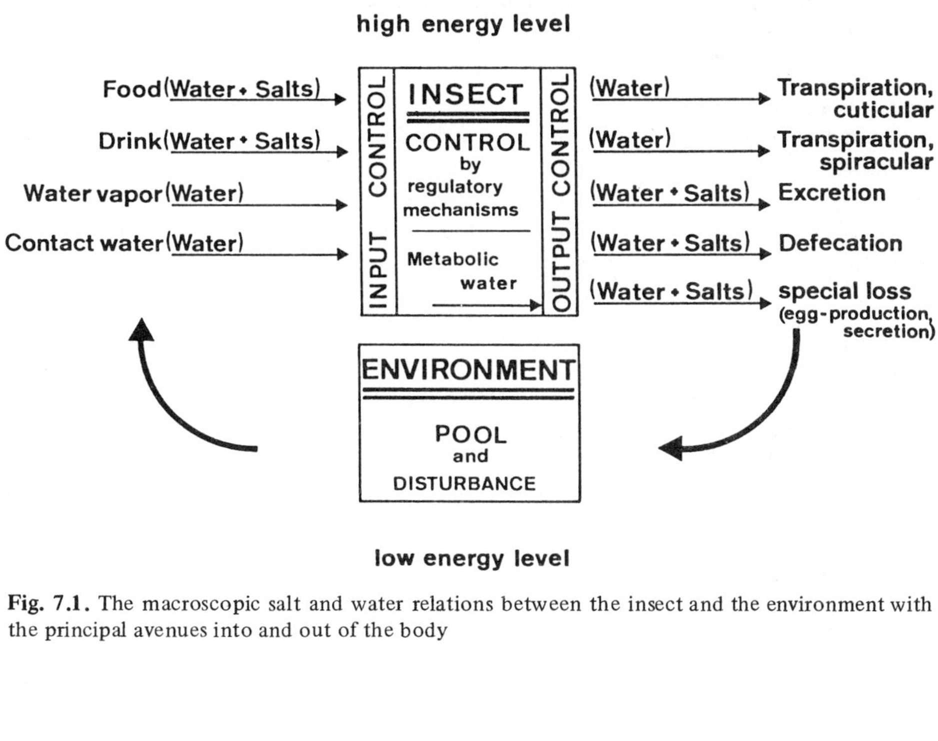

Fig. 7.1. The macroscopic salt and water relations between the insect and the environment with the principal avenues into and out of the body

and also the disturbance in this system, since the availability of water and ions is not constant.

However, this system becomes more complicated if we leave behind this simplifying physical approach and look at the living insect. It is certainly unsatisfactory to analyze only the input-output relations of a black box. The physiologist has to recognize that there are diverse avenues for water and ions into and out of the body and that there are diverse pathways and transports within the body. Most, maybe all, of these ways are either known, or believed, to be controllable by the insect (cf. Chap. 6). Although this statement is in opposition to Edney's (1980) opinion, who believes that the animal is not able to control water gain by food uptake, oxidative metabolism and water vapor absorption, it is important to understand the principle: the insect must be able to control the bulk of entering and leaving water and ions, thus, warranting over a period of time that all loss of water and ions is balanced by an equal gain or vice versa.

3 The Possible Habitats: Introductory Examples

Before now discussing the structures, mechanisms, and processes related to this control and regulation, it is helpful to get a general idea about the possible habitats and related problems in regard to gain and loss of water and salts by the insect.

Insects live very successfully in nearly every conceivable habitat with a view to the availability of water and ions. The range extends from water and ions in excess (I) (marine-like conditions) as in salt lakes and salt pits, to water in excess and ions in deficiency (II) (freshwater-like conditions), to water in deficiency and ions in dissimilar mixture (III) (land-like conditions), to complete absence of free water and deficiency of ions (desert-like conditions).

Table 7.1 gives some concrete examples and data to which we will refer in the further text.

A larva of *Aedes campestris* is able to thrive in salt waters with an osmolality in the range of seawater, maintaining an internal hemolymph concentration of 340 $mosmol \cdot l^{-1}$, but can also live at low salinities in the range of soft river water. Phillips et al. (1977) have demonstrated that these larvae can adapt to external concentrations that extend over a 500-fold range without changing the concentrations of the major ions within the hemolymph more than twofold. That means, the animal must be both a hyporegulator and a hyperregulator; regulatory mechanisms must be either switched on or off, if we do not suppose that one single pumping system will be able to operate in both directions.

Another kind of problem is found for the larvae of freshwater insects, for instance, *Sialis* with their high sodium concentration within the hemolymph which live permanently in river waters with low salinity, especially with extremely low sodium. The concentration factor for this ion in *Sialis* is over 270. A rough calculation of the chemical work (without considering permeabilities, Donnan potential, etc.) that would be necessary to sustain this gradient is given by

Table 7.1. Insect habitats with a view to the availability of water and ions

	Type of habitat		Examples	Composition							Author
	Water	Salts	Habitat / direction of gradient / insect (hemolymph) concentration factor	$mosmol \cdot kg^{-1}$	$mmol \cdot l^{-1}$						
				osmolality	Na^+	K^+	Ca^{++}	Mg^{++}	Cl^-	HCO_3^-	
I	Excess	Excess	Alkaline salt lake	800	500				50	450	Phillips et al. 1977
			Gradient	water: ↑	↓				↑	↓	
			Aedes campestris (larv.)	340	140				140	–	
			Factor		0.28				2.8		
II	Excess	Deficiency	River water	5	0.40	0.04	0.52	0.21	0.32	1.10	Livingstone 1963
			Gradient	water: ↑	↑	↑	↑	↑	↑	↑	
			Sialis lutaria (larv.)	339	109	5	7.5	19	31	15	Shaw 1955
			Factor		272.5	125.0	14.4	90.5	96.9	13.6	
III	Sufficiency	Dissimilarity	Fermenting fruits	Variable	4.5	42			17		Hevert 1974
			Gradient	water: ↕	↑	↓			↑		
			Drosophila hydei (larv.)	229	56	31	9		36		
			Factor		12.4	0.74			2.1		

$$W_{CHEM} = R \cdot T \cdot \ln \frac{(Na^+)out}{(Na^+)in}$$

and comes to the considerable value of 14 kJ · mol^{-1}.

Insects which thrive in habitats that can be described with regard to the osmotic and ionic conditions between land and brackish water, for instance in mushrooms, fermenting fruits, or decomposing leaves, are confronted with the third kind of problem: water content is in most cases sufficient, but rather unsteady; osmolality is very variable (consider what would happen, for example, to a fermenting pear during a long period of rain or after some days with hot and dry weather) and the major ions are in opposite ratio, namely, there is too much potassium and too little sodium. An example for this are the larvae of *Drosophila*. They are permanently running risk to lose sodium and to be deluged by potassium, consequently, they are forced to countertransport continuously these two ions.

As a last example in this compilation, the group of insects should be mentioned which lives under desert-like conditions. These are not only the real desert insects, such as some locusts, but also several beetles living in dry foodstuffs. For instance, the mealworm, *Tenebrio molitor*, is able to develop and complete its life cycle from egg to adult on flour with a water content of only 15% in equilibrium with 50% relative humidity without drinking free water at any time (Murray 1968); all water comes from this dry food either directly or indirectly by oxidative metabolism.

The above compilation of more or less extreme examples shows the wide range of environmental problems to which insects are exposed. In the context of this article it is important to point out that the insects must also have developed a wide range of mechanisms to overcome these problems, this certainly being one of the reasons of their great biological success.

But the picture is still incomplete; *Rhodnius*, such as other bloodsucking insects too, can survive many weeks or even months without any external supply of fluid. During such periods, they excrete, if at all, very little urine which is characterized by high potassium concentration and low sodium concentration. After a blood meal, the situation changes dramatically: they are able to increase the rate of fluid excretion 1000-fold (Wigglesworth 1931) and to switch within some minutes to a urine which is very rich in sodium (Maddrell 1962, 1963). The giant blood meal brings to the animal a large amount of water and deluges the body with sodium, which has to be eliminated as fast as possible.

Or consider an insect entering its flying phase, when it lowers its specific weight. Adults of *Pieris brassicae* lose nearly 50% of their total body weight within 3 h after emergence, corresponding to a loss of nearly 75% of their hemolymph volume (Nicholson 1976).

These two examples illustrate that insects are not only able to continuously tolerate extreme environments (*Aedes campestris* in $NaHCO_3$-waters) or to overcome slow fluctuations of salinity (*Drosophila* in fermenting fruits), but also to give a very fast response to sudden changes.

4 Hemolymph as a Regulated Tissue

Recently, Jungreis (1980) discussed the question whether insect hemolymph is a static (= passive) or dynamic (= active) tissue with regard to the metabolism of trehalose, amino acids, and ecdysone (cf. Chaps. 1 and 4). He concluded that the older view of Mellanby (1939) and other physiologists that "... the function of the hemolymph is to serve as a reserve ... which can be drawn upon as needed ... " is too static and has to be replaced by a more dynamic view which regards the hemolymph not as a reservoir or "sink" for metabolites, but compares it with mammalian blood as a highly dynamic tissue. Transferring this idea to the problem of water and salt balance we must put the question whether the hemolymph is a pool for water and ions from which the tissues can help themselves to maintain their own equilibrium against the environment or whether it is regulated itself. The question is whether the hemolymph is an extension of the environment into the body or an extension of the intracellular fluid compartment.

The literature describes many experiments with artificial external osmotic stress, which were primarily not carried out to answer this question, but can help to discriminate the two possibilities. The Cl^--concentration within the hemolymph of *Drosophila hydei*, for instance, responds in a characteristic way when the larvae are externally stressed either with low salinity (e.g., bathing the larvae in aqua bidest.) or with high salinity (e.g., bathing the larvae in 3.5% KCl). In both cases, oscillations can be observed which are very similar (Fig. 7.2) (Hevert 1974).

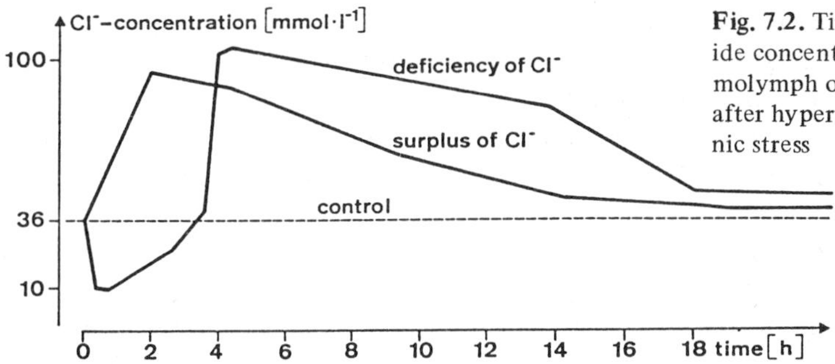

Fig. 7.2. Time course of chloride concentration in the hemolymph of *Drosophila hydei* after hypertonic and hypotonic stress

a) Low salinity stress, external deficiency of Cl^-.

1. In the first 10 min, the hemolymph chloride concentration falls off from 36 mmol $\cdot$ l^{-1} to 10 mmol $\cdot$ l^{-1}. This is likely due to diffusion, the organism loses Cl^- passively via integument, anal fields, gut, etc.
2. In the next 240 min, the chloride concentration rises again, not to the physiological value, but to about 100 mmol $\cdot$ l^{-1}. This is an overcompensation and doubtlessly an active regulation process. Since the external medium contains only the amount of chloride that has been released from the animal, the additional amount of chloride must have been released from inner depots.

3. In the next 17–20 h, the chloride concentration decreases slowly to about 45 mmol · l^{-1}. We assume chloride is both restored in depots and lost.
4. For several hours the larva maintains this new concentration of 45 mmol · l^{-1} which is significantly over the control.

b) High salinity stress, external surplus of Cl^-.

1. In the first 2 h, the hemolymph chloride concentration increases passively (inflow of Cl^-, loss of water) to about 100 mmol · l^{-1} [cf. with a (2)].
2. In the next 15–17 h, the chloride concentration decreases slowly to about 40 mmol · l^{-1} [cf. with a (3)].
3. For several hours the larva maintains this new concentration of 40 mmol · l^{-1} which is over the control.

Two conclusions can be drawn from this experiment: Firstly, hemolymph as a tissue is not passive, but reacts in a specific way. Secondly, hemolymph is regulated.

This second statement needs a little more explanation since one could believe that this increasing and decreasing of chloride concentration could also be an image of what happens with the cells. The most striking argument against this supposition is the fact that both the curves from the hypotonic and the hypertonic experiment come up to the same upper value of 100 mmol · l^{-1}. This finding indicates the existence of something like a receptor for Cl^- concentration (or a colligative property, such as osmolality, ionic strength, volume, hydrostatic pressure) within the hemolymph that triggers a regulation process if a certain threshold is reached. This type of regulation does not seem to be very effective since it does not react proportionally to the disturbance, e.g., wasting material and energy, but it enables the hemolymph to be a buffer between the changing environment and the "fine" regulation of the tissues.

5 The Role of Hormones

What really does regulate, what makes the chloride concentration increase or decrease? It is generally agreed that there are specialized organs, chiefly Malpighian tubules, gut with rectum, anal papillae, among others which are accountable for osmoregulation, but the answer to the above question is to be found at a lower complexity level, namely, at the cellular or even subcellular level. Ion pumps must be accelerated or retarded, possibly switched on or off, i.e., enzymes and/or metabolic pathways are to be influenced. From this, the questions arise, what is the messenger between these biochemical tools, localized within the cells of osmoregulatory organs and the ionic concentrations in the hemolymph, what is the connecting link between a "measuring receptor" and an "effecting regulation mechanism", are there hormones involved?

In comparison with the situation in vertebrates, our knowledge of this field is rather poor and goes back only some 20 years. Altmann (1956) and independent from him Nuñez (1956) were the first to demonstrate that water balance of insects were under some form of hormonal control. Altmann injected corpora cardiaca extracts into *Apis mellifica* and found that water excretion changed. Still clearer were the experiments of Nuñez who could demonstrate that the excretion in the beetle *Anisotarsus cupripennis* stops, if the circulation of hemolymph between the head and the rest of the body is interrupted, or if the nervous connections between the brain and the rest of the nervous system are cut, and that afterwards a rapid excretion began if he injected brain extracts or if he injected corpora cardiaca extracts.

From these pioneering experiments it was concluded that there must exist a diuretic hormone, released from the brain and/or the corpora cardiaca into the hemolymph and affecting the urine flow in the Malpighian tubules.

Recently, Maddrell (1980) gave a compilation of "insects for which there is evidence that the rate of fluid transport by their Malpighian tubules is regulated by hormones". There were until 1980, 19 species from Hymenoptera, Diptera, Lepidoptera, Coleoptera, Hemiptera, and Orthoptera and we may conclude that hormonal control of the activity of the Malpighian tubules is usual among insects. Only little is known of how this acceleration of fluid flow is achieved and about the nature of this diuretic hormone. However, the tubules of many insects are affected in in vitro preparations by 3',5'-cyclic adenosine monophosphate (cAMP) and Aston (1975) was able to show that the intracellular level of cAMP increased for a short time of some minutes from 0.6 pmol per tubule to 2.5 pmol per tubule when the tubules were treated with the diuretic hormone. In our laboratory it was shown that the addition of 5-hydroxytryptamine, which is known to stimulate the adenylate cyclase, to in vitro tubules is followed by an acceleration of urine flow (Wessing et al. 1978; Wessing and Rönnau unpubl. results). This leads to the possible concept of a hormone in which the intrinsic activity is mediated by a second messenger.

In locusts, the diuretic hormone is synthesized in cerebral neurosecretory cells and stored in special cells of the corpora cardiaca. From there it is released after a stimulus, e.g., feeding: starved *Schistocerca* have an average fluid secretion of 125 $nl \cdot min^{-1}$, after feeding for 2.5 h, fluid secretion increases to 460 $nl \cdot min^{-1}$. The injection of one pair of corpora cardiaca into starved locusts has nearly the same effect: the fluid secretion increases to 380 $nl \cdot min^{-1}$ (data calculated from Mordue 1972).

Meanwhile, this hormone (and this of *Glossina*, too) has been partially purified and seems to be a small polypeptide with a molecular weight of 1000–2000 Daltons. It is soluble in water and methanol and withstands boiling for 2 min (Aston and Hughes 1980). In *Rhodnius*, not only the rate of fluid secretion is stimulated by the hormone, but the ionic composition changes as well. In a bathing medium containing 100 $mmol \cdot l^{-1}$ Na and 50 $mmol \cdot l^{-1}$ K, the tubules secrete a fluid with 20 $mmol \cdot l^{-1}$ Na and 150 $mmol \cdot l^{-1}$ K when they are unstimulated. After stimulation with the hormone, they secrete a fluid with 100 $mmol \cdot l^{-1}$ Na and 85 $mmol \cdot l^{-1}$ K (after Madrell 1980).

In order to understand the biological meaning of this feature one must remember Wigglesworth's and Maddrell's observations (see Sec. 3 of this Chap.) that starved *Rhodnius* excrete very little, potassium-rich, urine and after one of their giant blood meals they excrete large amounts of a sodium-rich urine. The blood meal is the stimulus (we do not know exactly whether volume receptors are stimulated); the hormone is released from the corpora cardiaca. Urine flow is accelerated dramatically and the ion pumps are tuned to produce the sodium-rich urine. This leads to a rapid elimination of the high water and sodium load which comes from the blood plasma. Afterwards, *Rhodnius* begins to digest the erythrocytes, bringing potassium to the insect's hemolymph. In the meantime, the stimulus has disappeared, fluid secretion is again at its low level and the pumps are again tuned to excrete more potassium.

There is only little evidence for the existence of an antidiuretic hormone. A substance (so-called antidiuretic factor) has been isolated from the glandular lobes of the corpora cardiaca which is antidiuretic in as far as it increases the water uptake from the rectal lumen, but it has no effect on the Malpighian tubules. It is perhaps identical with the chloride transport stimulating hormone (CTSH), recently described by Phillips et al. (1982a,b) (see also Sec. 7.1.2 of this Chap.). Probably there is no need for the existence of an antidiuretic hormone to function with the Malpighian tubules since in the total absence of the diuretic hormone, the fluid production is yet at a very low level or even zero. For an extensive review of the literature on diuretic hormone in insects, the reader is referred to Phillips (1983).

6 The Structures and Principle Functions

6.1 The Malpighian Tubules

The Malpighian tubules are the main excretory organs of insects and thus, are chiefly involved in salt and water balance. They are tubular structures which lie either free in the hemocoel or in a particular connection with the rectum (cryptonephric complex), extending through the whole body cavity of larvae or only through the abdomen of adults. The distal end of each tubule is closed, the proximal end discharges, often via a short ureter, into the gut at the junction of midgut and hindgut. The terms "distal" and "proximal" are topographical in regard to the gut since the tubules are derivatives of the gut.

Among the diverse insect groups they are rather variable with regard to their number and length. The so-called oligonephria possess three to eight long tubules (e.g., *Drosophila*, such as most Diptera has two pairs, each tubule with a length ca. two times the whole body length); the so-called polynephria possess more than eight tubules (e.g., *Carausius*, such as other Orthoptera and Hymenoptera too has about 150 short tubules). Despite this variability, the relation of the overall tubulus surface and the insect's body mass seems to be rather constant. At the oligonephria, the tubules are often divided into two, three, or four morphologically distinguishable segments; for instance, *Drosophila* has a dilated distal segment (so-called initial segment, a short bent transitional segment and a proximal segment with a small lumen (so-called main segment), and a short ureter. This segmentation in *Drosophila* changes within the pupa and has disappeared after emergence of the adult fly (Wessing and Eichelberg 1978).

Instead of this segmentation, the polynephria often show polymorphism, i.e., they have different kinds of tubules (e.g., the "white tubules" and the "yellow tubules" in *Gryllotalpa* (L'Honoré 1972)). The epithelium of the tubules consists of one single cell layer, in most cases of transport active cells only, in several insects in association with some muscle fibers and tracheoles. The cells of the tubulus are characterized by a strongly marked polar structure as it is typical for transporting epithelia. Each cell is divided into (1) a basal region, facing the hemolymph and equipped with a ramifying network of deep infoldings of the basal plasmalemma and coated by a basement lamina; (2) an intermediate region with the nucleus, Golgi fields, and a distinct rough endoplasmatic reticulum, often in connection with vesicles and storage vacuoles and often in connection with the channels of the basal infoldings; and (3) an apical region, facing the urine with a brush border of more or less numerous and elongate microvilli. The mitochondria are situated mainly in the basal region between the infoldings and in the apical region, sometimes extending into the microvilli. In most insects, the cells are joined by septate junctions which are mostly pleated, but sometimes smooth, too.

Since the tubules are closed towards the hemolymph and since there is no hydrostatic pressure gradient to drive an ultrafiltration, the fluid production is achieved by secretory processes, i.e., the flow of water and solutes is consequent upon ion movements and we shall, therefore, first consider these ion transports.

Ramsay demonstrated in his pioneering papers on the tubules of the stick insect *Dixippus* (now *Carausius*) that ion potassium plays the predominant role among these ion transports (Ramsay 1953, 1954, 1955, 1958). He found in this insect and later in *Locusta migratoria, Pieris brassicae, Dytiscus marginalis, Tenebrio molitor, Aedes aegypti, Rhodnius prolixus*, and a tabanid larva (undet.) that the urine, when entering the hindgut has a potassium concentration which is 4–20 times as high as the potassium concentration within the hemolymph. Since he found the lumen of the tubules to be electrically positive in regard to the hemolymph he concluded that the translocation of potassium from hemolymph to urine is thermodynamically uphill. However, the potassium transport is active and electrogenic and we shall consider it to be the "prime mover" in fluid flow. Ions with a negative charge can follow without requiring primary energy, thus, achieving a transport of salt, mainly of KCl towards the lumen. In consequence upon this salt transport and mediated by osmotic coupling, a transport of water takes place and this water transport is able to bring by solvent drag small molecules into the lumen. All this results in a more or less isosmotic "ultrafiltrate" of the hemolymph, which is very rich in potassium. Some new ideas about the nature of the molecular mechanisms which are involved in this transport system and about the realization of the salt to water coupling will be discussed in Secs. 7.1 and 7.2 of this Chapter. The structural and functional principles described here are summarized in Fig. 7.3.

6.2 The Rectum

It was shown in Sec. 6.1 that the fluid coming from the Malpighian tubules and entering the hindgut is in most cases isosmotic to the hemolymph and considerably rich in potassium. If this urine would leave the body, the insect would have lost most of its water and potassium within some hours. The duty of the hindgut and especially of the rectum is, therefore, to reabsorb the bulk of this water and potassium. We, thus, can consider the excretion and related osmoregulation to be a two-step process with partial recycling as shown schematically in Fig. 7.4a,b.

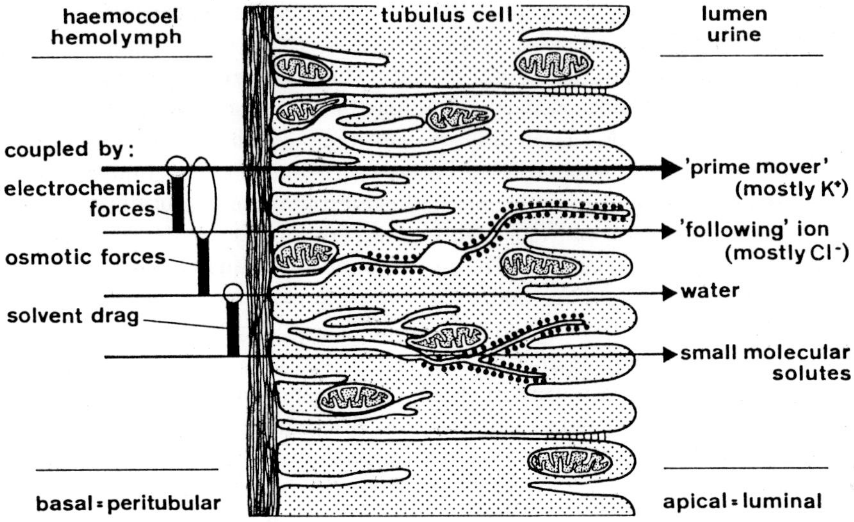

Fig. 7.3. The structural and functional principles of a Malpighan tubulus cell

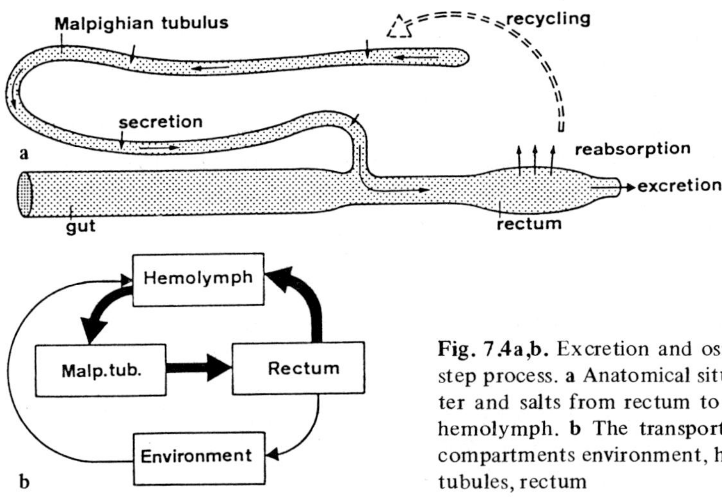

Fig. 7.4a,b. Excretion and osmoregulation as a two-step process. **a** Anatomical situation, recycling of water and salts from rectum to Malpighian tubules via hemolymph. **b** The transport relations between the compartments environment, hemolymph, Malpighian tubules, rectum

The morphology of the rectal epithelium shows a wide range of variability among the insects and the complexity of the morphological structures depends on the extent to which water has to be absorbed to concentrate the feces, hence, in the environmental situation (remember for instance *Tenebrio*, Sec. 3). The range extends from the complete absence of any specialized structure (*Drosophila* larvae) to the so-called rectal pads in the Dermaptera, Orthoptera, and some carabid Coleoptera to the "most complex ... transporting epithelia yet studied" (Berridge and Oschman

1972), namely, the rectal papillae in some Diptera and adult Lepidoptera (not to be confused with the anal papillae of mosquito larvae, see Sec. 6.4).

The absorptive cells of both the pads and the papillae are quite different from the transport active cells forming the Malpighian tubules. Their basal membrane is rather plain, lacking infoldings and borders a hemolymph sinus between the rectum and a muscle layer; the apical membrane is well equipped with a regular brush border lined by the rectal cuticle, but the predominant structure is an extensive membrane elaboration on the lateral membrane. This lateral membrane is highly folded into series of stacks of parallel membranes interdigitating with the neighboring cells and these membrane stacks are closely associated with numerous mitochondria. Both ends of the intercellular compartment formed by the interdigitations are closed by long septate desmosomes. Beside these interdigitations other deep intercellular spaces exist, often with a tracheole and open towards the hemolymph sinus, but closed towards the rectal lumen.

The idea that the rectal epithelium reabsorbs potassium and water is a very rough description of what really happens. Phillips (1964) demonstrated 20 years ago that as a result of selective reabsorptions in the locust rectum, the levels of potassium, chloride, and sodium can vary from 1–10 mmol $\cdot$ l^{-1} in water-loaded individuals to 200–600 mmol $\cdot l^{-1}$ in salt-loaded individuals. This points out clearly that the rectal transport activity is highly regulated by mechanisms depending on the environmental load and that (at least in the desert locust, *Schistocerca gregaria*) the rectum is the most important osmoregulatory organ. There is evidence that this regulation is under control of both neural and endocrine mechanisms involving the CTSH (see Secs. 5 and 7.1). From in vitro preparations of recta (sheet preparations in Ussing chambers) we know today that there are active reabsorptive mechanisms for Cl, Na, K, H/HCO_3, phosphate, acetate, and neutral amino acids (Andrusiak et al. 1980; Phillips 1980; Baumeister et al. 1981; Hanrahan and Phillips 1983) and that opposite to the situation in the Malpighian tubules, the prime mover is an active electrogenic transport of Cl^- from the gut lumen into the rectal pad cell in combination with an ATPase-mediated antiport of Na^+ and K^+ at the lateral membrane stacks. A more detailed description of the appropriate molecular mechanisms will be given in Secs. 7.1 and 7.2, the structural and functional principles are summarized in Fig. 7.5.

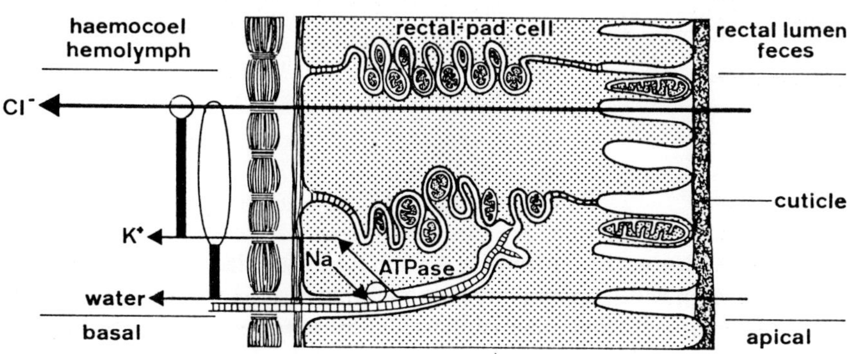

Fig. 7.5. The structural and functional principles of a rectal pad cell

6.2.1 The Cryptonephric Complex

The larvae of some Lepidoptera and mainly some Coleoptera living in extremely dry foodstuffs, such as flour or bran, show a very complex structure to reabsorb water with high efficiency from the feces. With their cryptonephric complex, they are enabled to produce fecal pellets resembling a "bone dry powder" (Wigglesworth 1934). The cryptonephric complex is a compound structure formed by the rectum and the distal ends of the Malpighian tubules. Both transporting tissues are intimately associated and enclosed by a cryptonephric membrane within a special chamber, the perinephric space (Fig. 7.6).

For a detailed description of the anatomy and the fine structure, the reader is referred to Grimstone et al. (1968) and Wall and Oschman (1975).

This anatomical peculiarity causes by the addition of a further compartment, the cryptonephric (= perirectal) space, an enhancing of the physiological complexity. The above described recycling of potassium and water (Sec. 6.2) is facilitated by short-circuiting the pathway from the rectum over the hemolymph back to the tubules. The schematic presentation in Fig. 7.6 is to be compared with Fig. 7.4.

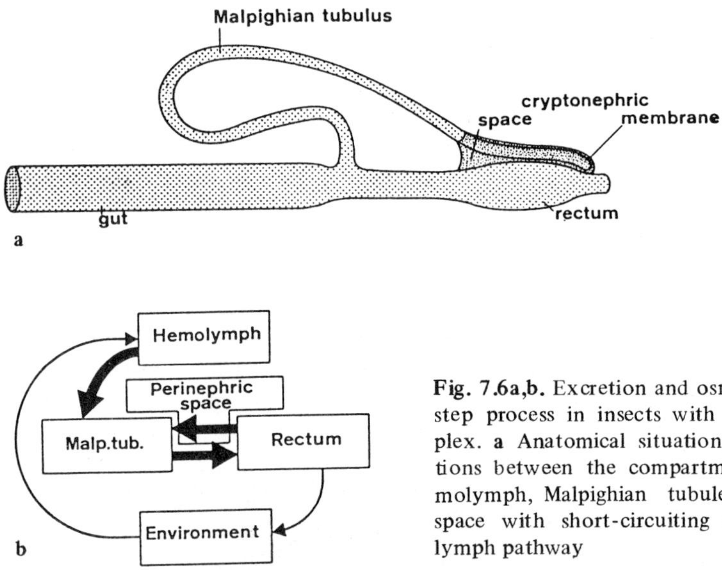

Fig. 7.6a,b. Excretion and osmoregulation as a two-step process in insects with a cryptonephric complex. **a** Anatomical situation; **b** the transport relations between the compartments environment, hemolymph, Malpighian tubules, rectum, perinephric space with short-circuiting of the rectum-hemolymph pathway

6.3 The Midgut

Recently, Harvey (1980) resummarized Sir Arthur Ramsay's classical view of osmomineral regulation in insects: "Ramsay ... proposed that Malpighian tubules function like vertebrate glomeruli in that their wall is permeable to most molecules smaller than proteins and that such small molecules are swept into the tubular lumen from the blood along with the fluid, which follows active potassium and sodium transport.

He proposed that the rectum functions like kidney tubules in that it selectively reabsorbs certain molecules and like kidney collecting ducts in that it has the capacity to reabsorb water or let is pass out depending on specific adaptations and conditions. According to this view the Malpighian tubule – rectal system is the primary regulator of the composition of the blood."

This concept has been generally agreed upon by insect physiologists and much proof has been accumulated. However, this feature is to be completed. Today, there is evidence that in some insects the tubules both secrete and reabsorb and that in some insects other additional organs, mainly the midgut, are involved in both secretion and reabsorption processes.

However, when considering the transport activity of the midgut epithelium, one must sharply discriminate between the normal "columnar cells" and the "goblet cells".

6.3.1 The Columnar Cells

The main cell type of insect midgut (= ventriculus) are the columnar cells with highly folded membranes on all sides. The apical microvilli are very uniform and are arranged in a close hexagonal array like in a honeycomb. Both basal and lateral membranes are so folded that they are difficult to distinguish; they form a basolateral membrane as is found in the vertebrate nephron. These columnar cells are absorbtive cells. They transport, as the cells of the vertebrate intestine, firstly, amino acids and sugars into the hemolymph, but also ions (sometimes coupled to the sugars) and water. Consequently, they are involved in osmomineral balance.

6.3.2 The Goblet Cells

Between the above described normal columnar cells, the midgut of larval Lepidoptera and Trichoptera has the remarkable goblet cells. They are characterized by a large inner cavity, the goblet, formed by a deep invagination of the whole apex with its microvilli, which nearly reaches the basal region of the cell. They connect their neighboring columnar cells at the apical side by both gap junctions and septate desmosomes, whereas the lateral membranes towards the basal region lack any junctions and form together with the basolateral membranes of the columnar cells a "third comparment" between the hemolymph and the cell.

These goblet cells transport potassium from the hemolymph to the lumen of the gut, thus, excreting abundant potassium from the plant diet of these phytophageous larvae, i.e., they function similar to the epithelial cells of the Malpighian tubules. Harvey and Nedergard (1964) found this second evidence for an active potassium pump in insects in the midgut of the larva of *Hyalophora cecropia.* Later it was measured and calculated at the midgut of *Manduca* that this potassium pump seems to be the most effective ever present in insect epithelia. It transports potassium against a high electrical gradient of more than 100 mV equivalent to a short circuit current of 1,200 $\mu A \cdot cm^{-2}$ (compared to stimulated salivary glands of *Calliphora*: 400 $\mu A \cdot cm^{-2}$; Harvey 1980 and Malpighian tubules of *Drosophila*: 880 $\mu A \cdot cm^{-2}$;

Hevert 1984, unpubl.). This potassium pump resembles in some respect the pump in the Malpighian tubules, but in other respects, it has some strange properties: it is strictly electrogenic in that it does not require any other ion to be transported with or against, it is not linked to a water transport, and it is not affected by the inhibition of a Na/K-ATPase by ouabain. However, we will discuss later in Sec. 7.1 the idea that these two pumps are not really different, but probably only different parts of a compound potassium pumping system in insect epithelia.

6.4 The Anal Papillae

In Sec. 2 we have discussed the capability of insects to live in waters with extreme low salinity, e.g., soft river water. Like a freshwater fish or crayfish, they need structures and mechanisms to replace a permanent passive loss of salts to the dilute environment, namely, salt absorbing organs. Such an organ can be on the one hand, the whole integument or gills (see Chap. 6) and on the other hand, specialized abdominal appendices, such as the anal papillae. They have been studied best in the larvae of *Aedes* and *Culex* (Copeland 1964; Stobbart 1967; Phillips et al. 1977). The ultrastructural findings of the transporting cells are very similar to those in crustacean gills and also show a certain analogy to the Malpighian tubulus cells: both the basal membrane and the apical membrane are highly folded and mitochondria are associated with these foldings. The basal membrane is coated by a considerable thin lamina basalis facing the hemolymph, the apical membrane is lined by a delicate cuticle which functions as a molecular sieve, and contacts directly the environmental water.

The physiological function of the salt absorbing cells is not yet as clearly understood as it is with the Malpighian tubules, the rectum, and the midgut, and the literature is still a little confusing. This may be explained by the fact that there are strictly freshwater inhabitants (*Aedes aegypti*) with papillae and others which can either live in freshwater or saline water (*Aedes campestris*).

The most probable transport system seems to be an antiport of sodium and ammonium or perhaps sodium and hydrogen ions at the apical side, likely not requiring primary energy, but driven by an active sodium uptake at the basal membrane (Na/K-ATPase). In this context it is of interest to look at a finding of Phillips (1977): "When saline-water larvae (add: of *Aedes campestris*) are transferred from hyperosmotic media to fresh water, no significant influx of ^{22}Na via the anal papillae is observed initially. However, over a period of 5–15 days, a large influx of this isotope through these organs develops ...". This could be explained by an induction of a "sleeping" ATPase activity or, because of the rather long delay time, more likely by a de novo synthesis.

7 The Mechanisms

It has been discussed (Secs. 1 and 2, this Chap., Chap. 4) that all mechanisms being accountable to produce the desirable discrepancy between the environment and the milieu interieur require energy in that they work against the dissipation of gradients. The connecting links between metabolic energy and transport of substances are the "pumps", molecules or molecule complexes, associated with membranes and having affinity sites for both energy delivering molecules (e.g., ATP) and ions to be translocated (e.g., Na^+ and K^+). A molecule complex with such properties is for instance the ubiquitous Na^+/K^+-(Mg stimulated)-adenosine-triphosphatase (Na/K-ATPase). Beside these pumps, which are called primary active since they use up directly metabolic energy, there are "carriers" which are called secondary active because they are also able to transport substances against gradients, but not by means of primary energy, but by spending electrochemical gradients generated by the pumps.

Moreover, there are passive pathways, namely, "conductivities" and diverse physical coupling forces. An extensive compilation and analysis of all these transports in terms of the thermodynamics of irreversible processes is given by Heinz (1978).

Most of our knowledge on these transport mechanisms comes from vertebrate tissues, mainly from the mammalian nephron, but it has been revealed that the basic molecular processes in as far as we understand them at the present time are not related to a species, but are principal evolutionary adaptations of living cells to maintain their balance with regard to the environment.

7.1 Ion Pumps and Transport Models

Insect transporting epithelia have four principal primary active ion transport systems:

1. the potassium pump (= the electrogenic potassium pump = the electrogenic alkali metal ion pump);
2. the sodium-potassium exchange pump;
3. the chloride pump;
4. calcium, magnesium pumps (?).

Before considering how these different pumps are involved and how they are composed to form the above described functions (Sec. 6) of the regulatory organs, it is advantageous to give a brief insight in their general localization among insects, their main operation, and biochemical characterization.

1. The potassium pump is present in the apical membrane of the goblet cells, on the apical membrane of Malpighian tubule cells, on the apical membrane of the secretory cells in salivary glands, and with uncertain or unknown localization in the rectum, in some labial glands, in the integument during molting and in some sensory cells. This pump is the most characteristic one for insects.

It is electrogenic in that it transports one single monovalent ion with positive charge out of the cell without being followed by a co- or counter-ion, thus, leaving the inside of the cell electrically negative. This ion is in most cases potassium, but it can also be under certain *natural* conditions sodium (in *Rhodnius,* see Sec. 5, and also in the tsetse fly, *Glossina*) and it has been shown that it takes under *experimental* conditions all other alkali metal ions depending on their availability. If the alkali metal ions are given in competition, the sequence of affinity is: K > Rb > Cs > Na > Li (Zerahn 1978). The biochemical nature of this pump is unknown at the present time, but there is growing evidence that it could be a potassium activated, ouabain insensitive, ATPase (Wieczorek 1982) somehow correlated to the so-called portasomes.

2. It has been discussed for some time whether this classic biological pump is present at all in insect transporting epithelia or not, this is due to the unusual difficulties to show oubain sensitivity of ion transports which we find in insects.

However, it has now been demonstrated on the basal or basolateral membrane of the cells of the rectum, salivary glands, labial glands, and in some, but not all Malpighian tubules.

It is also electrogenic in that it countertransports the two monovalent ions sodium and potassium in stoichiometric relation not equal to 1. Probably in all cases, three Na leaving the cell are replaced by two K entering the cell, thus, leaving the inside of the cell electrically negative as it does the potassium pump, too.

The biochemical nature is rather well-known. It is identical with the Na/K-ATPase, an enzymatic complex within the cell membrane with a binding site for inorganic phosphate and a second binding site for either Na or K depending on a conformational state. It is selectively inhibitable by ouabain (= g-strophantine).

3. In vertebrates, chloride is mostly cotransported with sodium or potassium or with sodium and potassium by means of secondary active carriers. There is evidence that such a symport also exists in insects, but there is also an additional primary active chloride pump which is after our present knowledge a "... novel Cl^--pump, different from models for vertebrates" (Phillips 1983). This electrogenic pump was first clearly demonstrated and described by Hanrahan and Phillips (1982) in the locust rectum.

It is located at the apical membrane, it is stimulated by low concentrations of potassium, but not affected by Na, Mg, Ca, HCO_3, and pH 6-8, and it has a high selectivity for Cl^-, only Br^- can partially substitute. Both the source of the primary energy and the biochemical nature are unknown.

4. There are disperse communications in the literature dealing with a possible active transport of calcium and magnesium. We know nearly nothing about these pumps in insects. Wood and Harvey (1976) speculate that since possibly each living cell has the capacity to extrude calcium, only the area-asymmetry of apical and basal membrane cause an asymmetry of pumping sites, thus, leading to a net Ca^{++} transport through the epithelium.

Beside the described pumps there are hints for active transports for sulfate, phosphate, and several organic anions (reviewed by Maddrell 1977).

The following Secs. 7.1.1 and 7.1.2 shall deal with two concrete examples and describe two models to show how such ion pumps are put together at the cellular level in order to realize a certain transport task.[2]

2 The experimental background for 7.1.1 comes from our laboratory in Giessen, FRG.

7.1.1 The Composed Potassium Transport in the Malpighian Tubules of *Drosophila*

As described in Sec. 6.1, the excretory and regulatory activity of the Malpighian tubules is consequent upon the transcellular movement of the ion potassium. The following model explaining how this movement could be achieved has been developed on the basis of experimental data from *Drosophila hydei* (larv.), but it is likely to be valid for others also. The measurements on potassium activities and electrical potentials and the calculated electrochemical potentials and related thermodynamic gradients are compiled in Table 7.2.

Table 7.2. Compilation of electrical and chemical parameters with regard to potassium transport at a cell from the proximal segment of the Malpighian tubules of *Drosophila hydei*

Compartment / Parameter[a]	Hemolymph		Proximal cell		Lumen
a_{K^+} [mmol · l^{-1}]	27		150		240
PD [mV]		–42		+ 91	
ED_{K^+} [mV]		+ 2.2		+102.8	
Therm · grad · K^+		Uphill		Steep uphill	

[a] a_{K^+} = potassium activity in a compartment; PD = electrical potential difference across a membrane; ED_{K^+} = electrochemical potential difference across a membrane for potassium; therm · grad · K^+ = direction and slope of the thermodynamic gradient for potassium

The crucial experimental findings are briefly summarized in the following:

1. Potassium is concentrated from hemolymph to cell approximately five times and from cell to urine again two times.
2. The resulting tenfold transepithelial K^+-gradient is affected by ouabain, the gradient does not completely disappear after ouabain treatment, but drops to half.
3. The electrical potential difference across both the basal and the apical membrane are depolarized by ouabain at low concentrations. The depolarization is biphasic in both cases.
4. Fluid secretion after ouabain treatment does not stop, but decreases to the half.

From these results it is confirmed that potassium is transported actively from hemolymph to urine, indeed, not only the step out of the cell, but also the step into the cell must be active. Moreover, it is evident that a ouabain sensitive Na/K-ATPase is involved.

Since this ATPase is located basally, it brings potassium actively from the hemolymph into the cell and since it is electrogenic it contributes to the electrical potential. The electrogenic metal ion pump is located on the apical membrane and pumps potassium out of the cell, thus, contributing also to the generation of the electrical potential.

A transport model for the movement of potassium must, of course, take into consideration both pumps. A sole cooperation of these two pumps could indeed explain the potential profile and the K^+-concentration profile under control conditions, but not the fact that potassium is still transported after blocking the Na/K-ATPase, since the electrogenic pump would not get fresh supply.

From other experiments we know that the entrance step of potassium into the cell is somehow coupled to an entrance of sodium and chloride. This leads to the idea of a carrier-mediated secondary active potassium transport into the cell, which is driven by a sodium gradient generated by the Na/K-ATPase. But since it is coupled to the Na/K-ATPase it also stops after ouabain treatment and is also not suited to explain the resting potassium flux. That means one is forced to suppose a third way for potassium transport at the basal membrane, which must be passive, namely, a conductivity. Under physiological conditions this conductivity causes a slight outflow of potassium, back to the hemolymph along its electrochemical gradient. After inhibition of the Na/K-ATPase, when this gradient becomes inverse, this leak becomes an entrance gate for potassium which can sustain the luminal electrogenic pump.

The first reaction after ouabain application is that the Na/K-ATPase stops; the basal electropotential begins to depolarize. But the sodium gradient is still persistent and potassium still comes into the cell by the carrier. Since the driving sodium gradient is no longer maintained, it leads to an additional depolarization of the potential difference.

Now, as a second reaction, the carrier stops, transepithelial resistance increases, and potassium entry via this gate is no longer possible. This leads to the observed repolarization of the potential difference. Meanwhile, the activities inside and outside the cell have shifted to a new equilibrium, thus, the electrochemical gradient for potassium becomes inverse and potassium begins to enter passively into the cell (see Fig. 7.7).

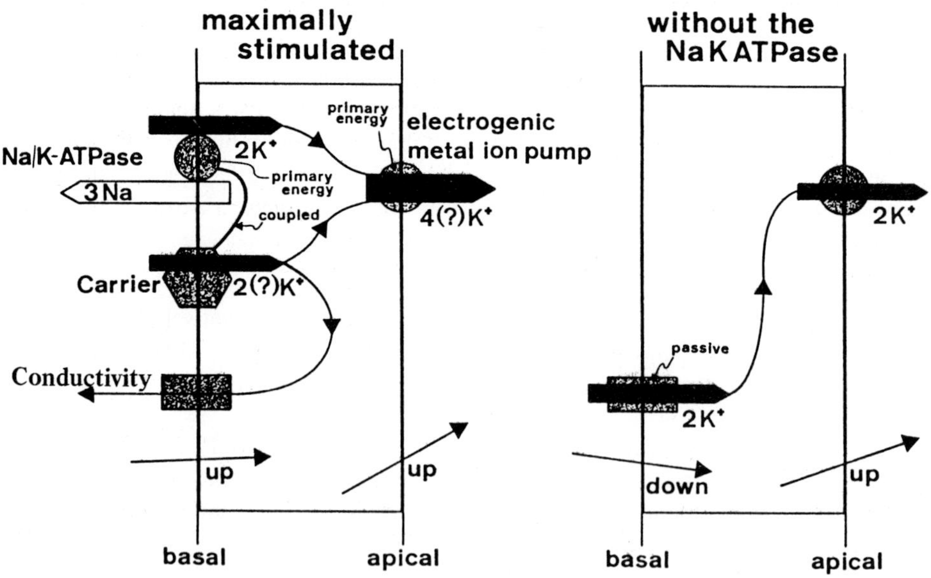

Fig. 7.7. Model for the movement of potassium through a cell of the Malpighian tubuels of *Drosophila*. All explanations are given in the text

We have described a compound potassium pumping system which is redundant at the input side. It is capable of translocating potassium when the apical electrogenic metal ion pump works together with the Na/K-ATPase, or when the apical pump works together with a conductivity (or, of course, with both of them).

Such a "more way"-system would be very appropriate from a biological point of view, since it has a large adaptability to varying conditions. We have learned that there is a possibility to regulate the transport rate by using different pathways through the cell.

7.1.2 The Active Transport of Chloride in the Rectum of *Schistocerca*

The rectum of terrestrial insects transports ions and water from the feces to the hemolymph, i.e. they reabsorb. A basally located Na/K-ATPase is involved which extrudes sodium out of the cell and generates an electrical potential difference with a positive hemolymph side. This gradient favors the passive outflux of chloride by a conductivity pathway. But how does chloride enter the cell from the rectal lumen? Since the cell is negative with respect to the lumen, there is an opposing electrochemical gradient which must be overcome. Phillips and co-workers failed in demon-

Ion Activities (mM)			
	lumen	cell	hemolymph
Na^+	75	9	75
K^+	7.2	70	7.2
Cl^-	82	47	82
mV	+64	0	+34

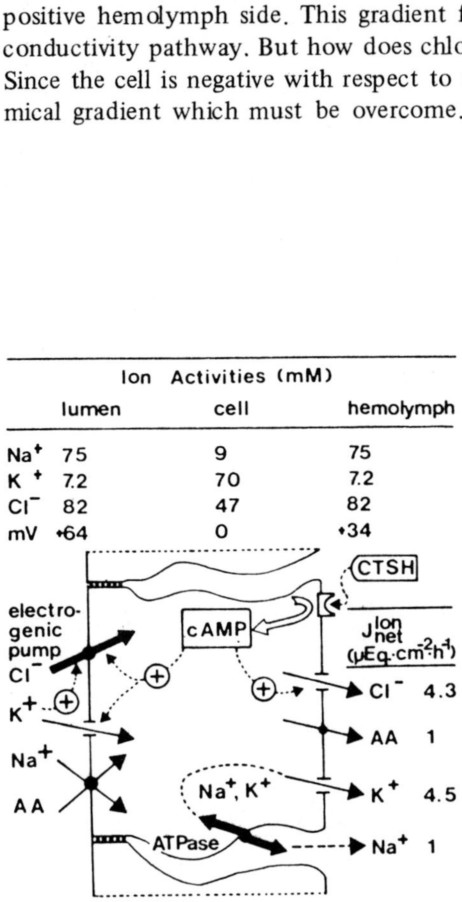

Net Electrochemical Potentials (mV)		
	apical	basolateral
Na^+	127 (favoring)	127 (opposing)
K^+	12 (favoring)	20 (favoring)
Cl^-	50 (opposing)	20 (favoring)

Fig. 7.8. Model for the movement of ions through a locust rectal pad cell. Structural complexity of the basolateral cell borders is ignored. An electrogenic Cl^- pump moves this anion actively into the cell from the lumen, while a typical Na/K-ATPase is the only primary transport mechanism in the basolateral membrane (*thick solid arrows*). Carrier-mediated processes (i.e., dissipative or secondary active transport) are indicated by *thin arrows* and *solid circles* at the membrane. Diffusive net movements through conductive pathways (possibly ion-selektive membrane channels) are indicated by *arrows through gaps* in the membrane. The specific points of stimulation by luminal K^+ and by intracellular cAMP, which acts as a second messenger for CTSH hormone, are indicated by *dotted lines*. This model is based on radiotracer flux studies (J^{ion}_{net}) and measurements of intracellular ion activities and electrical potentials (*upper table*) from which net electrochemical differences *(lower table)* were calculated favoring or opposing the observed net flux across apical and basolaternal cell borders. Data are for recta bathed bilaterally in a sline resembling locust hemolymph and in the open-circuit state, except for Na^+ (short-circuit state). Further explanations are given in the text. (From Phillips 1983)

strating any secondary active process in the rectum of *Schistocerca*, i.e., they were able to exclude an apical coupling to an electrochemical gradient. Therefore, they were forced to take into consideration a primary active transport, involving an anion-stimulated and anion-transporting ATPase (Fig. 7.8).

However, suggestions on the existence of a Cl^--ATPase are still controversial (critically reviewed by Gerencser and Lee 1983).

7.2 Water to Salt Coupling

In general, the transport of water in animal systems is consequent upon primary transport of a solute with water following due to osmotic forces. (There are two exceptions: water transport by hydrostatic pressure differences as in vertebrate glomeruli and water vapor condensation at especially synthesized organic compounds.) There is no problem, if water follows the osmotic gradient from one compartment to another, but there are problems if water is transported isotonically (as it is in most Malpighian tubules) or even against an osmotic gradient (as it is in recta). At least the latter transport is phenomenologically uphill. Two models have been developed to explain the apparent contradiction.

7.2.1 The Curran Hypothesis

The Curran hypothesis (Curran 1960), also called the three-compartment theory, is based on high structural complexity, that means if is based on the implication of more than one membrane (mainly two) with different partitioning properties between the compartments of interest, thus, forming more than two (mainly three) compartments. It will be best understood by considering an artificial model (Fig. 7.9).

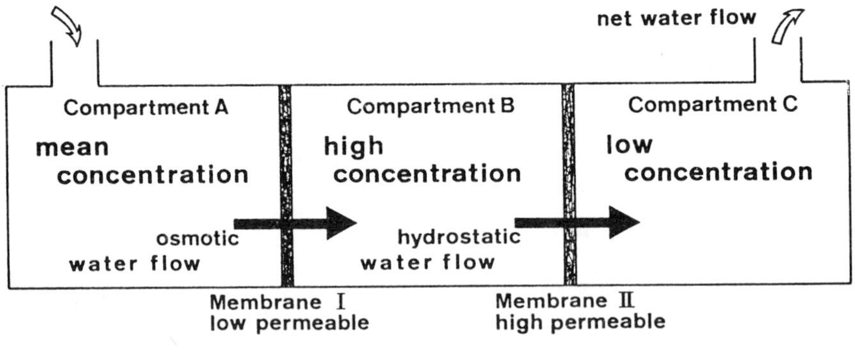

Fig. 7.9. Model to explain water transport from high to low concentration. Two compartments **A** and **B** are separated by a water permeable membrane; since the osmotic concentration in **B** is higher as in **A**, water is transported osmotically from **A** to **B**. A third compartment **C** is separated from **B** by a membrane that is more permeable to water as the membrane between **A** and **B**. Thus, water is transported from **B** to **C** by means of hydrostatic pressure

This model can easily explain the water transport in the cryptonephric complex (Sec. 6.2.1), and as shown by Diamond and Bossert (1967) with their "standing-gradient osmotic flow" theory, it may be also valid to explain the fluid flow through the cells of the Malpighian tubules. The additional compartment is in this case the space between the microvilli and the basal infoldings, respectively, where the higher ionic concentration is generated and sustained by ion pumps.

However, the Curran hypothesis is not suited to explain uphill water transport through membranes lacking microvilli or similar structures.

7.2.2 The Electroosmosis Hypothesis

This hypothesis is also based on a structural peculiarity. The electroosmosis needs a membrane with inside-charged ion channels. (For a detailed mathematical description see Hill 1975; for a concrete application see Küppers and Thurm 1980.)

The passage of ions with different charge through such channels results in a diffusion potential due to mutual effects between the charged particles and the charged channel wall. This diffusion potential is the force which drives water through these channels, either isosmotically or even against the osmotic gradient. The electroosmosis seems to be realized in insects living in extreme dry habitats.

8 Conclusions

We have started with a review on some extreme habitats with respect to the ion and water balance to which insects are adapted or can acclimate. This is due to a great many well-regulated and manifold cross-linked transporting processes related to a variety of morphological structures. From the evolutionary point of view, we can confirm that this is the material which has enabled the successful insects to make use of nearly every conceivable ecological niche. At the present time we are unable to decide whether the picture is going to become still more complicated or whether it is ripe for greater exploitation.

However, insect physiology and physiology of transport in particular with regard to ecological problems is and will be in the future a challenge to uncover more details about the linkage between animal and environment.

Acknowledgements. I am grateful to A. Wessing, K. Rönnau, and other colleagues for references to partially unpublished data and for helpful discussions. The model in Sec. 7.1.2 is reprinted from Phillips (1983) with kind permission of Dr. J. Phillips and Gustav Fischer Verlag, Stuttgart and New York.

Chapter 8

Color and Color Changes

Klaus H. Hoffmann[1]

Contents

1 Introduction

The description of insect coloration has long been an attractive field for entomologists. Coloration includes both color, which describes the absorption and reflectance of pigments and specific cuticular structures, and pattern, which describes the distribution of chemical and physical colors (Sec. 2). The colors of animals are mostly determined either by the requirements of camouflage and deception or advertisement. Many insects have patterns of colors and tones that so match the background that they are indistinguishable even when in the immediate foreground. In others,

1 Allgemeine Zoologie (Biologie I), Universität Ulm, Oberer Eselsberg, 7900 Ulm, FRG

Environmental Physiology and Biochemistry of Insects (ed. by K. H. Hoffmann)

strongly contrasting colors break up the continuity of surface and recognizable outline so that the appearance of form is destroyed. Sometimes both color and form exactly resemble inedible objects in the environment, such as leaves, twigs, and stones, so that the insects pass unnoticed by predators. Other animals escape attack because they mimic the bright color patterns of dangerous species and so "falsely trade upon an evil reputation" (Hinton 1976).

An extremely high variability in coloration can be found among acridoid grasshoppers in temperate regions (for review see Rowell 1971) and also in many cicadellid leafhoppers (Müller 1974). Such a variability may be due to differences in the genotype or to phenotypic differences (modificability) derived by the interaction of a given genotype with varying environments (e.g., color of the surroundings, temperature, humidity, diet) (Ibrahim 1974; Helfert 1978).

In contrast, animals that pass their existence concealed from view and from environmental variations in the soil, or as endoparasites, are more or less colorless. Color, like any other feature, will be selected against and in time disappears when it ceases to have a selective advantage (see also Sec. 4.2).

The holometabolous insects show a marked metamorphosis and different development stages which may occur in quite different environments. These insects often change color during development simply because at different times during development they are subjected to different selective pressures. If their color change is only part of the normal development it will be called an ontogenetic color change (Bückmann 1974a) (Sec. 3.4). Many insects, however, can also change their coloration in response to external stimuli by changes in pigment concentrations, that means as the result of morphological color changes (see Secs. 3.2 and 3.3). In only a few insect species can relatively rapid physiological color changes be observed, which are associated with the site of colored granules within the cell (Sec. 3.1). Insects do not possess the same type of specialized "pigment cells" as vertebrates. In insects the functions of pigment cells are carried out by the common epidermal cells or underlying tissues, which are capable of pigment formation, pigment storage, and pigment deposition. The structure, distribution, and chemistry of insect pigments have been recently reviewed by Needham (1978) and Kayser (1982, 1984), and will, therefore, be described only insofar as they concern visible coloration and adaptiveness of insect colors.

2 Pigments and Visible Coloration

The colors of immature and adult insects may be grouped into three classes: (1) pigmentary or chemical colors, (2) structural or physical colors, and (3) combination of chemicophysical colors (Richards and Davies 1977).

2.1 Pigmentary Colors

Pigmentary colors are due to chemical substances that absorb some wavelengths of the incidental light and reflect others. Insects synthesize, or acquire from their food members of all the known chemical classes of biochromes (Needham 1978): e.g., carotenoids, chromanes, flavonoids, aurones, ternary quinonoids, tetrapyrroles, including porphyrines and bilins, indolic melanins, ommochromes, papiliochromes, pteridins, and isoalloxazines. Those acquired from the food are metabolized, and often chemically modified in order to become "zoochromes". They may be present in the cuticle, epidermis, or in subepidermal tissues [usually fat body, gut (content), or hemolymph]. The following factors seem to be the most important in producing colors in living integument: (1) the absorption spectrum of the pigments, (2) the ultrastructure of the surroundings, (3) the site of the molecules, (4) the reflected wavelengths, and (5) the state, size, and shape of the pigments (Fuzeau-Braesch 1972). A color pattern often consists of an epidermal or subepidermal ground color and overlying areas of cuticular pigmentation.

2.1.1 Melanins and Cuticular Coloration

Most insect cuticles undergo drastic changes in properties shortly after ecdysis. Similar changes can also occur in parts of the pharate cuticle before ecdysis or during the transformation, e.g., of dipteran larval cuticle into the hard and dark puparium (Andersen 1980). As early as 1902, Dewitz studied the cuticular darkening that accompanies the formation of puparium in blowflies. He demonstrated that the darkening results from an enzymatic process for which oxygen is needed and which is similar to the darkening of insect hemolymph when shed. Gortner (1911) reported that localized darkening, such as the black markenings on the elytra of the Colorado potato beetle, *Leptinotarsa decemlineata*, results from an oxidase distributed in wing cuticle, and that formation of color is determined by the specific localization of a chromogen, the amino acid tyrosine, present in the hemolymph. Melanin can be formed from tyrosine by the enzyme tyrosinase with dehydrophenylalanine (DOPA) as an intermediate. Observations by Pryor and Fraenkel (1940) and Rudall (1947) completely changed the ideas about hardening. They found that a protein, an o-diphenolic compound, and an oxidase are both involved in the hardening. Further progress was made when Karlson and co-workers (e.g., Karlson and Sekeris 1962) found that during puparium formation tyrosine is transformed via DOPA and dopamine to N-acetyldopamine, which is incorporated into the cuticle.

Tanning of the cuticle is under humoral control, as has first been shown in adult flies (for review see Raabe 1982). Despite the importance of ecdysone in inducing the de novo synthesis of DOPA-decarboxylase and perhaps other enzymes associated with cuticular tanning (for review see Highnam and Hill 1977) it is not the only hormone involved. Tanning of the cuticle of many species is controlled by a hormone from the cerebral neurosecretory cells and released from the ganglia of the ventral nerve cord, for example, in flies (Cottrell 1962; Fraenkel and Hsiao 1962) or in cockroaches (Srivastava and Hopkins 1975), and called bursicon. Bursicon may act in concert with ecdysone and c-AMP, first by producing the substrate for the action of DOPA-decarboxyalse, and second by influencing the formation of protein which will react with the tanning chinone to produce sclerotin in the cuticle. The specific tanning of the puparium of the higher Diptera is not controlled by bursicon, but by a puparium tanning factor PTF (cf. Raabe 1982).

With regards to cuticle pigmentation, it should be borne in mind that most of the sclerotized cuticles can be described as amber. Cuticles of many insects, however, possess brown or black patterns. Brown and black cuticles have been considered as "melanized" long before good evidence was found that the coloration is due to true melanin (Fuzeau-Braesch 1972). Melanins are highly stable, dark brown or black pigments, which are insoluble in most solvents. For the chemical nature of melanins, see Needham (1978). Nicolaus (1968) classified such black pigments as eumelanins (indole-type) and allomelanins (catechol-type). Most melanins from insects are eumelanins which usually form microscopic granules (see industrial melanism in *Biston betularia*; Sec. 4.2). Melanins are often bound to protein to form melanoproteins (Hackman 1974). Where melanin pigmentation is discontinuous (e.g., spots, etc.) this is apparently the result of a localized distribution of the substrate tyrosine, the tyrosinase also occurring in areas which remain unpigmentated (Richards and Davies 1977). Generally, melanins are synthesized only at the time of ecdysis, associated with the molting process (Sec. 3.4). However, sclerotization and melanization are two distinct processes, the former capable of occurring in the absence of the latter (Raabe 1982), even when data are available which demonstrate that bursicon intervenes in melanization. Das and Gupta (1977) have shown that treatment of German cockroach nymphs, *Blatella germanica*, with juvenile hormone 12 h before ecdysis, leads to an excessive black coloration of the resulting nymphs or adults and that the excessive coloration is caused by melanin. In other species melanin formation during the molting cycle is inhibited by juvenile hormone (see Sec. 3.4).

In some species (e.g., in *Locusta migratoria*), new melanin formation can be induced unassociated with the molt, by rearing the animals on a black background (Sec. 3.3) (Fuzeau-Braesch 1966). Deposition of melanin also takes place when a cuticle is damaged mechanically or by a parasite. Since quinones are powerful oxidizing agents and rapidly destroy enzymes, a melanin-producing system may be one of the most primitive, nonspecific defence systems (Hackman 1974).

2.1.2 Carotenoids

The distribution and origin of carotenoids in insects has been recently reviewed by Kayser (1982). In insects the presence of carotenoids and their structural types depend on specific feeding habits. To date there is no convincing report on any de novo formation of carotenoids in animals. The carotenoid pattern in insects reflects group – or even species-specific capacities in selective absorption and metabolic transformation of dietary carotenoids. The occurrence of fungal carotenoids in some insects reflects the presence of microbial symbionts or complex food chains. The colors of most carotenoids are yellow, orange, or red. Together with a blue bile pigment (see Sec. 2.1.4) carotenoids produce the green color of many insects. Many carotenoids occur as chromoproteins in the integument of insects. Besides their role in pigmentation and in photoreception, carotenoids are suggested to play a part in taste, smell (for mimicry see Sec. 4.1), and even in sound reception (Rothschild 1973).

2.1.3 Pteridines

Pigments with a pteridine nucleus were first discovered in the wings of butterflies, but are also known to occur in other insect tissues. In contrast to melanins, integumental pteridines are located in the epidermal cells or in the interlamellar spaces in wings, scales, and hairs as ellipsoidal granules (Hackman 1974). The colors of pteridines are white, yellow, and red. In the integument as well as in the compound eye of insects pteridines frequently occur together with ommochromes (2.1.5). Chemistry and metabolism of pteridines have been reviewed by Cochran (1975) and Needham (1978).

2.1.4 Tetrapyrrole Pigments (Bile Pigments)

Linear tetrapyrroles, the true bile pigments, are of common occurrence in insects and are synthesized de novo. They are generally blue or green compounds and considered to be of the biliverdin type. Biliverdins of the IX_{α} structure are typically encountered in hemimetabolous insects, but have also been identified in some holometabolous species, e.g., in the Neuroptera *Chrysopa carnea* (Rüdiger and Klose 1970; see p. 217), and in the moth *Thaumetopoea pityocampa* (Choussy et al. 1973). In Lepidoptera, however, IX_{γ} type pigments are usually present (Kayser 1984). Until recently bile pigments had not been reported from any coleopteran species (Barbier 1981). Recently, however, Kayser and Dettner (1984) identified biliverdin IX_{γ} in the water beetle *Laccophilus minutus*. The two other biliverdin isomers IX_{β} and IX_{δ} have not yet been found in insects. Bilirubin, a red-brown diene, e.g., in the bug *Rhodnius* and in various Orthoptera, is formed from biliverdin by intramolecular rearrangement (Needham 1978).

2.1.5 Ommochromes (Phenoxazines)

The trivial name indicates that this class of pigments was first found in the ommatidia of insect eyes. Linzen (1974) proposed a classificatin of ommochromes in the order of increasing complexity: ommatins, ommidins, and ommins. The tryptophan → ommochrome pathway and chemical properties of ommochromes have been extensively described by Linzen (1974). Xanthommatin, a typical ommatin, is ubiquitous in insect eyes and integuments. In the integument of several lepidopteran genera (e.g., *Vanessa, Cerura, Anagasta*) it is the only pigment. Cinnabarinic acid is known only from hemolymph and excreta of the silkworm, *Bombyx mori*, among insects. Ommins are almost equally widespread in eyes and integuments, except in some Orthoptera and Diptera. Ommidins have been found in the epidermis of grasshoppers, e.g., in *Romalea microptera*.

Ommochromes are usually bound to protein, in 0.4–0.6 μm intracellular granules surrounded by a membrane (Linzen 1974). Only rhodommatin and ommatin D are "usually in solution, non-bound in vacuoles, and never in granules" (Needham 1978). The characteristic properties of ommochromes, redox behavior, absorption of ultraviolet and visible light, and low solubility, enable them not only to act as authentic functional pigments (eyes, integument), but also as an electron accepting or donating system (respiratory function) (see Sec. 3.5) and as metabolic end products. However, in the case of epidermal xanthommatin in larva of *Cerura vinula*, which changes its color under oxygen deficiency, Bückmann (1965) has doubted such a respiratory function. The role of ommochromes as pattern pigments and in morphological color change will be discussed in later sections. Some data are available on the relation of ommochromes to other pigments. It appears that hypodermal ommochromes can be associated with melanin, localized in the overlying cuticle (e.g., in *Carausius,* Dustmann 1964 or *Cerura,* Bückmann 1965). In some insects ommochrome pigmentation may supplement melanin pigmentation to make a species look entirely dark. This is the case in *Gryllus bimaculatus* (Fuzeau-Braesch 1960). With regard to pigment function, pteridines and ommochrones may also supplement each other in patterns of external coloration (Fuzeau-Braesch 1972). In the white mutants of *Drosophila* and in the mutant chalky of *Calliphora* both ommochrome and pteridine biosynthesis are affected (Linzen 1974).

2.1.6 Some Rare Pigments

Mention should also be made of the presence of some minor insect pigments occurring either in small amounts or only in a few species. Among such are flavonoids, polycyclic chinones, and purines. [For details on the chemistry and distribution see Needham (1978) and Kayser (1984)].

2.2 Structural Colors

Structural colors differ from those due to pigments (Sec. 2.1) in that they are changed or destroyed by physical changes in the cuticle, such as a result from shrinkage, swelling, distortion, or permeation with liquids of the same refractive index as the cuticle (Richards and Davies 1977). Four types of structural coloration may be distinguished: (1) structural white, which is caused by the scattering reflection and refraction of light by microscopic particles large in comparison with the wavelength of visible light and which, in themselves, are usually transparent; (2) interference colors, which are produced by optical interference between reflections from a series of superimposed laminae or ribs; (3) diffraction colors, which result from the presence of closely spaced striae, 0.5–3 μm apart; and (4) Tyndall blue which, although uncommon, occurs in some Odonata and Orthoptera and is due to the scattering of the shorter wavelengths by particles with dimensions of the same size as the wavelengths of light (see also p. 214 of this Chapter).

The desert beetle *Cryptoglossa verrucosa* exhibits distinct color phases that range from light blue to black when subjected to extremes of low and high humidity, respectively (see Sec. 3.6). The morphological basis for this humidity-induced color change was determined by Hadley (1979) by scanning electron microscopy. In black beetles an amorphous material (wax secretion) is squeezed from the tips of miniature tubercles that cover the cuticle surface. This condition is typical for beetles maintained at or near saturation and results in "reflection of incident light producing a

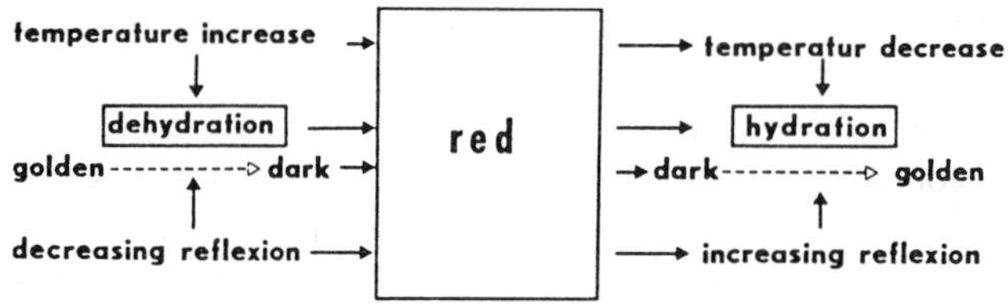

Fig. 8.1. Chemicophysical coloration (combination colors) in *Metriona trisignata.* The golden iridescence is produced by a film of moisture beneath the surface cuticle. (After Koch 1982)

dark or black coloration'' (Hadley 1979). When black beetles are transferred to low humidity, the amorphous wax secretion organizes into numerous slender (0.14 μm) filaments, forming a basket-like mesh that covers the entire surface of the beetle. This surface network is now able to refract incident light to produce the characteristic blue phase. After the return to high humidity the filaments simply "dissolve" into the epicuticular wax and contribute to its thickness. The meshwork that accumulates at low humidity reduces transcuticular water loss and may lower the rate at which body temperature rises under a radiation load by increasing reflectance.

Much more common than purely structural colors are combination colors which are produced by a structural modification in conjunction with a layer of pigments. Spectacular examples include the golden iridescence of *Cassida* (Coleoptera) and its allies, which is produced by a film of water beneath the surface cuticle. These insects rapidly lose their color when dried, but it returns after soaking in water (Fig. 8.1; Koch 1982) (see also Sec. 3.1). Similarly, the Hercules beetle, *Dynastes hercules*, changes color of the elytra due to humidity changes from black to greenish-yellow and vice versa. In several species of the genus *Dynastes* the epicuticle of the elytra is transparent and about 3 μm thick. Below this is a yellowish spongy layer about 5 μm thick. The cuticle below the yellow sponge is black. When the spongy layer is full of air it reflects yellow light back through the transparent layer. However, when the yellow sponge is filled with liquid, light passes through it and is absorbed by the layer below so that the elytra now appear black (Hinton and Jarman 1973).

3 Changes in Coloration

The resemblance of insects in shape and color to their background is often striking, but what is even more striking is the existence of internal mechanisms that enable a given species to adapt to environmental conditions in the course of its life. In the following sections attention is drawn to particular examples to illustrate the many and very different ways in which insects change color. The selective advantages of the color changes will be noted even when sometimes we can do no more than speculate about these.

3.1 Physiological Color Change

Physiological color changes in insects are rapidly reversible and occur by two different processes: by the expansion and concentration of chromatophores (chromatocytes) or by the surface migration of dark pigments within the epidermis cells. There are also some other kinds of rapidly reversible color changes, which will be briefly described at the end of this section (see also Sec. 2.2).

It is only in larvae of the genus *Chaoborus (= Corethra)* that chromatocytes are responsible for reversible color change (Hinton 1976). Large chromatocytes containing black ommochromes cover the dorsal face of two pairs of metathoracal and abdominal tracheal air sacs (Fig. 8.2). By varying the size of these air sacs the larva can vary its buoyancy and, hence, the depth at which it floats in the water. Darkly pigmented chromatocytes (jointed hexagonal cells) prevent the transparent larvae from being made conspicuous over a dark bottom. When larvae move back over a light background, the chromatocytes round up and move away from another. The change in color can occur in 10–30 min. As the larva grows, the number of chromatocytes increases with an increase in the surface area. Based on ligature and eye blinding experiments, it has been shown that chromatocyte function is humorally controlled and ordered by the head (for review see Raabe 1982). In the absence of hormone (chromactivating factor), the light-adapted condition is maintained.

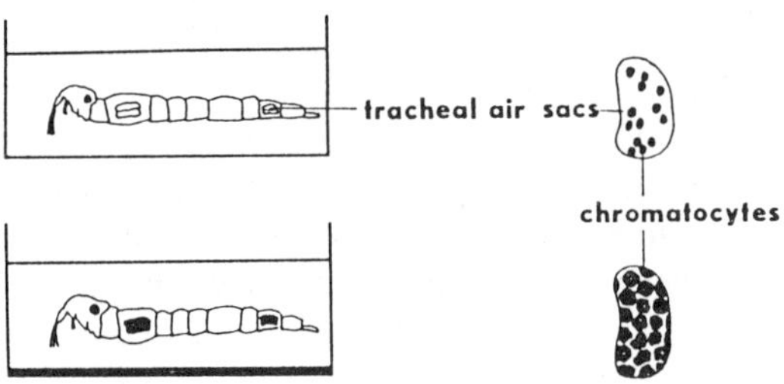

Fig. 8.2. Physiological color change in *Chaoborus (=Corethra)* larvae. Tracheal sacs with chromatocytes in the light-adapted and in the dark-adapted condition. (After Dupont-Raabe 1957)

Among the insects, physiological color change has been most intensively studied in the common stick insect, *Carausius morosus* (e.g., Schleip 1910; Giersberg 1928; Atzler 1931; Mothes 1960; Berthold 1977, 1980, 1982). *Carausius* epidermis cells contain yellow, orange, and red lipochromes, brownish ommochromes, and green pigments (five carotenoids and a bile pigment; Willig 1969). The stick insect displays a circadian color change that affects the entire body and occurs at dusk in varying degrees, depending on the pigmentation of the epidermis. Green insects remain unchanged, but beige insects turn brown and gray insects become completely black (Fig. 8.3). The distribution of ommochromes, carotenoids, and pteridines in light and dark adapted animals has been already analyzed by Schleip (1910). In light adapted animals the ommochrome granules concentrate at the proximal cell membrane. During dark adaptation they move to the distal cell membrane, dispersing over the whole cell and forming a shield of dark pigments. The carotenoid granules behave similarly. The red-shaped pteridine granules are concentrated in the distal half of the epidermis cells and show no daytime dependent movements. Apparently,

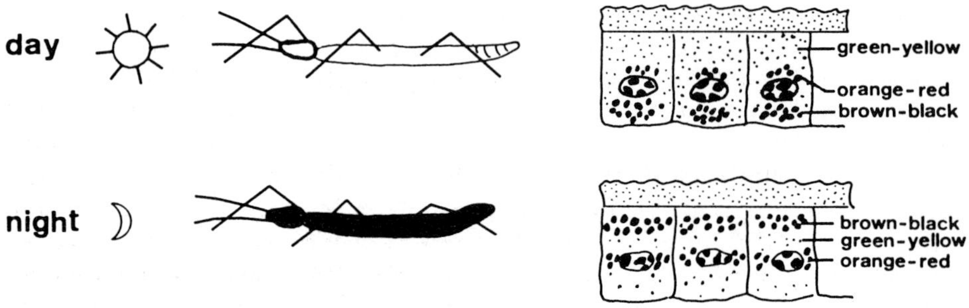

Fig. 8.3. Physiological, circadian color change in the common stick insect, *Carausius morosus*. *Pale position* (during day) with brown-black pigments below nucleus. *Dark position* (during night) with melanin granules at the distal cell membrane

biliverdin is not attached to granules, but is dissolved in the epidermis cells. The endogenous circadian rhythm of color change can continue for several weeks in complete darkness (Schleip 1910). It is possible to reverse the rhythm by keeping the animals in illumination by night and in darkness by day, and subsequently, the new established rhythm will also continue in constant darkness. The effect of light as "Zeitgeber" appears to be mediated only by the eyes and can be eliminated by cutting the optic nerves or by blackening the lower halves of the eyes. The effect of light seems to be determined by the amount of contrast in the background (ratio of light striking the dorsal and the ventral halves of the eyes).

Utilizing the fact that high humidity also produces body darkening, Giersberg proved as early as 1928 that the effect of the stimulus is indirect; via afferent nervous pathways, the brain, an endocrine source, and finally a blood-borne hormone. Further evidence that the physiological color change in *Carausius* is controlled by hormones has come from transplantations of portions of the skin of one insect to another (Janda 1936). The transplanted tissue begins to show color changes entirely paralleling those of the host within 2–3 days. The source of the chromactivating factor was investigated in the brain by the selective removal of different parts of this organ, and by the injection of their extracts (Raabe 1982). It is probably neurosecretory cells, identified in the tritocerebrum, which produce the chromactivating factor. The neurohemal structures which release the cerebral factor, however, have not yet been identified. There is also no direct evidence for the source of rhythm control, although it has been demonstrated that the chromactivating factors, synthesized in the ganglia, are stored by day in any neurohemal organs, before release at dusk. Besides the circadian color change, *Carausius* also reponds to background when illumination and humidity are kept constant. It turns dark colored on black and red surfaces and light colored on white and yellow ones.

Kusciuscola tristis is the only grasshopper known for changing color rapidly and reversible (Key and Day 1954). When it is cold males are nearly black, but if it is warmed, they change to a bright sky blue within minutes. The blue color is due to Tyndall scattering of light by small, white granules within the epidermis cells, and the blue results from being seen against a dark background. When the insect is cooled, the change to a dark color is produced by large blackish-colored granules in the epidermis migrating distally together with the small granules, and thus, "quenching" the light scattering. Changes of color occur without central nervous control and can also be demonstrated in isolated pieces of integument (Hinton 1976).

Physiological color change in dragonflies (e.g., in the damselfly *Austrolestes annulosus*) also depends on temperature and involves two phases: a light phase in which the pigment granules gather at the base of the epidermis cells (when warm by day), and a dark phase in which they are distributed throughout the cell (when cool by night). Veron (1973) showed that a darkening factor is located throughout the central nervous system, and in particular in the terminal abdominal ganglion. Temperature has the dual role of regulating the production of the darkening factor as well as directly regulating the rate of proximal pigment migration. For many Australian dragonflies in the suborders Zygoptera and Anisoptera it has been shown that the temperature threshold for this color change is higher in species from warmer regions than in those from colder regions (Hinton 1976). The selective advantage of such a color change may lie in the slightly greater speed with which the dark phase warms up in the early morning, and in the greater resistance to overheating in the bright phase (May 1976; see Sec. 4.3). It is more likely, however, that the chief selective value of the color change in dragonflies lies in its effect on predators. At night it is more difficult for predators to see the dark phase, whereas by day they find the bright phase more difficult to see (Hinton 1976).

Why are there only few insects with a physiological color change? Insects are as a group typical terrestrial animals. Terrestrial life, however, requires thick protective outer layers which would obscure pigment movement from external view. Therefore, there is a tendency in both vertebrates and arthropods, for physiological color change to be largely restricted to aquatic forms (Bückmann 1984a).

3.2 Morphological Color Change

Many insects change color during development by actual changes in the quantity of pigments within the animal or its integument. This may be related to the notion that in insects most developmental changes are matched to gross changes in environmental conditions (Bückmann 1977, 1984a). A true morphological color change, as a true physiological color change, however, should also be inducible by external factors (photoperiod, temperature, humidity, gregariousness, oxygen and carbon dioxide partial pressure; Fig. 8.4), independent of the phase of molting cycle, and should be reversible. If the pigmentation is irreversibly determined by environmental factors during a certain sensible period, this kind of controlled pigment deposition would be termed a color adaptation (Sec. 3.3). For color changes which are merely a part of the normal development and independent of environmental changes, the term ontogenetic color change has been proposed (Sec. 3.4) (for nomenclature cf. Bückmann 1974a, 1984a).

In the stick insect, *Carausius morosus*, a genuine, stage independent morphological color change occurs in response to external factors (illumination, background, humidity, temperature) (Bückmann 1977) and appears to involve similar mechanisms of control as the physiological color change. Morphological color change, however, requires stimulation over a much longer period (Atzler 1931). Darkening the ventral

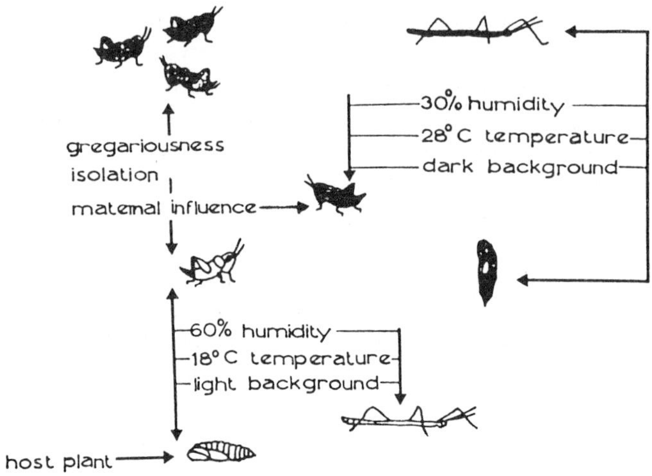

Fig. 8.4. Effects of environmental factors on pigmentation in some insects. (After Raabe 1983)

halves of the eyes, thus, imitating the visual situation of an animal in light on a dark ground causes, after a lag of 3–4 days, a continuous increase of epidermal ommochrome content by about 5 μg per day. Ommines and xantommatin contribute equally to the increase. Darkening the dorsal halves of the eyes, however, does not influence the ommochrome content (Bückmann 1977). Ommochromes once formed in the integument are never removed again (Dustmann 1964) and, therefore, this morphological color change is reversible only insofar as the animals can turn green again by "dilution" of integumental ommochromes, the ommochrome content remaining constant during further growth. Ommochrome synthesis is also enhanced at high temperature. Under these conditions it is much less dependent on the pattern and intensity of illumination.

The question arises as to how the insect adapts its tryptophan supply to its visual (or temperature) situation (Bückmann 1984b). Recently, Maisch and Bückmann (in prep.) proved the epidermal ommochrome formation dependent on the tryptophan content of the diet. The more tryptophan the diet contains, the more ommochrome is formed. In addition, according to Berthold and Bückmann (1975) an increase in ommochrome content correlates with a decrease in excretion of kynurenic acid, the normal end product of tryptophan metabolism in the stick insect. However, at any given tryptophan dose, those animals in which the eyes are blackened ventrally will always form more ommochrome (Bückmann unpubl.).

The independence of optically induced ommochrome synthesis of molting rhythm would indicate that it is also independent of the ecdysone/juvenile hormone system, which controls molting and metamorphosis (Bückmann 1977). Histological investigations and transplantation experiments indicate that pigmentation is under neurosecretory control (Raabe 1966). Since ommochrome production only depends on the relative magnitude of light stimuli in the dorsal and ventral halves of the compound eyes (a mechanism which would be required for an effective adaptation of the animals to the shade of the background independent of the amount of incident light), the central nervous system, which must accomplish the comparison and evaluation of dorsal and ventral stimuli, can control pigmentation by releasing a neurohormone (Bückmann 1979).

Besides ommochromes, other pigments also change during the morphological color change. An inverse relationship seems to occur between ommochrome and pterin contents (Berthold and Bückmann 1976). The synthesis of biliverdin is enhanced by darkness and low temperature, whereas the uptake of carotenoids is enhanced by warmth and light (Willig 1969). All these differences, however, are less striking than those of ommochromes.

Larvae of several species of dragonflies change their color in response to a change in background by decreasing and increasing concentrations of melanin in the cuticle (Krieger 1954). Since the amount of melanin and its distribution pattern can only be changed in the course of a molt, this morphological color change can only occur after a molt.

Another striking example of a facultative, morphological color change is that seen in some locusts, crickets, stick insects, and caterpillars according to population density. The relation of color to density is particularly clear in a few insects (e.g., *Locusta, Schistocerca, Podocanthus, Didymuria, Laphygma, Spodoptera)* which are cryptic when few and have bright aposematic colors when the population density is high (e.g., Key 1957; Staal 1961; Iwao 1962). In the Mediterranean field cricket, *Gryllus bimaculatus,* pigmentation is also influenced by crowding; crowding of the population produces a light coloration, whereas isolation leads to darkening, which is contrary to that observed in other species (Fuzeau-Braesch 1960). Carbon dioxide has recently been shown to be able to affect the phase dimorphism (dichroism) by strongly inducing the formation of solitaria colors, even when applied for only a few minutes per day to insects in the gregaria phase (Fuzeau-Braesch 1972, 1978, 1983). The endocrine control of phase dimorphism has been well studied in *Locusta migratoria.* Juvenile hormone is clearly the immediate cause of the differentiation of many of the solitary phase characteristics, and therefore, also induces the light pigmentation of the solitary phase (Joly and Joly 1974) (for review see Nijhout and Wheeler 1982). Similar observations have been made for various lepidopteran species (e.g., Ikemoto 1983). However, development of phase characters does not depend exclusively on juvenile hormone. Girardie (1970) has provided evidence for the existence of a neurosecretory factor, released from the corpora cardiaca, that stimulates melanization of the gregarious phase of *Locusta migratoria.* Another neurohormone termed melanization and reddish coloration hormone (MRCH) seems to control all the larval color changes corresponding to phase variations in various armyworm species (Matsumoto et al. 1981).

An obligatory morphological color change according to season is found in the order Neuroptera, e.g., in *Chrysopa carnea* (Honěk 1973). The autumnal color change from green to yellow-brown is caused by the increase of xanthommatin and the simultaneous decrease of biliverdin IX$_{\alpha}$ level (Rüdiger and Klose 1970). Since the winter coloration is not simply an external modification of diapause, the color change can be induced also in laboratory animals by various combinations of constant temperatures and photoperiods (Honěk 1973). Similar observations have been made by Michieli and Žener (1968) for color changes in adults of the bug *Nezara viridula.*

3.3 Morphological Color Adaptation

In many insect species the timing of a new pigmentation is part of the normal development program, but the resulting color is responsive to external influence, although for only a short sensitive period prior to the onset of the stage of variable pigmentation. The pigmentation, once determined, cannot be further changed, and therefore, should be called a morphological color adaptation (Bückmann 1974 a, 1984 a). Examples for such color adaptations are the many butterflies in which the pupae are not protected by a cocoon. Instead they are hidden from sight by a certain color adaptation to their background (Fig. 8.5) (Koch and Bückmann 1984). The sensitive period is shortly before pupation when the caterpillar has already settled at the place where it is going to pupate. The ocelli are involved in the control of color adaptation, but at least in some caterpillars an additional extraocular light receptor seems to exist (Angersbach 1975). The different color modifications are attributed not only to the variation of melanin, but also to that of carotenoids (Bückmann 1969), bile pigments (Kayser and Angersbach 1974), quinoid sclerotization (Kayser-Wegmann 1976), and even structural colors (Neville 1977).

The role of environmental factors in the determination of pupal color has been studied by numerous authors. In *Papilio polytes* and *Papilio demoleus* the curvature and texture of the pupational substrate is an important factor, brown pupae tending to be found on rough surfaces and green pupae on smooth ones (Smith 1978) (for other species see Wade and West 1979; Hondka 1981). Background color has only a minor effect in these species. Differences in the relative importance of texture and background color are considered to be related to the time of day a larva selects its pupation site. Texture is relatively more important when the selection of a pupation site occurs in the dark. In various species of the genus *Pieris* (which select pupation sites during daylight) light stimuli are the most important factors involved in the determination of pupal color (Oltmer 1968; Smith 1980). Yellow and blue spectral

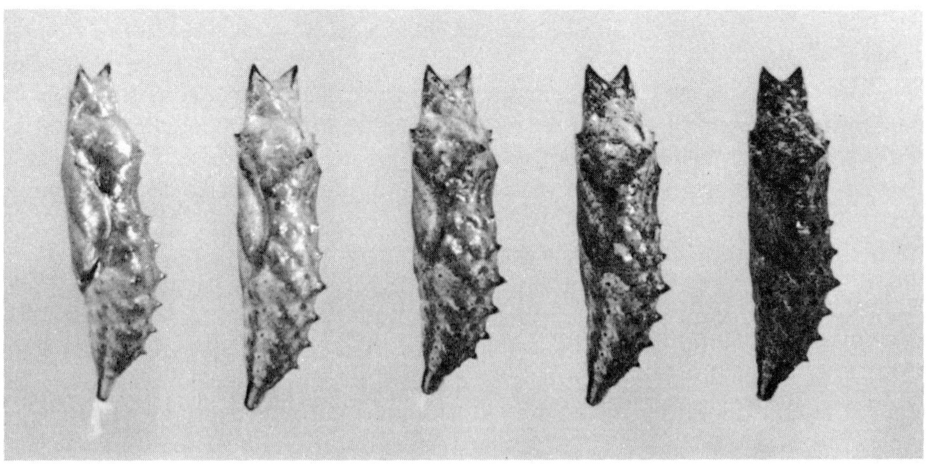

Fig. 8.5. Pupae of *Aglais urticae* (Nymphalidae), pupated under various illumination conditions (five classes of coloration). (Photography B. Koch, Ulm)

wavelengths determine the production of green and brown pupae, respectively, (Kayser and Angersbach 1974) and it is thought that this correlation results from yellow light being closely associated with green foliage and blue with sites away from green vegetation. Other nonoptical factors, such as relative humidity and photoperiod, may also influence pupal color determination. Diapausing pupae of *Pieris rapae,* reared under short day length, have a greater tendency to be brown rather than green, but the reversal is true in *Pieris brassicae.* In *Papilio zelicaon* short days (and cool temperature) also induce diapause brown pupae irrespective of pupation substrate (Sims 1983).

Results on the hormonal control of pupal color dimorphism in different genera of Lepidoptera are controversal. In *Papilio* brown cuticle is produced following the secretion of a hormone from a source in the head prior to the pupational molt (Smith 1978). However, in *Pieris* green pupal pigmentation requires the secretion of a hormonal factor from the head (Smith 1980). Ressin (1980) suggested that not only juvenile hormone, but also neurohormones and ecdysteroids, may be involved in the hormonal control of pupal pigmentation of *Pieris brassicae.* Recently, Awiti and Hidaka (1982) have proposed a neuroendocrine mechanism underlying the pupal color dimorphism of *Papilio xuthus.* A browning hormone is synthesized in the brain of the larva and transported to the prothoracic ganglion (PG), irrespective of the color of the future pupa. After the wandering stage, a larva settles either on the green twigs in the upper part of the food plant or on the dead brown trunks in the lower part for pupation. Such environmental conditions, as selected by the larva, are integrated and interpreted by the brain, which then sends a message via a neural path to the PG instructing it to release the browning hormone. Experiments on *Pieris brassicae, Aglais urticae,* and other Nymphalidae have also suggested that pupal color adaptation is controlled by a neurohormone, which, however, does not further darken pigmentation, but works by melanization inhibition (Bückmann 1984 a).

3.4 Ontogenetic Color Change

Ontogenetic color change is a step of the normal ontogenetic development of an insect and is usually linked with a larval molt or the metamorphosis, but it can also occur within a larval stage irrespective of molting (Bückmann 1974 a). In order to ensure a real biological adaptive effect of this color change the insects make use of the rigid and invariable insect instincts (Bückmann 1984 a, 1984 b), which at first sight may appear to be a disadvantage. These instincts, however, bring the insect at a particular stage of development reliably into a certain biological situation. For instance, in the great pussymoth, *Cerura vinula* (Fig. 8.6), the female always lays eggs on poplar or related plants. The young larva will, therefore, always find itself on those leaves, otherwise it would die of starvation. No further color adaptation to any other surrounding would help them to survive in a "wrong" environment. If one searches for these black caterpillars in their natural habitat one is bewildered by their resemblance to the black dried and furled edges of the poplar leaves. During the next molting the distribution of black cuticular melanin changes, the caterpillar now showing a black and green pattern. Again this pattern proves to be of protective value. The pattern has a "somatolytic" effect; it obscures the contours of the body among the quivering pop-

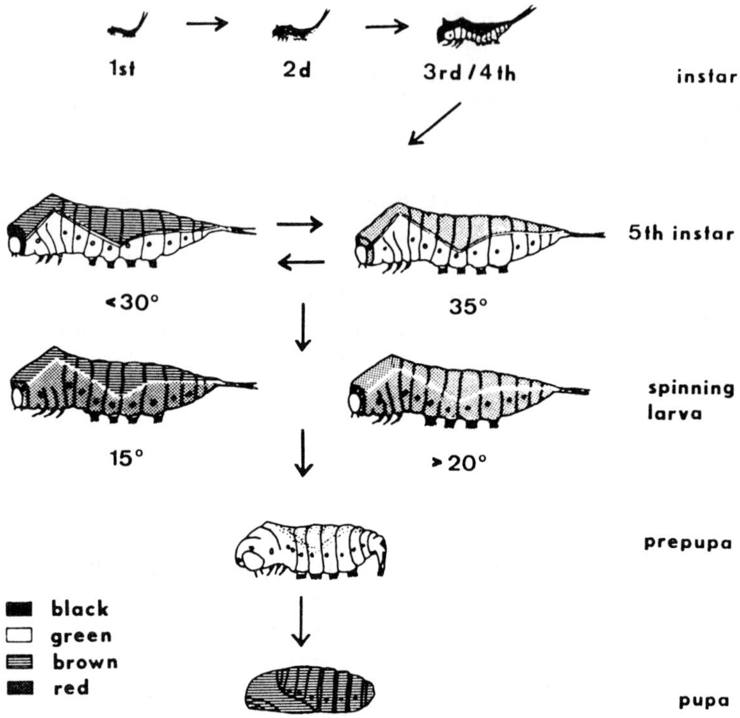

Fig. 8.6. Color patterns and color changes during the larval development of *Cerura vinula L.* For details see text. (From Bückmann 1984 b)

lar leaves. The shape of this pigment pattern remains the same until the last larval instar. In this stage the melanization becomes weaker, so that underlying brown or red epidermal pigments can be seen. The green as well as the red-brown parts of the body appear lighter towards the back, which has been called a "double inverse countershading" (de Ruiter 1956). As the normal position of the larva is upside down, this countershading may compensate for the shading of the larva on the leaves by natural illumination from above. If the caterpillar is nevertheless attacked, it suddenly shows its neck skin with a striking red and black pattern (warning coloration), from which it sprays a ray of formic acid against the predator. When the wandering larva leaves the green food, the parts of the body which are green or brown become dark red. This again appears to be a protective color change, the animal adapting in color efficiently to the bark of the tree. The colors disappear once the larva is sheltered in its cocoon.

As far as is known, all these ontogenetic color changes are controlled by the hormonal system which also regulates molting and metamorphosis (ecdysone and juvenile hormone; for details see Bückmann 1959, 1974 b, 1984 a; Ikemoto 1980). However, one cannot ascribe general pigmentary effects to juvenile hormone or ecdysone. In grown-up *Cerura* larvae juvenile hormone inhibits ommochrome formation, whereas in feeding *Mamestra brassicae* last instar larvae, both melanization hormone from the subesophageal ganglion and juvenile hormone from the corpora allata are necessary for normal epidermal pigmentation (Hiruma et al. 1984). In *Cerura* ecdysone causes formation of ommochrome, but later on the same ecdysone evokes the pupal molt (polytropy of hormones; Bückmann 1983). In last larval instars of the oleander hawk moth, *Deilephila nerii,* a first release of ecdysteroid is thought to trigger melanization in the absence or near absence of juvenile hormone, while a second release initiates events leading to apolysis and ecdysis (Chang and Jang 1980). Two other species of Lepidoptera, *Cephonedes hylas* and *Pieris brassicae,* also undergo developmental color changes during the last larval instar, similar to those in *Cerura.* In *Pieris* larvae, the deposition of epidermal ommochromes and the excretion of xanthommatin are related to the hemolymph ecdysone titer (Beydon et al. 1980). Juvenile hormone I inhibits the process of xanthommatin accumulation, which is thought to be controlled by a neuroendocrine mechanism. In *Cephonedes* larvae body color change is also primarily controlled by ecdysteroids (Ikemoto 1980).

3.5 Biochemical Color Change

Some effects will be mentioned which external factors may exert by acting directly and locally on the pigment metabolism. Here only the onset of the color change is environmentally controlled and the resulting color is invariable (Bückmann 1984 a). In *Cerura* the reddening of the brown ommochrome in the last instar "saddle patch" can be prevented by low temperature (15°C). However, in the green larva the reddening of the brown saddle patch can be evoked at any time by high temperature (30°–35°C). Furthermore, it can also be caused by a lack of oxygen. In both cases the ommochrome is chemically reduced (Bückmann 1963).

3.6 Physical Color Change

The humidity induced color change in the desert beetle *Cryptoglossa verrucosa,* as described in Sec. 2.3, can be called a "physical" color change. A similar kind of environmentally induced change in structural colors has been described by Mandl (1931) for various species of *Cicindela* (Coleoptera).

4 Adaptiveness of Insect Colors

Few of the principles underlying the evolution of colors or color patterns of animals are understood. Hypotheses which explain coloration fall into three major categories: (1) coloration affecting the animal's own vision, (2) coloration affecting the animal's visibility to others-conspecific and nonconspecific, and (3) physiological/structural functions of biochromes (Burtt 1981).

4.1 Color and Patterns as Optical Signals

The colors of animals that are normally visible to others are determined either by the requirement for deception or advertisement. Most animals resort to deception for protection. As has been noted in previous sections, many insects have patterns of colors and tones that so match the background so that they are indistinguishable even when plainly in view (Hinton 1976). In some insects both color and form exactly resemble common inedible objects in the environment so that they pass unnoticed by predators.

A few insect species are distasteful or poisonous, and these advertise their presence by bright colors so that predators recognize them for what they are (Müllerian mimicry). Some harmless animals escape attack by mimicking the bright colors of dangerous species (Batesian mimicry) (Heal 1979 a, 1979 b).

The fact that insects can see into the ultraviolet means that there is a part of the spectrum invisible to vertebrates that they are able to exploit (Hinton 1976). In the Pieridae, for instance, a thin film interference (see Sec. 2.2) is produced in the ultraviolet part of the spectrum. When such a butterfly is flying, bright flashes will be continually emitted as the wings beat through the appropriate angle. There is some evidence for the hypothesis that this ultraviolet pattern of male *Pieris* plays a part in successful courtship.

When attacked many insects engage in displays that are sometimes effective in intimidating the attacker. This kind of behavior is called deimatic behavior (Hinton 1976). Deimatic displays are best known in mantids, moths, and butterflies which when attacked suddenly reveal a brightly colored pair of eyes (see also for *Cerura* on p. 220). Other "flash" colors may be very effective in animals that do not stand their ground, but flee when attacked. For instance, a moth may have a brightly colored hind wing that "catches" the eyes of a predator, but as soon as it lands the hind wing is covered by the cryptically colored fore wings and the moth effectively disappears.

In Sec. 2.2 the interference colors in Cassidinae have already been noted. In recent years mutch attention has been paid to the iridescence produced by a very different kind of structure, the diffraction gratings of some ground beetles which live among grass roots or concealed in other places (Hinton and Gibbs 1971). The gratings are usually on the elytra and pronotum, but in some species they are also on the underside and legs. In shade or in diffuse light the spectra from different parts of the gratings may overlap so much that little if any color can be seen. However, bright spectra are found when the gratings are illuminated by parallel light, such as sunlight. Insects with a diffraction grating have an "all or none" warning color. In dim light their color may be cryptic, but as soon as they are exposed to sun, warning colors appear. A pattern of warning colors formed by pigments in the cuticle would render them conspicuous even in the dim light around grass roots.

4.2 Industrial Melanism

True melanism, an increased deposition of dark pigments in the wings or other structures of insects, is usually genetically controlled. Industrial melanism is brought about by the indirect effects of industrialization through natural selection. Industrial melanics are rather common within the Lepidoptera, but have manifested themselves only in the last 140 years. They are characterized by rapid spread, by increased vigor, and, in the majority of species, by having no intermediate forms. They occur only among those Lepidoptera that depend on crypsis for their survival and that spend the hours of daylight sitting motionless on blackened trees and walls and are sought by birds as food. Their inheritance usually is one of simple Mendelian dominants (for reviews on distribution and genetics see Kettlewell 1961; Bishop et al. 1977). One of first and best documented black specimen occurring in previously light colored species is the all-black form of the peppered moth, *Biston betularia f. carbonaria,* in Britain. *Gonodontis bidentata* is another geometrid moth with a melanic dominant *(Gonodontis bidentata f. nigra)* to the nonmelanic. In the last 25 years clean air policies have markedly reduced smoke and sulfur dioxide pollution, although there is still no evidence of a fall in the frequency of *Gonodontis bidentata f. nigra* associated with the decline in air pollution.

Non-lepidopteran insects which also show increases in the proportion of melanics in populations inhabiting the inner urban areas are the coccinellid *Adalia bipunctata* (Honěk 1975; Bishop et al. 1977) and the psocid *Mesopsocus unipunctatus* (Popescu 1979). In the ladybug, *Adalia bipunctata,* the advantage conferred by melanism is not completely understood since this species is warningly colored and distasteful (see Sec. 4.1). Melanics may be at an advantage in the presence of certain pollutants or they may be favored by the lower amount of sunshine received by areas suffering air pollution. In the lady bug there has been a spectacular decline in the frequency of melanics in Britain since the late 1960s. This rapid response would be expected if the environment acts directly on the species rather than through changes in the background.

4.3 Physiological and Structural Functions of Biochromes and Color

Animals absorb radiant energy from two different regions of the electromagnetic spectrum, the solar (290–2600 nm) and the thermal (>4000 nm). Differently colored surfaces absorb solar radiation differently, but biological tissues, regardless of color, absorb all incidental thermal radiation (Burtt 1981) (see also Sec. 5, Chap. 1). Edney (1971) and Hamilton (1975) have argued persuasively for the importance of color as a thermal strategy in desert insects, where either black or white seem to be preferred according to the time of day when a particular species is active (see Hoffmann 1978). By using a reflectometer, Willmer and Unwin (1981) have shown that insects of higher reflectivity heat up more slowly than dark forms of the same size. Further evidence for the importance of color comes from those insects which have differently pigmented morphs (Roland 1978). In alpine and arctic *Colias* butterflies variation in ventral coloration seems closely correlated with gradients of both latitude and altitude, possibly indicating selection for thermoregulatory requirements. In some other butterflies (Watt 1969; Douglas and Grula 1978) and in the honeybee, dark forms also have a higher body temperature when compared to light morphs under the same conditions. In a few cases a temperature-controlled physiological color change has been documented (Sec. 3.1). For instance, in the dragonflies *Austrolestes* (Veron 1973) and *Diphlebia* (O'Farell 1963), body temperature rises faster in the dark than in the pale blue form. Certain grasshoppers and beetles show a similar effect (Willmer 1982). In each of these cases microclimate seems to have a direct triggering effect. Seasonal changes in coloration, such as those seen in aphids, may be similarly linked to the weather (Dixon 1972).

In a few special cases insects may have unexpectedly high transmittance in the infrared region of the spectrum, which may be a specific thermally adaptive strategy to permit warm-up at low sun angles (Henwood 1975 b).

In some of the insects which change color humidity appears to be the most important causative factor (Hadley 1979). Dark animals in humid, shady habitats may absorb a maximum of solar radiation to maintain overall energy balance (Burtt 1981).

Water retention is another problem for terrestrial insects (see Chapter 6). Quinones darken the exoskeleton of many insects, reducing its permeability to water. (Needham 1974). Further resistance is conferred by melanin. However, water retention as a selective force in insect coloration has been little explored. Some insects conserve water and create color by depositing nitrogenous wastes in the often shed integument (Cochran 1975). (For coloration and water loss see also Sec. 2.2.)

Animals shield from ultraviolet radiation by pigmentation that protects ultraviolet sensitive tissues (e.g., gonads, nervous system) or by seeking microhabitats protected from UV-light (Burtt 1981). However, UV-radiation shielding as a selective force in (terrestrial) insect colors has not been studied in detail.

5 Conclusions

No single hypothesis can explain the coloration of insects. One aspect is that color is a by-product of an animal's physiological need. A second idea relates color to its effect on the animal's own vision. The third group of hypotheses associates color and pattern with optical signals. Prediction from these hypotheses depends on the particular characteristics of the photoreceptors and on environmental variables that affect the appearance of an optical signal (Burtt 1981).

Many insects respond to environmental challenge by alterating the color of their bodies. Although only a few insects are known to exhibit rapidly, reversible physiological color changes, these few show color changes in many different ways.

Slow changes in color are usually not reversible and are known as morphological color changes, morphological color adaptations, and ontogenetic color changes.

There is no doubt that predation is the main selective influence leading to the evolution of mechanisms of color changes. Color change may, however, also be triggered by thermal and other physical stimuli. In some insects, color changes are related to mating behavior.

Acknowledgements. I am grateful to Prof. Dr. D. Bückmann, Ulm, Dr. N. Mathers, Ulm, and B. Koch, Ulm for critical reading of the manuskript. Figures 8.5 and 8.6 were kindly provided by B. Koch and Prof. D. Bückmann, respectively.

Chapter 9

Environmental Aspects of Insect Bioluminescence

Klaus H. Hoffmann[1]

Contents

1 Introduction

The emission of light by organisms has a long and colorful history. Investigators from various fields of biology, chemistry, and physics have made important contributions to our understanding of bioluminescence. A great diversity of plants and animals has

1 Allgemeine Zoologie (Biologie I), Universität Ulm, Oberer Eselsberg, 7900 Ulm, FRG

Environmental Physiology and Biochemistry of Insects (ed. by K. H. Hoffmann)

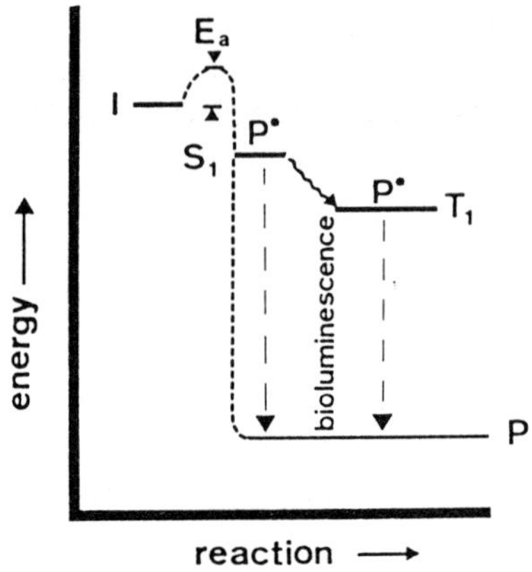

Fig. 9.1. Simplified energy level diagram for bioluminescence. *I* chemical intermediate; ***P**** product excited state; S_1 singulett state; T_1, triplett state; ***P*** product ground state; E_a Arrhenius energy of activation. (After Hoffmann 1981)

been found to posses the ability to emit light, and this literature has been reviewed extensively during past years (Herring 1978; Henry and Michelson 1978; Anctil 1979; Hoffmann 1981). Bioluminescence involves a chemical event followed by physical processes which convert the chemical reaction energy into light. Such luminescence is likely to occur at ordinary temperatures and without absorption of light from the environment. The energy necessary for light emission is always supplied by enzymatically catalyzed oxidation reactions ("luciferin-luciferase" reaction). The result of such light reaction pathways is a terminal chemical intermediate which can break down to form at least one excited state product (Fig. 9.1). Light is produced by the radiative transition of the product from the excited state to the ground state (for further details see, e.g., Mc Capra 1978; Wampler 1978). Of the various luminous forms that have been investigated none have received as much attention as the insects.

2 Distribution of Luminous Insects

Relatively few of the great number of insect species have evolved a light-producing mechanism. Insects positively known to be self-luminescent only occur in the families Poduridae and Onychiuridae (Collembola), Fulguridae (Homoptera) (?), Mycetophilidae (Diptera), and Lampyridae, Elateridae, Phengodidae, Homalisidae, and Telegeusidae (Coleoptera) (Lloyd 1983). A large number of cases in which luminescence has been reported for a particular insect species may be explained as being to infection by and growth of luminous bacteria (McElroy et al. 1974). Self-luminous insect species are widely distributed throughout the whole world. Since bioluminescence cannot compete with ambient sunlight intensities, its evolution in diverse species is, of course, associated with dimly lit environments and circumstances, such as nocturnal activity and subterranean life.

It is not possible within the close compass of this article to describe all of the different variations of luminescence that have developed in some insect families. Consequently, I will consider representative examples of various forms, and thereby emphasize the function and evolution of bioluminescence within the natural habitat of the species. Luminescent insects are particularly suitable for experiments relating to fundamental questions about communication in nonsocial insects, and the question then arises as to which aspects of their biology, especially their ecology, have been important in making their signals what they are today (Lloyd 1983).

2.1 Bioluminescence in Collembola (Springtails)

Self-luminous springtails are rather rare and, therefore, little work has been done on their physiology, biochemistry, and bioluminescent ecology (McElroy et al. 1974; Lloyd 1978, 1983). Light-emitting Collembola have been reported for two of the five families in the order (Poduridae, Onychiuridae). Luminescence is not known in juveniles and eggs, but only in adults. Actual mechanical stimulation of the animals is necessary for light emission, and the light emission usually lasts for 5-10 s. Oxygen is essential for light, and a very bright flash is observed when air is admitted to an anaerobic jar containing one of the animals. This indicates the possibility that a "luciferin" is accumulated under anaerobic conditions.

Collembola reproduce sexually only at time of environmental uncertainty or change and the appearance of luminescence in such forms is postulated to occur only during such a sexual phase, after appropriate ecological stimuli. Therefore, one can imagine that luminescence plays an important role in the mating behavior of Collembola (sperm transfer behavior) in the dark recesses and niches of leaf litter, soil, and rotting logs (Lloyd 1983). Adaptive bleeding occurs in some species of Collembola, and therefore, bioluminescence could also have a warning or defensive function (Lloyd 1983).

2.2 Bioluminescence in Homoptera (Lantern Flies)

Luminescence in lantern flies has often been reported, but there is little authenticated evidence for it. The only well-established luminous bug is found in the suborder Cicadina, known as the "lantern fly", *Fulgora laternaria,* from tropical South America (McElroy et al. 1974). There is some evidence that light emission in *Fulgora laternaria* is connected with season and with sexual activity. It is presumed that the head emits a bright, white light that is turned on only when males and females are flying together late at night (Lloyd 1971). No work has been done on the histology, physiology, and biochemistry of the luminous cells.

2.3 Bioluminescence in Diptera (Glowworm Flies, Fungus Gnats)

All of the 10 or 11 species presently known to be self-luminescent are in the family Mycetophilidae. They are typically found in wet forests, meadows, heaths, and caves. Only the larvae are luminescent and spin webs upon which they live. With these webs they snare prey which is presumed to be attracted to their light. The webs also aid in humidity control and locomotion (Lloyd 1978).

The scientific history of luminescence in fungus gnats began in 1849 with observations on larvae of *Keroplatus sesioides.* Meanwhile several other species of *Keroplatus* were found to be luminescent. Fat cells in the hypodermal fat layer emit the light and cessation of luminosity in the adult 2 days after eclosion is connected with the disappearance of the fat body. Some control over the light intensity has been observed, although the control mechanisms are as yet unclear. Luminous larvae of *Orfelia* (= *Platyura) fultoni* are usually found around wet, deeply shaded springs or streams. Light is associated with well-tracheated photogenic cells at the nonpigmented anterior and posterior ends of the larva (Bassot 1978) and is "continuous" and independent of stimulation. According to Fulton (1941), there is a day-night rhythm of luminescence that persists for several days in constant darkness. At night the larvae slide out on their web and expose their glowing bodies in order to attract tiny organisms which become entangled in the adhesive (Lloyd 1978).

Bioluminescent communication apparently occurs in the most famous fungus gnat, the New Zealand glowworm fly, *Arachnocampa luminosa.* All stages of the life cycle of *Arachnocampa,* except the egges, are luminous (Lloyd 1978). Immediately upon hatching, larvae glow brightly. In larvae, as well as in pupae and adults, the light organ is situated within the last abdominal segment. The luminous organ is actually an enlarged tip of four Malpighian tubules (see Chap. 7) which are well supplied with oxygen through large tracheae trunks. The biochemistry of *Arachnocampa* luminescence has been described by Anderson and Cormier (1976). The system resembles that of the fireflies (see Sec. 3 of this Chapter) in that light emission is stimulated by ATP, but not by other nucleoside-5'-triphosphates. Evidence has been presented that Mg^{2+} (but not Ca^{2+}) and oxygen are also required for luminescence. Stimulation is inhibited by simultaneous addition of 2 mM inorganic pyrophosphates. A luciferin-luciferase reaction, however, has not been characterized until today. The color of the emission is much bluer (λ_{max} = 488 nm) than that of Coleoptera. A lack of pH and temperature effects on the glowworm fly bioluminescence again suggests a different chemical structure for its "oxyluciferin" as against that in fireflies. The worm can emit $> 10^{15}$ photons in its lifetime, but at any one time possesses only about 10% of this total capacity (Lee 1976).

What has made this species so famous is a result of its development in caves. The eggs are deposited in a mucous glue on the walls and ceiling. Immediately upon hatching a larva will spin a silken sheath from which the animal hangs and starts to glow brightly. The tremendous network of glutinous threads serves to attract as well as to trap other small insects, mainly chironomid midges, on which the carnivorous larvae live. *Arachnocampa* larvae are also cannibalistic and neighboring larvae often lean out to their "nests" and fight. While fighting, larvae glow brightly and combatants are easy to recognize (Lloyd 1978). Male and female pupae hang from their silken threads, and turn their lights on for a few seconds when they are touched or rocked. A sexual significance of luminescence was discovered by Richards (1960). When a male fly lands on a female pupa she glows, and when a female is about to emerge one to three males may be clinging to her. If no males are present when females emerge, they apparently attract males to them with their glowing until finally they mate. Unfortunately, no experiments with artificial lights or decoys have been done to investigate the details of the signalling. Females usually stop glowing after oviposition begins, whereas males are capable of glowing throughout their entire life.

2.4 Bioluminescence in Coleoptera (Beetles)

Luminescent beetles belong to the superfamilies Elateroidea (Elateridae) and Cantharoidea (Telegeusidae, Homalisidae, Phengodidae, and Lampyridae). Virtually nothing is known about bioluminescence in Telegeusidae and Homalisidae.

2.4.1 Elateridae – The Click Beetles

Elaters with luminescent adults (fire beetles) probably use their light for mating communication. In addition, adults may also use their light for illumination and defense (Lloyd 1983). The lanterns generally occur at the posterior corner of the prothorax and between the thorax and abdomen (for further details see Lloyd 1978). The various light organs seem to have different mechanisms of control (Lloyd 1971). In some species the abdominal organ emits longer wavelengths (a more reddish light) than the thoracic organs. In other species, light organs are sexually dimorphic. Males of various species can be attracted to artificial light, captive individuals, and also to lighted cigarettes.

Juveniles, but not eggs, are also luminescent; the placement of larval light organs varies among the species. Larval luminescence appears only at several months of age, and in some cases only for a brief period before and after ecdysis (Lloyd 1978). It is generally suggested that they use their light defensively. Larvae of the species *Pyrearinus* live in tunnels and apparently attract prey with their lights (Lloyd 1983).

When pupae are stimulated by light, they glow brightly all over. Pupal duration is usually about 21 days, and after 10-14 days it is possible to see the outline of the adult light organs.

2.4.2 Phengodidae – The Glowworm Beetles

Although bioluminescent communication probably occurs in this family, in most cases it may play a role secondary to chemical communication (long-range attraction of males by pheromones). Light communication may especially occur during periods of low humidity and high temperature when diurnal attraction is prevented and the insects become active at dusk (Lloyd 1971; see also Chap. 1). Luminescence may then act as an amplifier to enhance the weakening reflected light cues. The amplification of reflected light signals by luminescence seems to be one possible evolutionary pathway leading to bioluminescent communication in animals (Lloyd 1978).

Females are larviform and brightly luminescent. Several interesting variations in their light patterns have been described, and spectacular examples include the starworm, *Diplocladon hasselti,* and the railroad worm, *Phrixothrix.* The latter has red headlights which perhaps emit a color to which the predators are not sensitive. Yellow-green abdominal lights serve perhaps as warning signals (Sivinski 1981). Larval luminescence often appears similar or identical to that of adult females. Its function, however, has not been experimentally determined. Only in some species are males luminescent. Males of *Dioptoma adamsi* and *Phrixothrix tiemanni,* for example, display brightly during mating, perhaps for seduction, perhaps for defense, or both (Lloyd 1983).

The light organs of phengoids so far as it is known are of two types. In *Phengodes* they consist of loose aggregations of oenocyte-like cells and they lack tracheal supply. Light control is very slow, with illumination or dousing requiring several minutes. The second type is found, e.g., in *Phrixothrix,* a compact organ supplied with a trachea that branches profusely between the cells. Since these animals are able to change light intensity very quickly, their luminescence may be under neural control.

2.4.3 Lampyridae – The Fireflies

In fireflies, all stages of the lifecycle can be luminescent. Freshly laid eggs commonly emit a faint, diffuse glow, which soon disappears. A few days before hatching lighting recurs and seems to be centered on the light organs of the developing embryo. As far as it is known larvae of all fireflies are luminescent, including those that develop into nonluminous adults (Lloyd 1978). Larval light organs occur, with minor exceptions, as a bilateral pair on the eighth abdominal segment (Case and Strause 1978).

The universal presence of larval luminescence in fireflies suggests that it is of considerable biological significance, but no one really knows why the glowworms glow. Sivinski (1981) has developed a model that explains the historical accumulation of conflicting observations of larval lighting. His model incorporates different defensive tactics, including hiding in darkness and glowing if discovered or when another's light is seen (Lloyd 1983). Larvae glow spontaneously and intermittently when active, but also when immobile. Some species glow when mechanically or photically stimulated, but sometimes they cannot be made to glow even with strong mechanical or photic stimulation (Lloyd 1978). Even under controlled, constant conditions, illumination patterns vary greatly among individuals. Observations on circadian patterns of larval luminosity, but with large inter- and intraspecific variations, have been made by Schwalb (1961) and Dreisig (1975).

In most fireflies the adult light organ is probably a new structure and not related ontogenetically to the larval organ. Adult firefly luminescence is of great significance in mediating sexual interactions, in illumination during landing and takeoff, in predation on males of other species, and in defense against predators (Lloyd 1978). Wynne-Edwards (1962) has suggested the flashes of males function in population regulation. This, however, has been strongly criticized by several authors (Lloyd 1971).

3 Chemical, Physical, and Morphological Aspects of Firefly Light Emission

In this chapter chemical, physical, and anatomical charcteristics of firefly bioluminescence are described so far as they concern the coded systems of flash exchange and their ecological adaption.

3.1 Chemistry of Light Emission

Firefly luminescence is characterized by the existence of a specific substrate, luciferin, which was isolated for the first time by Bitler and McElroy (1957). Its structure was determined and proven by total synthesis by White et al. (1961, 1963). Almost all of the detailed biochemistry has been performed using extracts from the North American species *Photinus pyralis.* Adequate research on other species, however, has indicated that the fundamental biochemistry is the same for all true fireflies (McElroy and DeLuca 1978; Pagel and Wulff 1980). The mechanisms of firefly luminescence have been recently reviewed by McElroy and DeLuca (1978) and DeLuca and McElroy (1981). The requirements for firefly luminescence are shown in the following equations:

$$\text{luciferin} + \text{luciferase} + \text{ATP} \xrightarrow{Mg^{2+}} \text{luciferyl adenylate} \cdot \text{luciferase} + \text{pyrophosphate} \qquad (1)$$

$$\text{luciferyl adenylate} \cdot \text{luciferase} + O_2 \longrightarrow \text{oxyluciferin} + \text{luciferase} + CO_2 + \text{AMP} + \text{light} \qquad (2)$$

In addition to reactions (1) and (2), luciferase will catalyze the formation of dehydroluciferyl adenylate as shown in reaction (3):

$$\text{luciferin} + \text{luciferase} + \text{ATP} \xrightarrow{Mg^{2+}} \text{dehydroluciferyl adenylate} \cdot \text{luciferase} + \text{pyrophosphate} \qquad (3)$$

The dehydroluciferyl adenylate • luciferase complex is capable of catalyzing the breakdown of ATP to AMP and pyrophosphate. It is suggested that ATP itself causes the hydrolysis of luciferase-bound luciferyl adenylate after which dehydroluciferin reacts with enzyme and ATP to release pyrophosphate. Dehydroluciferyl adenylate is a very potent inhibitor of the light reaction. This fact may be a possible cause of the slow decay in luminescence following a rapid peak in light intensity, as observed after light emission is initiated by the injection of ATP in the luciferin-luciferase reaction system, but since it is also observed in the light flash of living fireflies (Cope 1976). The first reaction (1) is an activation step analogous to the fatty acid and amino acid activation reactions. The oxidation of the activated substrate occurs in step (2). The detailed reaction mechanisms have been recently reviewed by McCapra (1982) from the standpoint of organic chemistry. The controversy centered on the detailed mechanisms (e.g., an involution of a dioxetane intermediate or not), however, has not as yet been resolved.

The quantum yield (number of photons produced for each molecule of substrate oxidized) for the enzymatic oxidation of firefly luciferin is nearly unit. Since the emission of most fireflies ranges from 500 to 630 nm, the energy requirement for the bioluminescence step should exceed about 240 kJ/mole as an absolute minimum. Different maxima in the light emission of various firefly species appear to be due to the fact that there are structural differences in the luciferase molecule, since the luciferin is identical for all fireflies that have been studied (McElroy et al. 1965). However, the stereochemistry of the nucleotide attachment is also of importance in determining the color of the light (McElroy and De Luca 1981). Studies of the substrate binding properties of firefly luciferase have shown that there are two binding sites each for luciferin and ATP per 100 000 mol. wt. of the enzyme. These results are consistent with the observation that luciferase is a dimer of 50 000 mol. wt. subunits. However, the enzymatically active substrate, the magnesium-ATP complex [see reaction (1)], is bound to only one site per 100 000 mol. wt. (McElroy and De Luca 1978). In vitro the luciferase activity as well as the overall bioluminescence emission spectrum can be altered also by changes in the enzyme assay conditions (temperature, pH, solvent system, and metal ion concentrations) (McElroy et al. 1974; Ugarova et al. 1981; Brovko et al. 1982; Kricka and DeLuca 1982). For the larval stage of *Photuris pennsylvanica,* an isozyme of luciferase, distinct from the adult form has recently been observed by Strause and DeLuca (1981).

3.2 Physical Characteristics of Luminescence

As already mentioned above, the lights of fireflies are green to yellow, and emission maxima lie between 545 and 595 nm. Spectra of males and females are usually the same (Nicol 1978). Measured luminous intensities range over about two orders of magnitude (*Photuris:* $1\text{-}7 \times 10^{-6}$ $\mu W \cdot cm^{-2}$ receptor surface at a distance of 1 m; *Lampyris:* 2×10^{-4} $\mu W \cdot cm^{-2}$). In order to detect such a low intensity bioluminescent signal which would further decrease by the inverse square law over the distance of flash communication, the firefly visual system must possess low threshold photoreceptors exhibiting action spectra which correspond with the peaks of bioluminescence emission of the respective species (Lall et al. 1980). There is some electrophysiological evidence for photon capture by dark adapted firefly eyes equal to that of the human eye (Nicol 1978). Such a high efficiency in capturing photons is ensured by the particular organization of the rhabdomers in the firefly eye (Nicol 1978). For details of the spectral sensitivity of firefly eyes see Sec. 4.

3.3 Light Organ Morphology and Pharmacology

Adult and larval firefly light organs have been extensively studied since 1948 (for details see Case and Strause 1978 and Hoffmann 1981). In adult fireflies *(Photuris)* the nearly 15 000 photocytes are arranged around a central tracheal supply which is also the access route of the innervation. The central regions of the photocytes are packed with membrane-bound photocyte vesicles. Hanna et al. (1976) found that photocyte vesicles have the characteristics of peroxisomes, exhibiting peroxidase activity. Density gradient fractionation of photocyte particles showed the presence of catalase and luciferase in the vesicles. The light organ innervation originates in the ganglia of the sixth and seventh abdominal segments. Larval light organs have important structural differences from that of adults (Peterson 1970). A lack of tracheal end cells in larval light organs may be the cause for lower light intensities and much longer latency and duration of luminescence in larvae than in adults. Observations made by Smalley et al. (1977) implicate the photocyte vesicles as a storage site for luciferin and suggest that luciferin might be present there in a bonded form which undergoes a change in state associated with luminescence. Since luciferase is also present the photocyte vesicles appear competent to serve as the subcellular light source. Several investigators suggested that the neural transmitter in the light organ is a catacholamine. However, subsequent biochemical investigations by Robertson and Carlson (1976) determined octopamine as the only transmitter present in firefly lanterns.

3.4 Luminescence Control Theories

Two possible mechanisms of firefly luminescence control are extensively discussed in the literature, namely, the oxygen and the neural control theories. According to both, neural commands initiate events leading to flashing. In the case of the oxygen control theory, neural activity causes oxygen liberation into an oxygen-poor luminescent system. The neural control theory calls for a direct neural action on the light emitting cells.

As early as 1932, Snell demonstrated that a decrease in oxygen tension decreases the intensity and amount of light in the flash of *Photuris pennsylvanica* (Fig. 9.2 A). Furthermore, a decrease in oxygen tension below 2.67 kPa (20 mmHg) caused an increase in duration of flash (Fig. 9.2 B). Below 0.53 kPa (4 mmHg) flashing never occurred and the luminous organs glowed continuously. Experiments on luminescence reaction kinetics (DeLuca and McElroy 1974) showed that only 60ms are required to reach the peak of luminescence in conditions where oxygen is injected into an anaerobic mixture of enzyme and substrate, whereas several hundred ms are required if the reaction is initiated by combination of luciferin with luciferase and cofactors. Unfortunately, there is no direct evidence at present concerning oxygen levels in photocytes, and also there has been no plausible mechanism for neurally controlled oxygen gating advanced (Case and Strause 1978).

Arguments for the neural theory suffer from a similar array of uncertainties. None of the ions commonly gated by membrane potential changes (Na^+, K^+, Ca^{2+}) seem likely candidates for controlling luminescence chemistry (Case and Strause 1978).

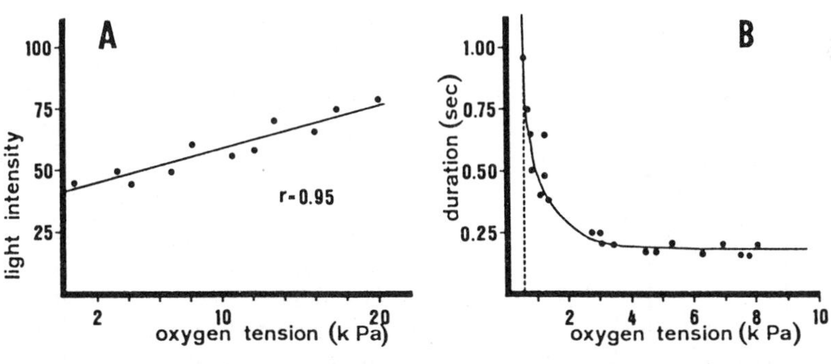

Fig. 9.2. A Maximum light intensity of brightest flashes of the firefly *Photuris pennsylvanica* at various oxygen tensions. **B** Duration of flashes in the firefly *Photuris pennsylvanica* at various oxygen tensions. (After Snell 1932)

Since in most mechanisms neural commands have to arise in the brain and nerve impulses have to be transmitted to the lanterns, several authors have investigated the organization of the neural structures in fireflies very closely. For example, in the rhythmically flashing males of *Luciola lusitanica* (see p. 238) the photomotor neurons of the brain are located in the deep protocerebral neuropile. Their rhythmic activity is the result of the interaction with an oscillator located in the optic lobes. The retina-lamina complex and the medulla may function as a modulator (Bagnoli et al. 1976). Although the females of *Luciola lusitanica* do not flash spontaneously, no fundamental difference seems to exist in the organization of the neural structures of both sexes.

4 Bioluminescence and Vision in Fireflies

In Fig. 9.3 histograms of the emission maxima for 55 species of North American fireflies are shown. The data form two broad groups: "early starting" species which begin flashing activity earlier than 30 min after sunset (dusk-active), and "late starting" species, which do not begin to flash before darkness (dark-active). Of 32 dark-active species, 23 emit green light ($\lambda_{max} \leqslant 558$ nm) and 21 of 23 dusk-active species emit yellow light ($\lambda_{max} \leqslant 560$ nm) (Lall et al. 1980). Even within closely related species, the earlier flashers exhibit the yellow color (e.g., *Photinus collustrans*), while late flashers emit green light (e.g., *Photinus tanytoxus*). For all species studied so far, the peak of the luminescence emission was identical with the peak of the visual spectral sensitivity of their compound eyes (Lall et al. 1980, 1981, 1982; Seliger et al. 1982a). In addition to the green (in dark-active species) or green-yellow (in dusk-active species) receptor type a near-UV maximum of sensitivity was found in firefly eyes. Near-UV radiation is more intense in open space than in dense foliage. Hence, the near-UV visual system in fireflies could faciliate orientation of the insects in open space. In the presence of only two photoreceptors (near-UV and green-yellow), wavelenght discrimination in the green-yellow region of the spectrum would be poor. With

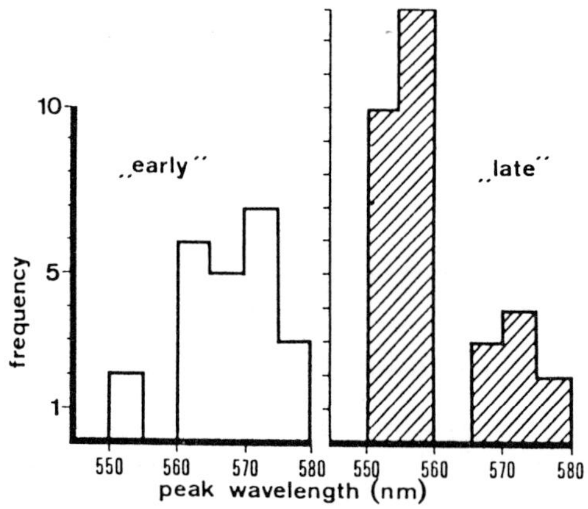

Fig. 9.3. Histrograms of the peak wavelength of emission spectra from 55 species of North American fireflies. The open area *(left side)* is representative of dusk-active (early-flashing) fireflies; the crosshatched area *(right side)* is representative of dark-active (late-flashing) species. (After Lall et al. 1980)

a third, violet receptor type, wavelenght discrimination across the entire spectrum is possible. Lall et al. (1982) actually demonstrated the presence of three (near-UV, violet, and green) receptor types in the compound eye of *Photuris lucicrescens,* and hence, a possibility of the existence of color vision as well. However, behavioral experiments have also shown that these animals respond essentially to the temporal pattern of the flash signals regardless of its color.

Based on the aforementioned results, Seliger et al. (1982b) developed an optimization model for the detection of firefly bioluminescence in nature during twilight. From this model it was proposed that in a green ambient light background, a shift to yellow visual sensitivity and yellow bioluminescence would increase the signal-to-noise ratio for detection of bioluminescence during twilight. The colors of firefly bioluminescence are, therefore, species-specific adaptations to optimize the detection of bioluminescence in the different photic environments in which the species have developed (Seliger et al. 1982a). A curve of biologically effective adaptation (BEA – a dimensionless ratio proportional to the relative advantage of a species to detect bioluminescence during twilight – calculated from distributions of ambient spectral radiances, visual spectral sensitivities, and bioluminescence emissions) like that in Fig. 9.4 may be used as a first approximation to predict the relative times of onset of flashing activity. The higher the value of BEA, the earlier a firefly species may initiate flashing activity.

Since fireflies frequently emit light in short or long flashes or pulses, sometimes respectively, these aspects of luminescence also have to be related to their visual performance. About 30 species of fireflies emit flickers. The flicker frequencies are usually 7-20 Hz, although one Florida *Photuris* (males) reaches 40 Hz (Lloyd 1978). The ability to distinguish individual flashes depends upon their frequency as well as on the critical flicker frequency of the observer (Nicol 1978). For *Luciola lusitanica* Brunelli et al. (1977) demonstrated discrete electroretinogram responses up to a flash repetition rate of 3.3 per second. In many fireflies, however, a photic stimulus exerts a long lasting inhibition on the photomotor centers, which determines the upper

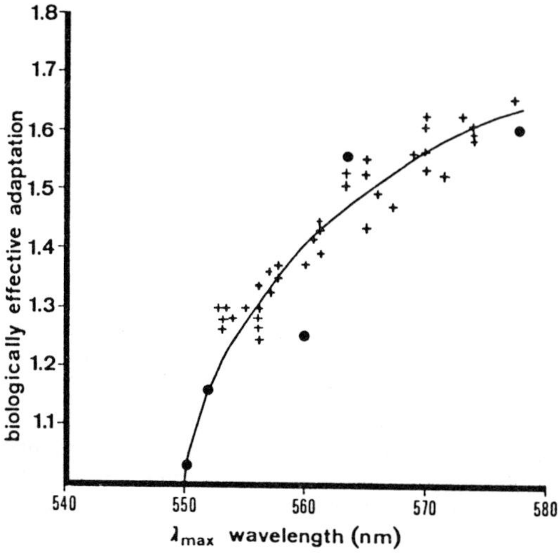

Fig. 9.4. Biologically effective adaptation, BEA, for North American fireflies as a function of peak wavelength of their emission spectra (λ_{max}). The five *circled* values represent the species *Photuris* "LR", *Photuris lucicrescens, Photinus collustrans, Photinus pyralis,* and *Photinus scintillans.* (After Seliger et al. 1982a)

limit of the rate of repetition. The duration of such an active inhibitory influence on the photomotor neurons is directly related to the stimulus intensity (for stimulus associated inhibitory effects see also p. 242).

5 Bioluminescence and Communication in Fireflies

5.1 General

Lampyrid sexual communication and communication ecology have been studied extensively. The sexual signals show considerable variation among different species, but the information being transferred is generally the same: "Broadcasters and correspondents inform potential and prospective mates of their sex, spatial location, specific identity, and quality" (Lloyd 1978). There are two fundamental patterns of signal interactions in fireflies (Fig. 9.5 A, B). In signal system I, one sex (usually the female, which presumably has the less risky sexual role) is stationary or sedentary, broadcasts a signal (sometimes continuously), and the other sex is attracted to it (e.g., in the European fireflies, *Lampyris noctiluca* and *Phausis splendidula*). In signal system II, one sex (usually the flying male) broadcasts a species-specific signal, the other sex responds in turn with a species-specific signal, and then the advertizer approaches the respondent (e.g., in *Photinus, Pyractomena, Photuris, Aspisoma* and some *Luciola spp.*). System I seems to be primitive, and the evolutionary precursor of system II (Lloyd 1978). It should be mentioned, however, that it is not possible to use this scheme to arrange, classify, and understand the signalling biology of all firefly species. The mating behavior of some fireflies is conspicuously complex. Synchronizing *Pteroptyx* fireflies of Southeast Asia (Sec. 5.3 of this Chapter; Fig. 9.5 C), for example, have prolonged sexual interactions involving a variety of luminescent emissions, and obscure dialogues. Such a complex mating program includes luminescent signals, signal systems, movements, and reaction chains, as well as specific sites of activity in the habitat, diurnal and seasonal times of occurrence, etc. (Lloyd 1978).

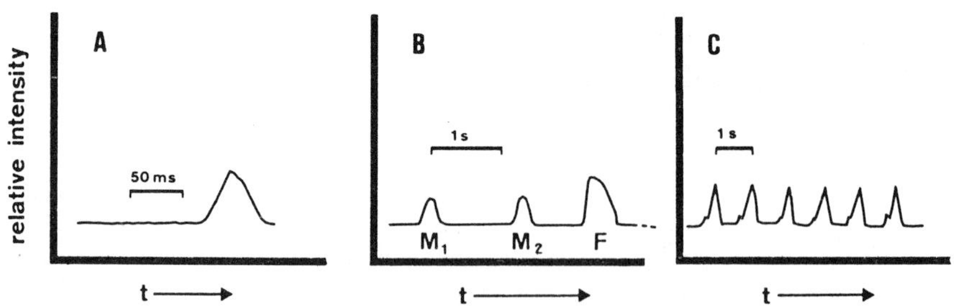

Fig. 9.5. A-C. Flash patterns of fireflies: **A** system I *(Lampyris noctiluca)*; **B** system II *(Photinus greeni; M_1, M_2,* male flash pair; *F* female flash response); **C** congregating, synchronous fireflies *(Pteroptyx malaccae)*. *Vertical axis,* relative intensity; *horizontal axis,* time. (After Hoffmann 1981)

5.2 Courtship Behavior in Individual Luminescent Fireflies

The elements that may be varied to produce distinctive, species-specific signals are color (see Sec. 4), brightness, and timing of the emission, shape, size, and movement of the light organ. Of course, there are limitations in the use of brightness and size for coding because their appearance varies with the distance from the emitter (Lloyd 1971). In experiments with the European fireflies, *Lampyris noctiluca* and *Phausis (= Lamprohiza) splendidula* (system I), however, Schwalb (1961) found differential responses to color and size of experimental decoys.

Most flashing fireflies have elaborated timing in their coding, and most use system II, because of its greater coding potential (Lloyd 1978). The importance of flash duration, number, and timing in the male flash pattern, and of the female delay, has been demonstrated in a huge number of species. Flash patterns of more than 130 species are known, and most of them have been electronically recorded. In *Photinus greeni,* for example, the female must be able to recognize as an information unit a pair of flashes separated by a particular range of intervals (1.1–1.7 s at 27°C) (Buck and Buck 1972). The two flashes of the male's stimulus phrase can be made multiple, or increased several-fold in duration, without preventing the female from responding, as long as the proper time relation between the start of the first flash and the subsequent one is preserved. The limit of the female's response latency is from 0.9–1.1 s (at 27°C), measured from the second of the male's paired flashes. Several other flash patterns of individual free-flying males are demonstrated in Fig. 9.6. Differences in flash pattern, such as those in Fig. 9.6, are important in reproductively isolated species that are sexually active in the same sites at the same time. Males of some species

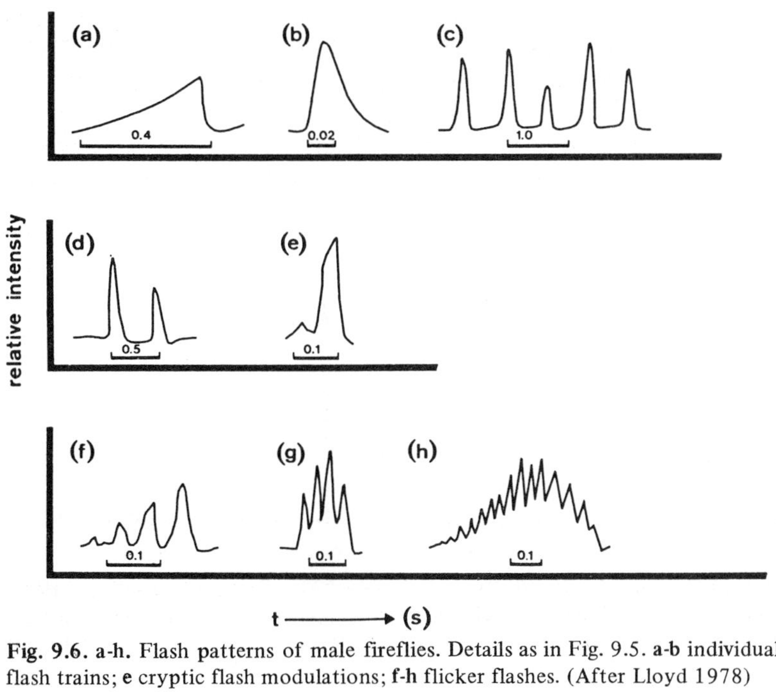

Fig. 9.6. a-h. Flash patterns of male fireflies. Details as in Fig. 9.5. **a-b** individual flash forms; **c-d** flash trains; **e** cryptic flash modulations; **f-h** flicker flashes. (After Lloyd 1978)

emit several flashes in quick successions even though the communicative unit is a single flash. This appears to enhance the changes of the male being seen by females that are perched in dense vegetation.

The few flashing patterns of females (system II) that are known are generally single, unmodulated flashes, although in some species (e.g., in *Luciola* and in a few *Photinus*), female patterns are also more complex (Lloyd 1978). The flashes of some females persist as dim glows that are probably used by males for orientation. Obviously, the time at which the female signal is emitted, and not the form or pattern of the signal, is the most important factor in females.

5.3 Synchronous Fireflies

Males of many firefly species emit their flashes in synchrony. The light emissions which are synchronized by the males vary from single flashes to multiple flickers and differ in frequency over a ten fold range (Buck and Buck 1978). Within a species, the synchrony is too perfect to be explained by a leader-follower relationship because nerve impulses do not travel fast enough for it to be possible (Buck and Buck 1968). One example of synchronous flashing is the (nonaggregating) harvester/searcher fireflies flying in high densities, such as *Photinus concisus* and *Photuris congener* in North America. Their mass synchrony may extend across the population for 30 m or more if neighboring males are able to maintain visible contact (Lloyd 1983). Synchrony, whether rhythmic or not, seems to be part of courtship in these species. By synchronizing his flashes with those of males near him, a male may be better able to detect the female's response which another male gets. Otte (1980) reviewed several theories which explain how synchrony might reduce the environmental noise level and improve sexual communication. Synchrony only occurs under conditions of close proximity between males. In *Photinus* and *Photuris* close spacing appears to be a passive consequence of local population density, whereas in many fireflies of Southeast Asia (genus *Pteroptyx*) close spacing is achieved through active congregation of males.

Reports of synchronous rhythmic flashing by fireflies in Southeast Asia have appeared for more than 200 years. The classical displays, as seen for example along the banks of the Chao Pyraya River below Bangkok, comprise "myriads of fireflies perched on the leaves of mangrove trees, flashing in unison, hour after hour, night after night, for weeks or even months" (Buck and Buck 1978) (Fig. 9.5 C). Although mangroves are important firefly perches, numerous other trees have also been mentioned as firefly trees. There seems to be no causal relationship between synchronous flashing, or even congregation, and particular kinds of trees (Buck and Buck 1966). Although only males are synchronously flashing, females are also present in the trees and mating pairs are common. Many males and females remain assembled in trees by day, and the display of the following evening is presumably begun by these fireflies. Lloyd (1973) recorded the flashed mating signals of 22 species in the genera *Pteroptyx* and *Luciola*. The results provided indirect evidence that information-coding parameters are similar to those of New World *Photinus* species (duration of male flashes, the period of flashes or flicker phrases in long series, and the period of flash and flicker modulation within phrases). However, a new element of potential coding significance, the intensity ratios of modulations within flashes, was revealed by the emission of three *Pteroptyx* species, including *Pteroptyx cribellata.* Of course, apparent brightness is a function of distance, and therefore, there are several limitations on coding with this parameter (see also p. 232). But a particular intensity ratio between two flashes could code information if the ratios were constant and not significantly altered by any change in distance that occurred during the time between flashes. In these three fireflies the modulation ratios are fairly constant and the modulation occurs so rapidly that the ratios are probably little, if at all, distorted.

Although mass synchronous flashing of fireflies is conspicuous and seems to be associated with mating behavior, its biological significance in reproduction is still disputed (Buck and Buck 1978; Otte 1980; Lloyd 1983). Recently, display, courting, and mating behavior in the aggregative, synchronously flashing species *Pteroptyx tener* were extensively analyzed (Case 1980). Males flash synchronously with other males in many phases of courtship, competition with other males and mating, but also during flight. The courting male perches on the back of the female and flashes his lantern directly into her eyes. Characteristic flash exchanges occur before copulation (but never prior to actual contact), which is accompanied by flashing of both sexes. The holding of the male's lantern over the female's head while flashing may reduce her dark adaptation and thereby limit her ability to see luminescence of other males. Another possibility is that the male's flashing might alter any possibly attractive flashing by the female to a less attractive or less obvious type to other searching males. Also, a strong photic stimulation might be a necessary initiator of female copulatory behavior. However, known differences on the behavior of different habitual synchronizers caution against assuming that all synchronizing species have the same photic reproductive adaptations.

Only recently have investigators attempted to study the physiological mechanisms underlying synchronous flashing. In many approaches synchronously flashing fireflies are examined as possible modes for pacemaker activity or multiple oscillator systems. Analogies are drawn with such phenomena as circadian rhythms (see Sec. 3.3 in Chap. 1 and Sec. 3.1 in Chap. 8). Males flash spontaneously at a regular species- and temperature-dependent frequency (Fig. 9.7). This free run rhythm will be timed by a

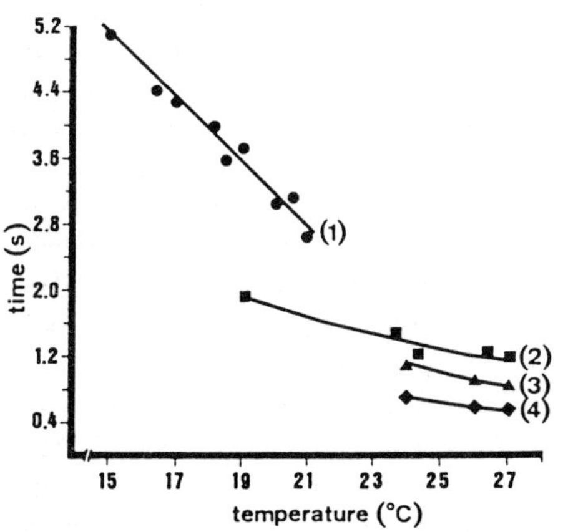

Fig. 9.7. Relationship between flash periods and ambient temperature in four species of Malanesian fireflies. *1 Pteroptyx cribellata; 2 Pteroptyx sp.; 3 Luciola pupilla; 4 Pteroptyx antennata.* (After Lloyd 1973)

pacemaker in the brain that periodically triggers a neural volley which runs down the cord to the light organ in the abdomen and there evokes a flash (Buck and Buck 1976). Data from display flashing of *Pteroptyx cribellata* under laboratory conditions (Buck et al. 1981) showed that the duration of the flash-to-flash interval can be used as a measure of endogenous pacemaker timing behavior. The cycling of the pacemaker runs continuously, does not require visual feedback, is dependent of the intensity of the flash, or even of its occurrence, and undergoes occasional spontaneous phase-shifting. Its periods are independent in duration, cycle-by-cycle, and hence, not subject to additional regulation by a reference oscillator. The spontaneous flashing of the female, which was observed in *Pteroptyx cribellata,* has a radically different period from that of the males, thus, arguing against the possibility that the same pacemaker may be used for control of flash frequency in the male and for frequency recognition in the female.

An interesting puzzle has been posed in a laboratory experiment (Buck and Buck 1978) where males of *Pteroptyx cribellata* were observed to synchronize with males of *Luciola pupilla.* Since *Pteroptyx cribellata* occurs in at least two localities also inhabiting *Luciola pupilla,* it seems possible that these species of different genera, but essentially identical flash period (1000 ms), could congregate and synchronize together.

The tendency for synchronizing spezies to live in areas that are densely vegetated (or were in preagricultural past), whereas roving species tend to inhabit relatively open sites, may be significant in terms of the evolutionary origin of this behavior. But since snychronous and roving types sometimes occur together today, the ecological correlation is by no means definite.

5.4 Interspecific Light Communication

5.4.1 Genera Discrimination

The two European fireflies, *Lampyris noctiluca* and *Phausis splendidula,* live in the same habitat. Males of *Lampyris noctiluca* distinguish the photophore pattern of their females from that of the females of *Phausis splendidula* (Schwalb 1961). *Phausis* males, however, do not discriminate between the two patterns, and *Phausis* males, thus, attracted by *Lampyris* females are eaten.

5.4.2 Aggressive Mimicry

Females of at least 12 species of *Photuris* (from temperate and tropical America) are specialized predators of male fireflies of several genera (Lloyd 1978). Apparently, only those species which are nonpredatory occur "alone" in ecological restricted areas (Lloyd 1983). The predatory females lure males by mimicking the mating signals of the males' own females. The males are then seized and devoured. Females of *Photuris versicolor* and *Photuris* "B", for example, mimic the reply of *Photinus macdermotti* females, thus, attracting *Photinus macdermotti* males, which they eat. While the predators are decoying *Photinus macdermotti* males, they even produce "extra" flashes that mimic the "competition" flashes used by the males when competing for their own females (Lloyd 1981). This is remarkable insofar as the responses are not imitations, but apparently unlearned answers. Predatory use of luminescence by females of several species of the genus *Photuris* is probably of much wider significance than mere nutrition (Buck 1978). Predation by *Photuris* females is probably the single most important influence on the communication of New World fireflies (Lloyd 1983). Aggressive mimicry is a New World phenomenon exclusively. The flashing species of Asia, Africa, and Europe have not had this sort of pressure on their communication. To avoid or reduce risk from predators, prey species must avoid risky air space, activity times, signal patterns, and flight patterns, use special tactics, including deceptive signals, hiding against background, and quick movements. For example, in summer when *Photuris* "D", an extreme hawker, is active, *Photinus macdermotti* males begin activity more than a half-hour later when daylight is greatly reduced, and they fly less (Lloyd 1983).

5.4.3 Defensive Strategies

The presence of luminescent organs almost guarantees that fireflies will constitute conspicuous targets for both vertebrate and invertebrate predators. The facts that fireflies are often rejected after being tasted by other terrestrial arthropods, lizards, frogs, and birds and are rarely preyed on by bats provides support for the proposal that firefly flashing is a warning of distastefulness (Müllerian mimicry; see also Sec. 4.1 in Chap. 8). As early as 1934, Crawford stated that many lampyrids would be unpalatable because of their odoriferous qualities. The reflex bleeding in various species of *Photinus* and *Photuris*, as described by Blum and Sannasi (1974) and Kloft et al. (1975), is a defensive adaptation suited to invertebrate predators. If a terrestrial arthropod predator seizes a lampyrid, the former will almost certainly encounter the primed battery of autohemorrhagic sites positioned around the elytral and pronotal margins of the firefly. The distribution of reflex bleeding points maximizes the probability that an aggressive arthropod will be engulfed by viscid blood. In addition, the enrichment of the blood with astrigent compounds makes this fluid a deterrent for at least some species of vertebrates as well (Blum and Sannasi 1974).

5.5 Illumination

Another apparent function of bioluminescence is seen in the marked temporary increase in frequency of flashing of certain lampyrid fireflies at the time the insect alights (Lloyd 1968). This flicker is presumed to help in choosing a perch and in avoiding hazards, such as water or spider webs (Buck 1978). Many fireflies, however, do not show this behavior.

6 Environmental Control of Luminescent Activity in Fireflies

The daily onset of activity in fireflies is a complicated process involving both a circadian rhythm and a triggering mechanism controlled by ambient illumination (see also p. 238). Dreisig (1971) observed the time of start and duration of glowing in females of the European firefly *Lampyris noctiluca* (system I) in Denmark. The light emission of the females is a behavior that is synchronized in all the individuals comprising the population. The start of activity is related to a certain low illumination (Fig. 9.8 A). On the average the first start was seen at 8.3 lux and the last one at 0.2 lux (which means an average critical illumination – the illumination at the time when 50% of the population has commenced activity – of 1.4 lux). The variation in the start of activity is probably due to individual differences in response to light and to differences in illumination at the hideouts of the animals. This assumption was confirmed by laboratory experiments under seminatural conditions (Dreisig 1975). If the jars, in which the animals were kept were not provided with vegetation and the animals then directly exposed to the illumination, the level of critical illumination was considerably lower. The average period of glowing was 62.5 min. However, the daily mean ranged from 27–118 min, being almost solely dependent on the number of males present and the number of matings taking place (during copulation females stop glowing and never resume it). The activity of the males started about 55 min later than that of the females (average critical illumination: 0.032 lux). It also stopped earlier, most often because the temperature dropped below 12°C, when the activity of the males decreased sharply. The activity of the males, and therefore, indirectly that of the females, was also inhibited by wind. Males completely suspended activity at wind velocities beyond 3–4 m/s.

In the Florida fireflies *Photuris congener* and *Photinus umbratus* the variation in onset of activity differs from that in *Lampyris noctiluca* (Dreisig 1975). During the long twilight of summer in Denmark, *Lampyris noctiluca* has a greater variation of

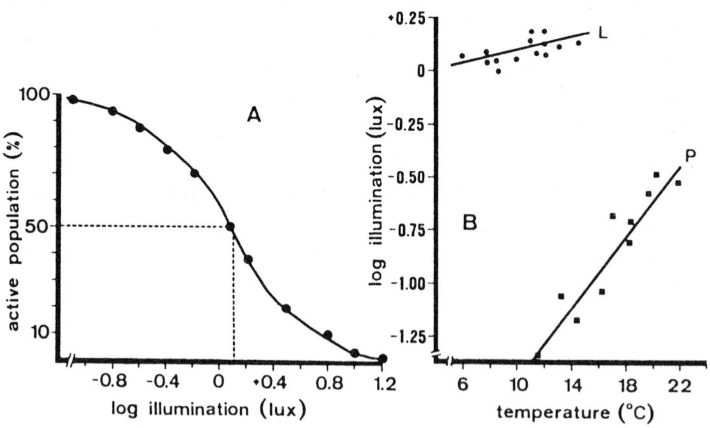

Fig. 9.8. A Effect of average illumination (log lux) on luminescence activity of *Lampyris noctiluca* (After Dreisig 1971). **B** Relationship between air temperature and critical illumination at the daily onset of luminescence activity in *Lampyris noctiluca (L)* and *Photuris congener (P)*. (After Dreisig 1975)

onsets than have the American fireflies during the short twilight of Florida, although the period of onset of activity occurs within approximately the same range of illumination. Buck as early as 1937 mentioned that *Photinus pyralis* starts to flash earlier at higher rather than lower temperatures. This was also seen in *Photuris congener* and in *Lampyris noctiluca*. In both species, the onset of activity occurred at a lower illumination in low rather than in high temperatures (Fig. 9.8 B). The results indicated that the light emission of *Lampyris* could be compared with other endogenously controlled activity rhythms that are free-running under constant conditions, but normally are entrained to a light-dark cycle (Dreisig 1975). Therefore, the release of luminescence activity depends on an endogenous circadian rhythm of sensitization (with a period from 19–23.5 h in *Lampyris*; Dreisig 1976). The animal must be in a certain state of receptivity, if an external signal (critical illumination) is to be effective.

The luminescence ecology of a West African firefly, *Luciola discicollis,* was extensively studied by Kaufmann (1965). Immediately after sundown, males rise from the herbage and fly hovering over the habitat. While on the wing, the males give short, quick flashes in rapid succession, each flash lasting an average of 0.35 s. The females remain in the grass and flash, halfway between the top and the root of the grass blade. The female light is weaker and the flashing rhythm is slower. No definite exchange of flashes between the sexes was found, as each sex flashed independently of the other. Yet there is no doubt whatsoever about the function of adult light organs in bringing the sexes together, since males failed to recognize females when the photogenic organs of the latter were coated with dark paint. Light intensity, height and density of vegetation, and soil conditions, in addition to temperature and moisture, were the important factors in determining their habitat. Areas which are otherwise habitable, but are exposed nightly to artificial lighting are not suitable for activity. Even those habitats where many fireflies congregate and display their lights may temporarily become unfavorable in moonlight, with the insects remaining hidden in the vegetation. The larvae of *Luciola discicollis* are also luminous, but they are less sensitive to weather conditions than adults. When the temperature falls to 14°–$15^{o}C$ at the grass level, larvae may still flash, but adults are totally inactive. The optimum temperature for flashing activity is between 19° and $26^{o}C$, with a relative humidity of 90% or higher. The duration of larval flash is between 0.8 and 2.0 s, depending upon the temperature and humidity. The function of larval flashing also remains uncertain in this species. Kaufmann (1965) suggested that larval flashing may contribute to the survival of the species. Since larvae feed on firefly eggs, and ovipositing females may recognize larval lights, just as they do male lights, it is quite probable that the females take off to migrate into less crowded areas when the number of larvae in a habitat exceeds a certain critical limit.

7 Conclusions

Luminescent insects have become the subject of choice for approaching fundamental questions about communication in nonsocial insects. The following subjects have been recently compiled by Lloyd (1983) as the most important factors influencing firefly communication: (1) Phylogenetic opportunities and constraints. A system II signal format (i.e., flash pattern – flash response), for example, cannot evolve quickly from a signal I format in a species that has nonluminous males. Those species might then switch to pheromone communication or change their activity in time or space. (2) Physical characteristics of the habitat, biotic and abiotic, including duration of twilight, ambient light quality, height and density of vegetation, wind, and temperature. (3) Predators, including aggressive mimics. Signal mimicry has probably caused some prey species to modify their signals and to incorporate codes and devices to which predators cannot adjust. (4) Sexual selection, including male congregation, female choice, and reproductive isolation.

The unique spectacle of mass congregation and flash synchrony in some Southeast Asian fireflies probably evolved when a number of males acting in concert attracted a greater per male quota of females than solitary males. The "noise problem" created by the crowd of males was then alleviated through the synchrony in flashing (Otte 1980).

Finally, it should be mentioned that bioluminescence, especially firefly bioluminescence, has been used routinely as an analytical tool for about 30 years.

Acknowledgement. I am grateful to Dr. N. Mathers, Ulm who critically read the mansucript.

Final Remarks

The uniqueness of living organisms lies in their ability to operate as an open thermodynamic system, through which energy is obtained from the environment and used to maintain life. The evolution of specific animals, however, required an ability to maintain a constancy in their organization and functioning (internal environment), more or less independent of the external environment. In previous chapters we have shown that many questions relating to operation and evolution of insect metabolism cannot be answered without consideration of animals' ecology. We have also presented a number of homoeostatic mechanisms designed to keep the internal environment of insects constant, even when insects' external environment is highly variable.

It is important to recognize that the diverse reactions in insects to environmental changes do not occur in isolation; rather they are components in a complex of integrated biochemical pathways which respond to various parameters in environmental and physiological conditions so as to promote the biological success of the individual and the species (Downer and Kallapur 1981). Each of the mechanisms shown may be part of a mosaic for optimization of insects' efficiency in the fluctuating environment. The idea that the neuro-endocrine system is a mediator between environmental factors and (sub)cellular responses together with the existence of neurohormonal multiregulation systems in insects may explain the complexity in their answer to environmental stimuli. Details in this highly complex neuro-endocrine integration must be studied in further detailed experiments.

References

Abul-Nasr S, Awadallah KT, Omar HM (1974a) Effect of aging of cotton bolls on diapause in the pink bollworm, *Pectinophora gossypiella* (Lepidoptera, Gelechiidae). Bull Soc Entomol Egypte 58:303–308

Abul-Nasr S, Awadallah KT, Omar HM (1974b) Oil and moisture contents of cotton bolls and seeds as factors inducing diapause in the pink bollworm, *Pectinophora gossypiella* (Lepidoptera, Gelechiidae). Bull Soc Entomol Egypte 58:405–413

Afzelius BA, Gonnert N (1972) Intramitochondrial tracheoles in flight muscle from the hornet *Vespa crabro*. J Submicr OSC Cytol 4:1–6

Ahearn GA (1970) The control of water loss in desert tenebrionid beetles. J Exp Biol 53: 573–595

Alexandrov VY (1977) Cells, molecules and temperature. Springer, Berlin Heidelberg New York

Allen JC (2976) A modified sine wave method for calculating degree days. Environ Entomol 5:388–396

Altmann G (1956) Die Regulation des Wasserhaushaltes der Honigbiene. Insects Soc 3:33–40

Anctil M (1979) Physiological control of bioluminescence. Photochem Photobiol 30:777–780

Anders G, Drawert F, Anders A, Reuther KH (1964) Über kausale Zusammenhänge zwischen der Zuchttemperatur, dem Aminosäure-Pool und einigen quantitativen morphologischen Phänen bei *Drosophila melanogaster*. Z Naturforsch 196:495–499

Andersen SO (1980) Cuticular sclerotization. In: Miller TA (ed) Cuticle techniques in arthropods. Springer, Berlin Heidelberg New York, pp 185–215

Anderson JM, Cormier MJ (1976) Transductive coupling in bioluminescence. Biochem Biophys Res Commun 68: 1234–1241

Anderson RL, Mutchmor JA (1971) Temperature acclimation in *Tribolium* and *Musca* at locomotory, metabolic and enzyme levels. J Insect Physiol 17:2205–2219

Andrewartha HG, Birch LC (1954) The distribution and abundance of animals. The University of Chicago Press, Chicago London

Andrewartha HG, Birch LC (1960) Some recent contributions to the study of the distribution and abundance of insects. Annu Rev Entomol 5:219–242

Andrusiak EW, Phillips JE, Speight J (1980) Phosphate transport by locust rectum in vitro. Can J Zool 58:1518–1523

Angersbach D (1975) The direction of incident light and its perception in the control of pupal melanization in *Pieris brassicae*. J Insect Physiol 21:1691–1696

Applebaum SW, Janković M, Birk Y (1971) Studies on the midgut amylase activity of *Tenebrio molitor* L. larvae. J Insect Physiol 7:100–108

Applebaum SW, Janković M, Grozdanović J, Marinković D (1964) Compensation for temperature in the digestive metabolism of *Tenebrio molitor* larvae. Physiol Zool 37:90–95

Ar A, Paganelli CV, Farhi LE, Rahn H (1980) Convective gas flow caused by diffusion explains positive hydrostatic pressure in hen's eggs. Physiology 23:178

Archer TL, Musick GL, Murray RL (1980) Influence of temperature and moisture on black cutworm (Lepidoptera: Noctuidae) development and reproduction. Can Entomol 112:665–673

Asahina E (1966) Freezing and frost resistance in insects. In: Meryman HT (ed) Cryobiologie. Academic, New York

Aston RJ (1975) The role of adenosine 3', 5' – cyclic monophosphate in relation to the diuretic hormone of *Rhodnius prolixus*. J Insect Physiol 21:1873–1877

Aston RJ, Hughes L (1980) Diuretic hormone extraction and chemical properties. In: Miller TA (ed) Neurohormonal techniques in insects. Springer, Berlin Heidelberg New York, pp 91–115

Atzler M (1931) Untersuchungen über den morphologischen und physiologischen Farbwechsel von *Dixippus (Carausius) morosus*. Z Vgl Physiol 13:505–533

Audit-Lamour C, Busson D (1981) Oögenesis defects in the ecd-1 mutant of *Drosophila melanogaster*, deficient in ecdysteroid at high temperature. J Insect Physiol 27:829–837

Awiti LRS, Hidaka T (1982) Neuroendocrine mechanism involved in pupal colour dimorphism in swallowtail *Papilio xuthus* L. Insect Sci Application 3:181–191

Bagnoli P, Brunelli M, Magni F, Musumeci D (1976) Neural mechanisms underlying spontaneous flashing and its modulation in the firefly *Luciola lusitanica*. J Comp Physiol A 108:133–156

Bailey E (1975) Biochemistry of insect flight. Part 2- Fuel supply. In: Candy DJ, Kilby BA (eds) Insect biochemistry and function. Chapman and Hall, London, pp 89–117

Baker JE (1983) Temperature regulation of larval size and development in *Attagenus megatoma* (Coleoptera: Dermestidae). Ann Entomol Soc Am 76: 752–756

Barbier M (1981) The status of blue-green bile pigments of butterflies, and their phototransformations. Experientia (Basel) 37:1060–1062

Barker JF, Herman WS (1976) Effect of photoperiod and temperature on reproduction of the monarch butterfly *Danaus plexippus*. J Insect Physiol 22:1565–1568

Barnes HF (1952) Studies of fluctuations of insect populations. XII. Further evidence of prolonged larval life in the wheat blossom midges. Am Appl Biol 39:370–378

Bartholomew GA (1981) A matter of size: An examination of endothermy in insects and terrestrial vertebrates. In: Heinrich B (ed) Insect thermoregulation. Wiley, New York, pp 45–78

Bari MA, Lange WH (1980) Influence of temperature on the development, fecundity, and longevity of the artichoke plume moth. Environ Entomol 9: 673–676

Bassot J-M (1978) Les corps noirs, cellules géantes du Diptère mycétophilide lumnieux *Platyura fultoni* et leur sécrétion mitochondriale. C R Acad Sci (Paris) Ser D 286:623–626

Baumeister T, Meredith J, Julien W, Phillips JE (1981) Acetate transport by locust rectum in vivo. J Insect Physiol 27:195–201

Baust JG (1972) Temperature-induced neural adaptations of motoneuron discharge in the Alaskan beetle *Pterostichus brevicornis* (Carabidae). Comp Biochem Physiol 41A:205–213

Baust JG, Lee RE (1982) Environmental triggers to cryoprotectant modulation in separate populations of the gall fly, *Eurosta solidaginis* (Fitch). J Insect Physiol 28:431–436

Baust JG, Miller LK (1970) Variations in glycerol content and its influence on cold hardiness in the Alaskan carabid beetle, *Pterostichus brevicornis*. J Insect Physiol 16:979–990

Baust JG, Lee RE, Ring RA (1982) The physiology and biochemistry of low temperature tolerance in insects and other terrestrial arthropods: A bibliography. Cryo-Letters 3:191–212

Beach RF, Craig GB (1979) Photo inhibition of diapause in field populations of *Aedes atropalpus*. Environ Entomol 8:392–396

Beament JWL (1964) Active transport and passive movement of water in insects. Adv Insect Physiol 2:67–129

Beament JWL (1965) The active transport of water: Evidence, models and mechanisms. Symp Soc Exp Biol 2:273–298

Bean DW, Beck SD (1980) The role of juvenile hormone in the larval diapause of the European corn borer, *Ostrinia nubilalis*. J Insect Physiol 26:579–584

Bean DW, Beck SD (1983) Haemolymph exdysteroid titres in diapause and nondiapause larvae of the European cornborer, *Ostrinia nubilalis*. J Insect Physiol 29:687–693

Beck SD (1980) Insect photoperiodism, 2nd edn. Academic, New York

Beck SD (1983a) Thermal and thermoperiodic effects on larval development and diapause in the European corn borer, *Ostrinia nubilalis*. J Insect Physiol 29:107–112

Beck SD (1983b) Insect thermoperiodism. Annu Rev Entomol 28: 91–108

Beckel WE (1958) The morphology, histology and physiology of the spiracular regulatory apparatus of *Hyalophora cecropia* (L.). Proc 10th Intern Congr Entomol 2:87–115

Beckwith RC (1982) Effects of constant laboratory temperatures on the Douglas-fir tussock moth (Lepidoptera: Lymantriidae). Environ Entomol 11:1159–1163

Beenakkers AMT (1965) Transport of fatty acids in *Locusta migratoria* during sustained flight. J Insect Physiol 11:879–888

Beenakkers AMT, van den Broek ATM, de Ronde TJA (1975) Development of catabolic pathways in insect flight muscles. A comparative study. J Insect Physiol 21:849–859

Beenakkers AMT, van der Horst DJ, van Marrewijk WJA (1978) Regulation of release and metabolic function of the adipokinetic hormone in insects. In: Gaillard PJ, Boers HH (eds) Comparative endocrinology. Elsevier, North-Holland, Amsterdam, pp 445–448

Beenakkers AMT, van der Horst DJ, van Marrewijk WJA (1981a) Role of lipids in energy metabolism. In: Downer RGH (ed) Energy metabolism in insects. Plenum, New York London, pp 53–100

Beenakkers AMT, van der Horst DJ, van Marrewijk WJA (1981b) Metabolism during locust flight. Comp Biochem Physiol 69B:315–321

Behrens W, Hoffmann KH (1982) Function of ecdysteroids in adult insects. Acta Endocrinol 99 (Suppl 146):31–32

Behrens W, Hoffmann KH (1983) Effects of exogenous ecdysteroids on reproduction in crickets, *Gryllus bimaculatus*. Int J Invertebr Reprod 6:149–159

Behrens W, Hoffmann KH, Ressin WJ (1980) Temperaturspezifische Unterschiede im Ecdysteroidtiter bei adulten Feldgrillen, *Gryllus bimaculatus* de Geer. Verh Dtsch Zool Ges 1980:313

Behrens W, Hoffmann KH, Kempa S, Gässler S, Merkel-Wallner G (1983) Effects of diurnal thermoperiods and quickly oscillating temperatures on the development and reproduction of crickets, *Gryllus bimaculatus*. Oecologia (Berl) 59:279–287

Bell WJ (1982) Pheromones and behaviour. In: Bell WJ, Adiyodi KG (eds) The American cockroach. Chapman and Hall, London New York, pp 371–397

Benestad E (1970) Laboratory experiments on the biology of *Syrphus corollae* (Fabr) (Dipt., Syrphidae). Nor Entomol Tidsskr 17:77–85

Bennettova B, Fraenkel G (1981) What determines the number of ovarioles in a fly ovary? J Insect Physiol 27:403–410

Bernays EA, Simpson SJ (1982) Control of food intake. Adv Insect Physiol 16:59–118

Berridge MJ, Oschman JL (1972) Transporting epithelia. Academic, New York London

Bertalanffy L von (1957) Quantitative laws in metabolism and growth. Q Rev Biol 32:217–231

Berthold G (1977) The migration of pigment granules in the epidermis during physiological colour change of *Carausius morosus*. J Insect Physiol 23:235–240

Berthold G (1980) Microtubules in the epidermal cells of *Carausius morosus* (Br.). Their pattern and relation to pigment migration. J Insect Physiol 26:421–425

Berthold G (1982) Histochemical localization of magnesium activated ATPase on microtubule bundles in epidermal cells of *Carausius morosus*. Possible significance to pigment migration. J Insect Physiol 28:137–142

Berthold G, Bückmann D (1975) Morphologischer Farbwechsel and Kynurensäureexkretion bei der Stabheuschrecke *Carausius morosus* Br. J Comp Physiol 100:347–350

Berthold G, Bückmann D (1976) Pterin- und Ommochromgehalt im Integument grüner und schwarzbrauner Imagines von *Carausius morosus* Br. J Comp Physiol 111:33–37

Bestmann HJ, Dippold K (1983) Temperaturabhängigkeit von Elektroantennogrammen bei Lepidopteren. Naturwissenschaften 70:47–48

Beydon P, Mauchamp B, Lhonoré J, Bouthier A, Lafont R (1980) The epidermal cell cycle during the last larval instar of *Pieris brassicae* (Lepidoptera). J Comp Physiol 136:21–29

Bhaskaran G (1981) Regulation of corpus allatum activity in last instar *Manduca sexta* larvae. In: Bhaskaran G, Friedman S, Rodriguez JG (eds) Current topics in insect endocrinology and nutrition. Plenum, New York, pp 53–81

Bishai FR (1960) On some aspects of respiration in locusts. PHD Thesis, Ain Shams University, Kairo

Bishai FR, Zebe E (1959) Enzymverteilungsmuster und Metabolitspiegel in der Beinmuskulatur der Heuschrecke. Verh Dtsch Zool Ges, Münster, Zool Anz 23, Suppl, p 314

Bishop JA, Cook LM, Muggleton J (1977) The response of two species of moths to industrialization in Northwest England. Part 1. Polymorphisms for melanism. Philos Trans R Soc Lond B Biol Sci 281: 489–516

Bitler B, McElroy WD (1957) Preparation and properties of crystalline firefly luciferin. Arch Biochem Biophys 72: 358–368

Bligh J, Cloudsley-Thompson JL, Macdonals AG (eds) (1976) Environmental physiology of animals. Blackwell Scientific Publications, Oxford

Block W, Sømme L (1983) Low temperature adaptations in beetles from the sub-antarctic Island of South Georgia. Polar Biol 2: 109–114

Block W, Young SR (1978) Metabolic adaptations of Antarctic terrestrial micro-arthropods. Comp Biochem Physiol 61A: 363–368

Blum MS, Sannasi A (1974) Reflex bleeding in the lampyrid *Photinus pyralis*: defensive function. J Insect Physiol 20:451–460

Blunck H (1914) Die Entwicklung des *Dytiscus marginalis* vom Ei bis zur Imago. 1. Teil: Das Embryonalleben. Z Wiss Zool 111:76–151

Boller EF, Prokopy RJ (1976) Bionomics and management of *Rhagoletis*. Annu Rev Entomol 21:223–246

Bonnemaison L (1946) Actions des températures constantes ou variables sur le développement d'un Hémiptère: *Eurydema ornatum* L. (Pentat). Ann Epiphyt (Paris) 12: 115

Bosquet G (1971) Evolution de la consommation d'oxygène au cours du jeûne chez *Bombyx mori*. C R Acad Sci (Paris) Ser D 273:765–759

Bracken GK (1982) The Bertha armyworm *Mamestra configurata* (Lepidoptera: Noctuidae): effects of dietary linolenic acid on pupal syndrome, wing syndrome, survival, and fat composition. Can Entomol 114:567–574

Bradshaw WE (1980) Thermoperiodism and the thermal environment of the pitcher-plant mosquito, *Wyeomyia smithii*. Oecologia (Berl) 46:13–17

Bradshaw WE, Phillips DL (1980) Photoperiodism and the photic environment of the pitcher plant mosquito, *Wyeomyia smithii*. Oecologia (Berl) 44:311–316

Braune HJ (1971) Der Einfluß der Temperatur auf Eidiapause und Entwicklung von Weichwanzen (Heteroptera, Miridae). Oecologia (Berl) 8:223–266

Braune HJ (1980) Ökophysiologische Untersuchungen über die Steuerung der Eidiapause bei *Leptopterna dolobrata* (Heteroptera, Miridae). Zool Jahrb Syst 107:32–112

Briers T, de Loof A (1981) Moulting hormone activity in the adult Colorado potato beetle, *Leptinotarsa decemlineata* Say in relation to reproduction and diapause. Int J Invertebr Reprod 3:145–155

Bridges RG, Watts SG (1975) Changes in fatty acid composition of phospholipids and triglycerides of *Musca domestica* resulting from choline deficiency. J Insect Physiol 21:861–871

Bridges CR, Kestler P, Scheid P (1980) Tracheal volume in the pupa of the saturniid moth *Hyalophora cecropia* determined with inert gases. Respir Physiology 40:281–291

Bridges CR, Scheid P (1982) Buffering and dissocation of CO_2 body fluids in the pupa of the silkworm moth, *Hyalophora cecropia*. Respir Physiology 48:183–197

Briers T, Peferoen M, de Loof A (1982) Ecdysteroids and adult diapause in the Colorado potato beetle, *Leptinotarsa decemlineata*. Physiol Entomol 7:379–386

Britz L, Höhne W (1955) Temperaturschwankung und Entwicklungsgeschwindigkeit bei *Anopheles atroparvus* (Diptera, Culicidae). Z Angew Zool 42:208–234

Brockway AP (1967) Water loss in Saturniid pupae. Am Zool 8:777

Brockway AP, Schneiderman HA (1967) Strain-gauge transducer studies on intratracheal pressure and pupal length during discontinuous respiration in diapausing silkworm pupae. J Insect Physiol 13:1413–1451

Brovko LY, Belyaeva EI, Ugarova NN (1982) Subunit interactions in luciferase from the firefly *Luciola mingrelica*. Their role in the manifestation of enzyme activity and the process of thermal inactivation. Biokhimiya 47:760–766

Brown JJ (1980) Haemolymph protein reserves of diapausing and nondiapausing codling moth larvae, *Cydia pomonella* (L.) (Lepidoptera: Tortricidae). J Insect Physiol 26:487–491

Brown JJ, Chippendale GM (1977) Ultrastructure and respiration of the fat body of diapausing and non-diapausing larvae of the corn borer, *Diatraea grandiosella*. J Insect Physiol 23:1135–1142

Brunelli M, Magni F, Pellegrino M (1977) Excitatory and inhibitory events elicited by brief photic stimuli on flashing of the firefly *Luciola lusitanica* (Charp.). J Comp Physiol A 119: 15–35

Buck J (1958a) Possible mechanism and rationale of cyclic CO_2 retention by insects. Proc 10th Intern Congr Entomol 2:339–342

Buck J (1958b) Cyclic CO_2 release in insects. IV. A theory of mechanism. Biol Bull (Woods Hole) 114:118–140

Buck J.(1962) Some physical aspects of insect respiration. Annu Rev Entomol 7:27–56

Buck J (1978) Functions and evolutions of bioluminescence. In: Herring PJ (ed) Bioluminescence in action. Academic, New York, pp 419–460

Buck J, Buck E (1966) Biology of synchronous flashing of fireflies. Nature (Lond) 211:562–564

Buck J, Buck E (1968) Mechanisms of rhythmic synchronous flashing of fireflies. Science (NY) 159:1319–1327

Buck J, Buck E (1972) Photic signaling in the firefly *Photinus greeni*. Biol Bull (Woods Hole) 142: 195–205

Buck J, Buck E (1976) Synchronous fireflies. Sci Am 234:74–85

Buck J, Buck E (1978) Toward a functional interpretation of synchronous flashing by fireflies. Am Nat 112:471–492

Buck J, Friedman S (1958) Cyclic CO_2 release in diapausing pupae. – III. CO_2 capacity of the blood: carbonic anhydrase. J Insect Physiol 2:144–163

Buck J, Keister M (1955) Cyclic CO_2 release in diapausing *Agapema* pupae. Biol Bull (Woods Hole) 109:144–163

Buck J, Buck E, Hanson FE, Case JF, Mets L, Atta GJ (1981) Control of flashing in fireflies. IV. Free run pacemaking in a synchronic *Pteroptyx*. J Comp Physiol A 144:277–286

Buck JB (1937) Studies on the firefly. II. The signal system and color vision in *Photinus pyralis*. Physiol Zool 10:412–419

Bückmann D (1959) Die Auslösung der Umfärbung durch das Häutungshormon bei *Cerura vinula* L. (Lepidoptera, Notodontidae). J Insect Physiol 3:159–189

Bückmann D (1963) Der Einfluß der Temperatur auf das Epidermispigment der Raupen von *Cerura vinula* L. Z Naturforsch Sect B 18:255–264

Bückmann D (1965) Das Redoxverhalten von Ommochrom in vivo. Untersuchungen an *Cerura vinula* L. J Insect Physiol 11:1427–1462

Bückmann D (1969) Die Biochemie der morphologischen Farbanpassung bei Tagschmetterlingspuppen. Zool Anz Suppl 32:636–640

Bückmann D (1974a) Die hormonale Steuerung der Pigmentierung und des morphologischen Farbwechsels bei den Insekten. Fortschr Zool 22:1–22

Bückmann D (1974b) Die Wirkung von Ecdyson und von Tryptophanüberschuß bei der Ommochrombildung von *Cerura vinula* L. vor der Verpuppung. Zool Jahrb Physiol 78:257–278

Bückmann D (1977) Morphological colour change: Stage independent, optically induced ommochrome synthesis in larvae of the stick insect, *Carausius morosus* Br. J Comp Physiol B 115: 185–193

Bückmann D (1979) Morphological colour change (2): The effects of total and partial blinding on epidermal ommochrome content in the stick insect, *Carausius morosus* Br. J Comp Physiol B 130:331–336

Bückmann D (1983) The phylogeny and the polytropy of hormones. Horm Metabol Res 15: 211–217

Bückmann D (1984a) Colour change in insects. Pigment cell, vol 7. University of Tokyo Press, Tokyo (to be published)

Bückmann D (1984b) Tryptophan metabolism and morphological colour change in insects. In: Schlossberger HG, Kochen W, Linzen B, Steinhart H (eds) Progress in tryptophan and serotonin research. Walter de Gruyter, Berlin New York, pp 733–742

Bünning E (1959) Zur Analyse des Zeitsinnes bei *Periplaneta americana.* Z Naturforsch Sect B 14:1–4

Bult T (1939) Over de beweging der vloeistof in de tracheolen der insecten. Thesis, Rijks-Univerteit de Groningen, Groningen. Van Gorcum, Assen

Burkett BN, Schneiderman HA (1974a) Roles of oxygen and carbon dioxide in the control of spiracular function in *Cecropia pupae.* Biol Bull (Woods Hole) 147:274–293

Burkett BN, Schneiderman HA (1974b) Discontinuous respiration in insects at low temperatures: intra-tracheal pressure changes and spiracular valve behavior. Biol Bull (Woods Hole) 147: 294–310

Burr MJ, Hunter AS (1970) Effects of temperature on Drosophila-VII. Glutamate-aspartate transaminase activity. Comp Biochem Physiol 37:251–256

Burrows M (1980) The tracheal supply to the central nervous system of the locust. Proc R Soc Lond Ser B 207:63–78

Bursell B (1957) Spiracular control of water loss in the tsetse fly. Proc R Entomol Soc Lond Ser A 32:21–29

Bursell E (1964) Environmental aspects: temperature. In: Rockstein M (ed) The physiology of insecta, vol I. Academic, New York, pp 284–317

Bursell E (1974a) Environmental aspects. In: Rockstein M (ed) The physiology of insecta, vol II, 2nd edn. Academic, New York, London, pp 2–41

Bursell E (1974b) Environmental aspects – humidity. In: Rockstein M (ed) The physiology of insecta, vol II, 2nd edn. Academic, New York London, pp 44–84

Bursell E (1981) The role of proline in energy metabolism. In: Downer RGH (ed) Energy metabolism in insects. Plenum, New York London, pp 135–154

Burtt EH Jr (1981) The adaptiveness of animal colors. Bioscience 31:723–729

Butler GD Jr, Foster RN (1979) Longevity of adult pink bollworm at constant and fluctuating temperatures. Ann Entomol Soc Am 72:267–268

Butler GD Jr, Lopez JD (1980) *Trichogramma pretiosum:* development in two hosts in relation to constant and fluctuating temperatures. Ann Entomol Soc Am 73:671–673

Candy DJ (1981) Hormonal regulation of substrate transport and metabolism. In: Downer RGH (ed) Energy metabolism in insects. Plenum, New York London, pp 19–52

Carlsen J, Herman WS, Christensen M, Josefsson L (1979) Characterization of a second peptide with adipokinetic and red pigment-concentrating activity from the locust corpora cardiaca. Insect Biochem 9:497–501

Carlson SD, Steeves HR, Vandeberg JS, Robbins WE (1967) Vitamin A deficiency: Effect on retinal structure of the moth *Manduca sexta.* Science (Wash DC) 158:268–270

Case JF (1957) The median nerves and cockroach spiracular function. J Insect Physiol 1:85–94

Case JF (1980) Courting behavior in a synchronously flashing aggregative firefly, *Pteroptyx tener.* Biol Bull (Woods Hole) 159:613–625

Case JF, Strause LG (1978) Neurally controlled luminescent systems. In: Herring PJ (ed) Bioluminescence in action. Academic, New York, pp 331–366

Casey TM (1981) Behavioral mechanisms of thermoregulation. In: Heinrich B (ed) Insect thermoregulation. Wiley, New York, pp 79–114

Champlain RA, Butler GD Jr (1967) Temperature effects on development of the egg and nymphal stages of *Lygus hesperus* (Hemiptera:Miridae). Ann Entomol Soc Am 60:519–521

Chang F, Jang E (1980) Endocrine regulation of melanization in late larvae of the oleander hawk moth, *Deilephila nerii.* Ann Entomol Soc Am 73:89–92

Chapman RF (1969) The insects – structure and function. English University Press, London

Chapman RF (1982) The insects. Hodder and Stoughton, London

Cheeseman P, Goldsworthy GJ (1979) The release of adipokinetic hormone during flight and starvation in *Locusta.* Gen Comp Endocrinol 37:35–43

Chino H (1958) Carbohydrate metabolism in diapause eggs of the silkworm, *Bombyx mori.* II. Conversion of glycogen into sorbitol and glycerol during diapause. J Insect Physiol 2:1–12

Chino H (1963) Respiratory enzyme system of the *Bombyx* silkworm egg in relation to the mechanism of the formation of sugar alcohols. Arch Biochem Biophys 102:400–415

Chippendale GM (1977) Hormonal regulation of larval diapause. Annu Rev Entomol 22:121–138

Chippendale GM (1978a) Behavior associated with the larval diapause of the southwestern corn borer, *Diatraea grandiosella*. Probable involvement of juvenile hormone. Ann Entomol Soc Am 71:901–905

Chippendale GM (1978b) The functions of carbohydrates in insect life process. In: Rockstein M (ed) Biochemistry of insects. Academic, New York, pp 1–55

Chippendale GM (1983) Larval and pupal diapause. In: Downer RGH, Laufer H (eds) Endocrinology of insects. Alan R Liss, New York, pp 343–356 (Invertebrate endocrinology, vol 1)

Chippendale GM, Turunen S (1981) Hormonal and metabolic aspects of larval diapause of the southwestern corn borer, *Diatraea grandiosella* (Lepidoptera: Pyralidae). Entomol Gen 7: 223–231

Choussy M, Larrouquère J, Barbier M, Lamy M (1973) Les protéines de l'hémolymphe de la processionaire du pin *Thaumetopoea pityocampa* Schiff (Lépidoptère). Etude des pigments associés aux chromoprotéins. C R Acad Sci (Paris) Ser D 277:365–368

Ciemior KE, Sehnal G, Schneiderman HA (1979) Moulting, growth and survival of *Galleria mellonella* L. (Lep., Pyralidae) treated with juvenoids. ZAngew Entomol 88:414–425

Claret J (1966) Mise en évidence du rôle photorécepteur du cerveau dans l'induction de la diapause chez *Pieris brassicae*. Ann Endocrinol 27:311–320

Claret J, Carton Y (1980) Diapause in a tropical species, *Cothonaspis boulardi* (parasitic Hymenoptera). Oecologia (Berl) 45:32–34

Clark MG, Bloxhamd DP, Holland PC, Lardy HA (1973) Estimation of the fructose diphosphatase-phosphofructokinase substrate cycle in the flight muscle of *Bombus affinis*. Biochem J 134:589–597

Clarke A (1980) A reappraisal of the concept of metabolic cold adaptation in polar marine invertebrates. Biol J Linn Soc 14:77–92

Clarke KU (1966) Histological changes in the endocrine system of *Locusta migratoria* L. associated with the growth of the adult under different temperature regimes. J Insect Physiol 12: 163–170

Clarke KU (1967) Insects and temperature. In: Rose AH (ed) Thermobiology. Academic, London New York, pp 293–352

Clay ME, Vernard CE (1972) Larval diapause in the mosquito *Aedes triseriatus:* effects of diet and temperature on photoperiodic induction. J Insect Physiol 18:1441–1446

Cloudsley-Thompson JL (1953) The significance of fluctuating temperatures on the physiology and ecology of insects. Entomologist (Lond) 86:183–189

Cloudsley-Thompson JL (1958) Studies in diurnal rhythms- VIII. The endogenous chronometer in *Gryllus campestris* L. (Orthoptera: Gryllidae). J Insect Physiol 2:275–280

Cochran DG (1975) Excretion in insects. In: Candy DJ, Kilby BA (eds) Insects biochemistry and function. Chapman and Hall, London, pp 179–281

Coenen-Staß D, Kloft W (1976) Transpirationsmessungen an den beiden Schabenarten *Periplaneta americana* und *Blaberus trapezoideus*. J Insect Physiol 22:945–950

Cohet Y, David J (1978) Control of the adult reproductive potential by preimaginal thermal conditions. Oecologia (Berl) 36:295–306

Colbo MH, Porter GN (1981) The interaction of rearing temperature and food supply on the life history of two species of Simuliidae (Diptera). Can J Zool 59: 158–163

Colhoun EH (1958) Acetylcholine in *Periplaneta americana* L. – I. Acetylcholine levels in nervous tissue. J Insect Physiol 2:108–116

Connolly J (1981) Analysis of data in changing temperature experiments. J Therm Biol 6: 341–347

Conradi-Larsen EM, Sømme L (1973) The overwintering of *Pelophila borealis* Payk. II. Aerobic and anaerobic metabolism. Nor Entomol Tidsskr 20: 325–332

Cope FW (1976) Kinetics of the light flash of the living firefly: A supramolecular process. Physiol Chem Phys 8:343–347

Copeland DE (1964) A mitochondrial pump in the cells of the anal papillae of mosquito larvae. J Cell Biol 23:253–263

Cottrell CB (1962) The imaginal ecdysis of blowflies. The control of cuticular hardening and darkening. J Exp Biol 39:395–411

Crabtree B, Newsholme FA (1975) Comparative aspects of fuel utilization and metabolism by muscle. In: Usherwood PNR (ed) Insect muscle. Academic, London New York, pp 405–500

Crawford SC (1934) The habits and characteristics of nocturnal animals. Q Rev Biol 9:210–214

Cunningham RE, Williams RJJ (1980) Diffusion in gases and porous media. Plenum, New York London

Curran PF (1960) Na, Cl and water transport by rat ileum in vitro. J Gen Physiol 43:1137–1148

Cymborowski B, Bogus M, Beckage NE, Williams CM, Riddiford LM (1982) Juvenile hormone titers and metabolism during starvation induced supernumerary larval molting of the tobacco hornworm, *Manduca sexta.* J Insect Physiol 28:129–136

Dadd RH (1973) Insect nutrition: Current developments and metabolic implications. Annu Rev Entomol 187:381–420

Dahlman DL (1969) Cuticular pigments of tobacco hornworm *(Manduca sexta)* larvae: Effect of diet and genetic differences. J Insect Physiol 15:807–814

Dallman SH, Herman WS (1978) Hormonal regulation of hemolymph lipid concentration in the monarch butterfly, *Danaus plexippus.* Gen Comp Endocrinol 36:142–150

Danilevsky AS (1965) Photoperiodism and seasonal development of insects. Oliver & Boyd, Edingburgh

Danks SM, Tribe MA (1979) Biochemical changes in blowfly flight muscle mitochondria following temperature acclimation. J Therm Biol 4:183–195

Das AB, Singh SP (1972) Thermal adaptation of insects: Physiology and biochemistry. Indian Biol 4:19–24

Das AB, Singh SP (1974) Thermal acclimation in the rate of endogenous oxygen consumption of the fat body and the coxal muscle of *Periplaneta americana* (Linn.). Environ Physiol Biochem 4:139–146

Das AK, Das AB (1982a) Patterns of the enzymatic compensations for temperature in the fat body of male and female *Periplaneta americana.* Proc Indian Natl Sci Acad Part B Biol Sci 48: 44–54

Das AK, Das AB (1982b) Compensations for temperature in the activities of digestive enzymes of *Periplaneta americana.* Comp Biochem Physiol 71A:255–263

Das YT, Gupta AP (1977) Nature and precursors of juvenile hormone-induced excessive cuticular melanization in German cockroach. Nature (Lond) 268:139–140

Davey KG, Treherne JE (1964) Studies on crop function in the cockroach *Periplaneta americana.* III. Pressure changes during feeding and crop-emptying. J Exp Biol 41:513–524

Davies HC, Goldsworthy GJ (1982) Studies on the control of gluconeogenesis in *Locusta migratoria.* Gen Comp Endocrinol 46:379

Davison TF (1971) The effect of temperature on oxidative phosphorylation in isolated flight muscle sarcosomes. Comp Biochem Physiol 38B:21–34

DeLuca M, McElroy WD (1974) Kinetics of firefly luciferase catalysed reactions. Biochemistry 13:921–925

DeLuca M, McElroy WD (eds) (1981) Bioluminescence and chemiluminescence: Basic chemistry and analytical applications. Academic, New York London

Deb DC, Chakravorty S (1980) Effect of a juvenoid on the growth and differentiation of the ovary of *Corcyra cephalonica* (Lepidoptera). J Insect Physiol 27:103–111

Dehnel PA, Segal E (1956) Acclimation of oxygen consumption to temperature in the American cockroach *(Periplaneta americana).* Biol Bull (Woods Hole), 111:53–61

Dejours P (1981) Principles of comparative respiration physiology, 2nd edn. Elsevier, Amsterdam

Denlinger DL (1974) Diapause potential in tropical flesh flies. Nature (Lond) 252:223–224

Denlinger DL, Campbell JJ, Bradfield JY (1980) Stimulatory effect of organic solvents on intiating development in diapausing pupae of the flesh fly *Sarcophaga crassipalpis* and the tobacco hornworm *Manduca sexta.* Physiol Entomol 5:7–16

Destouches ML (1921) Prolongation de la vie chez les *Galleria mellonella.* C R Acad Sci (Paris) Ser D 172:998

Dewitz J (1902) Untersuchungen über die Verwandlungen der Insektenlarven. Arch Anat Physiol Abt Physiol 2:327–340

Diamond JM, Bossert WH (1967) Standing-gradient osmotic flow – a mechanism for coupling of water and solute transport in epithelia. J Gen Physiol 50:2061–2083

Dixon AFG (1972) Control and significance of the seasonal development of colour forms in the sycamore aphid, *Drepanosiphum platanoides* (Schr.). J Anim Ecol 41:689–697

Dortland JF (1978) Synthesis of vitellogenins and diapause proteins by the fat body of *Leptinotarsa,* as a function of photoperiod. Physiol Entomol 3:281–288

Douglas MM, Grula JW (1978) Thermoregulatory adaptations allowing ecological range expression by the pierid butterfly, *Nathalis iole* Boisduval. Evolution 32:776–783

Dow JAT (1981) Counter current flows, water movements and nutrient absorption in the locust midgut. J Insect Physiol 27:579–585

Downer RGH (1978) Functional role of lipids in insects. In: Rockstein M (ed) Biochemistry of insects. Academic, New York, pp 57–92

Downer RGH (1979) Induction of hypertrehalosemia by excitation in *Periplaneta americana.* Insect Physiol 25:59–63

Downer RGH, Kallapur VL (1981) Temperature-induced changes in lipid composition and transition temperature of flight muscle mitochondria of *Schistocerca gregaria.* J Therm Biol 6: 189–194

Dreisig H (1971) Control of the glowing of *Lampyris noctiluca* in the field (Coleoptera: Lampyridae). J Zool (Lond) 165:229–244

Dreisig H (1975) Environmental control of the daily onset of luminescent activity in glowworms and fireflies (Coleoptera: Lampyridae). Oecologia (Berl) 18:85–99

Dreisig H (1976) Phase shifting the circadian rhythms of nocturnal insects by temperature changes. Physiol Entomol 1:123–129

Duman JG (1979) Thermal-hysteresis-factors in overwintering insects. J Insect Physiol 25: 805–810

Duman JG, Horwath KL (1983) The role of haemolymph proteins in the cold tolerance of insects. Annu Rev Physiol 45:261–270

Duman JG, Horwath KL, Tomchaney A, Patterson L (1982) Antifreeze agents of terrestrial arthropods. Comp Biochem Physiol 73A:545–555

Dupont-Raabe M (1957) Lex mécanismes de l'adaptation chromatique chez les insectes. Arch Zool Exp Gen 94:61–262

Dustmann JH (1964) Redoxpigmente von *Carausius morosus* und ihre Bedeutung für den morphologischen Farbwechsel. Z Vgl Physiol 49:28–57

Edney EB (1980) The components of water balance. In: Locke M, Smith DS (eds) Insect biology in the future "VBW 80". Academic, New York London, pp 39–58

Edney EB (1971) The body temperature of tenebrionid beetles in the Namib desert of Southern Africa. J Exp Biol 55:253–272

Edney EB (1977) Water balance in land arthropods. Springer, Berlin Heidelberg New York (Zoophysiology, and Ecology, vol 9)

Edney EB, McFarlane J (1974) The effects of temperature on transpiration in the desert cockroach, *Arenivaga investigata,* and *Periplaneta americana.* Physiol Zool 47:1–12

Edwards LJ (1970) A thermal conductivity device for detecting the discontinuous release of carbon dioxide from insects. Ann Entomol Soc Am 63:627–629

Edwards LJ, Patton RL (1967) Carbonic anhydrase in the house cricket *Acheta domesticus.* J Insect Physiol 13:1333–1341

Egwuatu RI, Ajibola Taylor T (1977) The effects of constant and fluctuating temperatures on the development of *Acanthomia tomentosicollis* Stal (Hemiptera, Coreidae). J Nat Hist 11: 601–608

El-Salhy M, Falkmer S, Kramer KJ, Speirs RD (1983) Immunohistochemical investigations of neuropeptides in the brain, corpora cardiaca, and corpora allata of an adult lepidopteran insect, *Manduca sexta* (L.). Cell Tissue Res 232:295–317

Elbert A (1979) Ecological studies on the regulation of larval diapause of *Trogoderma variabile* (Coleoptera: Dermestidae). Z Angew Entomol 88:268–282

Engelmann F (1970) The physiology of insect reproduction. Pergamon, Oxford

Engelmann F (1979) Insect vitellogenin: identification, biosynthesis, and role in vitellogensis. Adv Insect Physiol 14:49–108

Engelmann F (1983) Vitellogenesis controlled by juvenile hormone. In: Downer RGH, Laufer H (eds) Endocrinology of insects. Alan R Liss, New York, pp 259–270 (Invertebrate endocrinology, vol 1)

Engle CE, Barnes MM (1983) Developmental threshold temperature and heat unit accumulation required for egg hatch of navel orangeworm (Lepidoptera: Pyralidae). Environ Entomol 12: 1215–1217

Englisch H (1981) Konzentration verschiedener Metabolite bei den Larven der Mücke *Chaoborus crystallinus* während Anaerobiose und nachfolgender Erholung. Staatsexamensarbeit, Universität Münster

Englisch H, Opalka B, Zebe E (1982) The anaerobic metabolism of the larvae of the midge *Chaoborus crystallinus*. Insect Biochem 12:149–155

Eubank WP, Atmar JW, Ellington JJ (1973) The significance and thermodynamics of fluctuating versus static thermal environments on *Heliothis zea* egg development rates. Environ Entomol 2:491–496

Farley RD, Case JF, Roeder KD (1967) Pacemaker for tracheal ventilation in the cockroach, *Periplaneta americana* L. J Insect Physiol 13:1713–1728

Fast PG (1970) Insect lipids. In: Holman RT (ed) Progress in the chemistry of fats and other lipids, voll II, part 2. Pergamon, New York, pp 181–242

Fleming JE, Miquel J (1983) Effects of temperature on the metabolic rate of young and old *Drosophila*. Experientia (Basel) 39:267–268

Foley DH (1981) Pupal development rate of *Heliothis armiger* (Hübner) (Lepidoptera: Noctuidae) under constant and alternating temperatures. J Aust Enotmol Soc 20:13–20

Forster RE (1966) Diffusion as a limiting factor in oxygen transport in cells. In: Proceedings of the international symposium of cardiovascular respiration effects of hypoxia. Karger, Basel New York, pp 16–28

Foster DR, Crowder LA (1980) Diapause of the pink bollworm, *Pectinophora gossypiella*, related to dietary lipids. Comp Biochem Physiol 65B:723–726

Fraenkel G (1976) Molting and development in undersized fly larvae. In: Hepburn HR (ed) The insect integument. Elsevier, Amsterdam Oxford, pp 323–338

Fraenkel G, Blewett M (1946) The dietetics of the clothes moth, *Tineola bisselliella* Hum. J Exp Biol 22:156–161

Fraenkel G, Hsiao C (1962) Hormonal and nervous control of tanning in the fly. Science (NY) 138:27–29

Fraenkel G, Rudall KM (1947) The structure of insect cuticle. Proc R Soc Lond B Biol Sci 134: 113–143

Frank P, Mises R von (1935) Die Differential- und Integralgleichungen der Mechanik und Physik, vol II Physikalischer Teil, 2nd edn. Viehweg, Braunschweig Dover New York

Fulton BB (1941) A luminous fly-larva with spider traits (Diptera, Mycetophilidae). Ann Entomol Soc Am 34:289–302

Furunishi S, Masaki S, Hashimoto Y, Suzuki M (1982) Diapause response to photoperiod and night interruption in *Mamestra brassicae* (Lepidoptera: Noctuidae). Appl Entomol Zool 17: 398–409

Fuzeau-Braesch S (1960) Étude biologique et biochemique de la pigmentation d'un insecte *Gryllus bimaculatus* de Geer (Gryllide, Orthoptère). Bull Biol Fr Belg 94:527–627

Fuzeau-Braesch S (1966) Étude du noircissement cuticulaire nonexuvial chez *Locusta migratoria*. J Insect Physiol 12:1363–1368

Fuzeau-Braesch S (1972) Pigments and color changes. Annu Rev Entomol 17:403–424

Fuzeau-Braesch S (1978) A new action of carbon dioxide. The change of subsocial species of insects by a chronic treatment. C R Seances Soc Biol Fil 172:427–431

Fuzeau-Braesch S (1983) Les changements de couleur chez les insectes: Déterminismes et rolês. Bull Soc Zool Fr 108:195–205

Gäde G (1970) Zur Aktivität der "Schlüsselenzyme" des Energiestoffwechsels im Muskel. Eine vergleichende Untersuchung an Arthropoden. Staatsexamensarbeit, Universität Münster

Gäde G (1975) Zur stoffwechselphysiologischen Spezialisierung von Insektenmuskeln. Verh Dtsch Zool Ges 1974:258–261

Gäde G (1983) Energy metabolism of arthropods and mollusks during environmental and functional anaerobiosis. J Exp Zool 228:415–429

Gäde G, Ellington WR (1983) The anaerobic molluscan heart: adaptation to environmental anoxia. Comparison with energy metabolism in vertebrate hearts. Comp Biochem Physiol 76A:615–620

Garen A, Kauvar L, Lepesant JA (1977) Roles of ecdysone in *Drosophila* development. Proc Nat Acad Sci US 74:5099–5103

Geankoplis CJ (1983) Transport processes and unit operations, 2nd edn. Allyn and Bacon, Boston London Sidney Toronto

Geer BW, Perille TJ (1977) Effects of dietary sucrose and environmental temperature on fatty acid synthesis in *Drosophila melanogaster.* Insect Biochem 7:371–379

Geldiay S (1967) Hormonal control of adult reproductive diapause in the Egyptian grasshopper, *Anacridium aegyptium* L. J Endocrinol 37:63–71

Gelman DB, Woods CW (1983) Haemolymph ecdysteroid titres of diapause- and nondiapause-bound fifth instars and pupae of the European corn borer, *Ostrinia nubilalis* (Hübner). Comp Biochem Physiol 76A:367–375

Gelperin A (1971) Regulation of feeding. Annu Rev Entomol 16:365–378

Gerencser G, Lee SH (1983) Cl^--stimulated adenosine triphosphate: existence, location and function. J Exp Biol 106:111–222

Gharib B, Legay JM, de Reggi M (1981) Potentiation of development abilities of diapausing eggs of *Bombyx mori* by 20-hydroxy-ecdysone. J Insect Physiol 27:711–713

Gharib B, de Reggi M, Connat JL, Chaix JC (1983) Ecdysteroid and juvenile hormone changes in *Bombyx mori* eggs, related to the initiation of diapause. FEBS Lett 160:119–123

Giersberg J (1928) Über den morphologischen und physiologischen Farbwechsel der Stabheuschrecke *Carausius morosus.* Z Vgl Physiol 7:657–695

Gilby AR, Rumbo ER (1980) Water loss and respiration of *Lucilia cuprina* during development within the puparium. J Insect Physiol 26:153–161

Gilmour D (1961) The biochemistry of insects. Academic, New York London

Girardie A (1970) Neurosecretions cérébrales chez les acridiens. Bull Soc Zool Fr 95:783–802

Gösswald K (1936) Physiologische Untersuchungen über die Einwirkung ökologischer Faktoren, besonders der Temperatur und Luftfeuchtigkeit, auf die Entwicklung von *Diprion (Lophyrus) pini* L. zur Feststellung der Ursachen des Massenwechsels. Z Angew Entomol 22:331–384

Goldson SL, Emberson RM, Bickerstaffe R (1981) Seasonal changes in the fatty acid composition of Argentine stem weevil, *Hyperodes bonariensis* (Coleoptera: Curculionidae). NZ J Zool 8:79–81

Goldsworthy GJ (1984) The endocrine control of flight metabolism in locusts. Adv Insect Physiol 17:149–204

Goldsworthy GJ, Gäde G (1983) The chemistry of hypertrehalosemic factors. In: Downer RGH, Laufer H (eds) Endocrinology of insects. Alan R Liss, New York, pp 109–119 (Invertebrate endocrinology, vol 1)

Goldsworthy GJ, Wheeler CH (1984) Adipokinetic hormones in locusts. In: Hoffmann JA, Porchet M (eds) Biosynthesis, metabolism and mode of action of invertebrate hormones. Springer, Berlin Heidelberg New York (to be published)

Gortner RA (1911) Studies on melanin-IV. The origin of the pigment and the color pattern in the elytra of the Colorado potato beetle *(Leptinotarsa decemlineata* Say). Am Nat 45:743–755

Graham T (1833) On the law of diffusion of gases. Phil Mag 2:175

Grigo F, Topp W (1980) Einfluß von Adaptationstemperatur und Photoperiode auf den Sauerstoffverbrauch bei Staphyliniden (Col.) in Diapause und Non-Diapause. Zool Anz 204:19–26

Grimstone AV, Mullinger AM Ramsay JA (1968) Further studies on the rectal complex of the mealworm *Tenebrio molitor,* L. (Coleoptera, Tenebrionidae). Philos Trans R Soc Lond B Biol Sci 253:343–382

Gruetzmacher MC, Keely LL (1982) Cytochrome degradation and synthesis in fat body mitochondria during diapause, diapause termination and metamorphosis of *Heliothis zea.* Insect Biochem 12:49–54

Grüner C, Sauer KP (1984) Aestivation of *Mamestra brassicae* (Lep.): an adaptive strategy. In: Engels W (ed) Advances in invertebrate reproduction, vol III, Elsevier, Amsterdam, p 588

Gunn DL (1933) The temperature and humidity relations of the cockroach *Blatta orientalis*. I. Desiccation. J Exp Biol 10:274–285

Gunn DL, Hopf HS (1942) The biology and behaviour of *Ptinus tectus* Boie (Coleoptera, Ptinidae), a pest of stored products. III. J Exp Biol 18:278–289

Guppy JC (1969) Some effects of temperature on the immature stages of the armyworm, *Pseudaletia unipuncta* (Lepidoptera, Noctuidae), under controlled conditions. Can Entomol 101: 1320–1327

Hackman RH (1974) Chemistry of the insect cuticle. In: Rockstein M (ed) The physiology of insecta, vol VI. Academic, New York London, pp 216–270

Haderspeck W, Hoffmann KH (1983) Temperaturanpassungen bei Verdauungsenzymen eines antarktischen Käfers. Verh Dtsch Zool Ges 1983: 285

Hadley NF (1972) Desert species and adaptation. Am Sci 60:338–347

Hadley NF (1979) Wax secretion and color phases of the desert tenebrionid beetle *Cryptoglossa verrucosa* (LeConte). Science (Wash DC) 203:367–369

Hagstrum DW, Hagstrum WR (1970) A simple device for producing fluctuating temperatures, with an evaluation of the ecological significance of fluctuating temperatures. Ann Entomol Soc Am 63:1385–1389

Hagstrum DW, Leach CE (1973) Role of constant and fluctuating temperatures in determining development time and fecundity of three species of stored-products Coleoptera. Ann Entomol Soc Am 66:407–410

Hagstrum DW, Silhacek DL (1980) Diapause induction in *Ephestia cautella*, an interaction between genotype and crowding. Entomol Exp Appl 28:29–37

Hamilton AG (1959) The infrared gas analyser as a means of measuring the carbon dioxide output of individual insects. Nature (Lond) 184:367–369

Hamilton AG (1960) The relationship between the output of carbon dioxide and sexual maturation in two species of locusts (*Schistocerca gregaria* and *Nomadacris septemfasciata*). XI. Proc Int Congr Entomol 1:701–706

Hamilton AG (1964) The occurrence of periodic or continuous discharge of carbon dioxide by male desert locusts (*Schistocerca gregaria* Forskal) measured by an infrared gas analyser. Proc R Soc Lond Ser B 160:373–395

Hamilton AG (1972) The combined use of a twin channel null-balance paramagnetic O_2 analyser and in infra-red CO_2 analyser for measuring respiration in insects. Lab Pract 21:807–809

Hamilton WJ (1975) Coloration and its thermal consequence for diurnal desert insects. In: Hadley NF (ed) Environmental physiology of desert organisms. Dowden, Hutchinson and Ross, London Stroudsberg, pp 67–89

Hammond RB, Poston FL, Pedigo LP (1979) Growth of the green cloverworm and a thermal unit system for development. Environ Entomol 8:636–642

Handel E van (1966) The thermal dependence of the rates of glycogen and triglyceride synthesis in the mosquito. J Exp Biol 44:523–528

Handel E van (1973) Temperature dependence of caloric expenditure and mortality in the starving mosquito. Comp Biochem Physiol 44:1321–1323

Hanegen JL, Heath JE (1970) Control of body temperature in moth *Hyalophora*. J Exp Biol 53: 349–362

Hanna CH, Hopkins TA, Buck J (1976) Peroxisomes of the firefly lantern. J Ultrastruct Res 57: 150–162

Hanrahan JW, Phillips JE (1982) Electrogenic K^+-dependent chloride transport in locust hindgut. Philos Trans R Soc Lond B Biol Sci 299:585–595

Hanrahan JW, Phillips JE (1983) Mechanism and control of salt absorption in locust rectum. Am J Physiol 244:R131–R142

Hanski I (1976) Assimilation by *Lucilia illustris* (Diptera) larvae in constant and changing temperatures. Oikos 27:288–299

Hanski I (1977) An interpolation model of assimilation by larvae of the blowfly *Lucilia illustris* (Calliphoridae) in changing temperatures. Oikos 28:187–195

Harms E (1972) Alkohol-Dehydrogenase aus Larven der Zuckmücke *Chironomus tentans*. Isolirung und Charakterisierung. PHD Thesis, Freie Universität Berlin

Harnisch O (1938) Studien zum anaeroben und Erholungsstoffwechsel der Larven von *Chironomus thummi*. I. Wechsel im Glykogen-, Fett- und N-Gehalt. Z Vgl Physiol 26:200–220

Harnisch O (1939) Studien zum aneroben und Erholungsstoffwechsel der Larven von *Chironomus thummi*. II. Säurebildung und "Notoxybiose". Z Vgl Physiol 27:275–303

Harnisch O (1942) Die Bedeutung der Säureabscheidung bei der Anaerobiose der *Chironomus*-Larven. Naturwissenschaften 30:147–148

Harnisch O (1953) Bestimmung der Trockensubstanz bei Larven von *Chironomus thummi* und *Prodiamesa oblivacea* in Aerobiose und Anaerobiose. Biol Zentralbl 72:518–530

Hartman JH, Hynes CD (1980) Embryonic diapause in *Tipula simplex* and the action of photoperiod in its termination (Diptera: Tipulidae). Pan-Pac Entomol 56:207–212

Harvey GT (1967) On coniferophagous species of *Choristoneura* (Lepidoptera: Tortricidae) in North America. V. Second diapause as a species character. Can Entomol 99:486–503

Harvey WR (1980) Water and ions in the gut. In: Locke M, Smith DS (eds) Insect biology in the future "VBW 80". Academic, New York London, pp 105–124

Harvey WR, Nedergaard S (1964) Sodium independent active transport of potassium in the isolated midgut of the *Cecropia* silkworm. Proc Nat Acad Sci US 51:757–765

Harvey WR, Williams CM (1961) The injury metabolism of the *Cecropia* silkworm. I. Biological amplification of the effect of localized injury. J Insect Physiol 7:81–99

Harwood RF, Takata N (1965) Effect of photoperiod and temperature on fatty acid composition of the mosquito *Culex tarsalis*. J Insect Physiol 11:711–716

Hassan AAG (1944) The structure and mechanism of the spiracular regulatory apparatus in adult Diptera and certain other groups of insects. Trans R Entomol Soc Lond 94:103–153

Hazel JR, Prosser CL (1974) Molecular mechanisms of temperature compensation in poikilotherms. Physiol Rev 54:620–677

Hazelhoff EH (1926) Regeling der ademhaling bij insecten en spinnen. Proefschrift, Rijks-Universiteit te Utrecht, Utrecht (Translation loaned by courtesy of Dr. John Buck)

Headlee TJ (1940) The relative effects on insect metabolism of temperatures derived from constant and variable sources. J Econ Entomol 33:361–364

Heal J (1979a) Color patterns of Syrphidae. 1. Genetic variation in the drone fly *Eristalis tenax*. Heredity 42:223–236

Heal J (1979b) Color patterns of Syrphidae. 2. *Eristalis intricarius*. Heredity 43:229–238

Heath JE (1970) Behavioural regulation of body temperature in poikilotherms. Physiologist 13: 399–410

Heath JE, Heath MS (1982) Energetics of locomotion in endothermic insects. Annu Rev Physiol 44:133–143

Hegdekar BM (1979) Epicuticular wax secretion in diapause and nondiapause pupae of the Bertha armyworm. Ann Entomol Soc Am 72:13–15

Hegel JR, Casey TM (1982) Thermoregulation and control of head temperature in the sphinx moth, *Manduca sexta*. J Exp Biol 101:1–15

Heimbach F (1973) Einfluß von tagesperiodischen Wechseltemperaturen auf Wachstum und Körpergröße des Kohlweißlings *Pieris brassicae*. Diplomarbeit, Universität Köln

Heinrich B (1971) Temperature regulation of the sphinx moth, *Manduca sexta*. I. Flight energetics and body temperature during free and tethered flight. J Exp Biol 54:141–152

Heinrich B (1974) Thermoregulation in endothermic insects. Science (Wash DC) 185:747–756

Heinrich B (1975) Thermoregulation in bumblebees. II. Energetics of warm-up and free flight. J Comp Physiol 96:155–166

Heinrich B (1977) The physiology of exercise in the bumblebee. Am Sci 65:455–465

Heinrich B (1981) Ecological and evolutionary perspectives. In: Heinrich B (ed) Insect thermoregulation. Wiley, New York, pp 236–302

Heinz E (1978) Mechanics and energetics of biological transport. Springer, Berlin Heidelberg New York

Helfert B (1978) Environmental effects on color variation and ommochromes in *Chortippus biguttulus* (Orthoptera, Acrididae). Zool Anz 200:41–52

Henry JP, Michelson AM (1978) Bioluminescence. Photochem Photobiol 27:855–858

Henwood K (1975a) A field-tested thermoregulation model for two diurnal Namib Desert tenebrionid beetles. Ecology 56:1329–1342

Henwood K (1975b) Infrared transmittance as an alternative thermal strategy in the desert beetle *Onymacris plana*. Science (Wash DC) 189:993–994

Herford GM (1938) Tracheal pulsation in the flea. J Exp Biol 15:327–338

Herman WS (1981) Studies on the adult reproductive diapause of the monarch butterfly, *Danaus plexippus*. Biol Bull (Woods Hole) 160:89–106

Herring PJ (ed) (1978) Bioluminescence in action. Academic, London

Hertz G (1922) Ein neues Verfahren zur Trennung von Gasgemischen durch Diffusion. Z Physik 23:433–434

Hertz G (1923) Über Trennung von Gasgemischen durch Diffusion in einem strömenden Gase. Z Physik 19:35–42

Heslop JP, Price CM, Ray JW (1963) Anaerobic metabolism in the housefly *Musca domestica* L. Biochem J 87:35–38

Hevert F (1974) Zur Regulation der Ionenkonzentration von *Drosophila hydei:* Na^+, K^+, Ca^{++}, and Cl^-. J Insect Physiol 20:2225–2245

Hew CL, Kao MH, So YP, Lim KP (1983) Presence of cysteine-containing antifreeze proteins in the spruce budworm *Choristoneura fumiferana*. Can J Zool 61:2324–2328

Highnam KC, Hill L (1977) The comparative endocrinology of the invertebrates, 2nd edn. Arnold, London

Hilbert DW, Logan JA (1983) Empirical model of nymphal development for the migratory grasshopper, *Melanoplus sanguinipes* (Orthoptera: Acrididae). Environ Entomol 12:1–5

Hill AE (1975) Solute-solvent coupling in epithelia: an electroosmotic theory of fluid transfer. Proc R Soc Lond Ser B 190:115–134

Hinton HE (1976) Colour changes. In: Bligh J, Cloudsley-Thompson JL, Macdonald AG (eds) Environmental physiology of animals. Blackwell Scientific Publications, Oxford London Edinburgh Melbourne, pp 390–412

Hinton HE, Gibbs DF (1971) Diffraction gratings in gyrinid beetles. J Insect Physiol 17:1023–1035

Hinton HE, Harman GM (1973) Physiological colour change in the elytra of the Hercules beetle, *Dynastes hercules*. J Insect Physiol 19:533–549

Hiruma K, Matsumoto S, Isogai A, Suzuki A (1984) Control of ommochrome synthesis by both juvenile hormone and melanization hormone in the cabbage armyworm, *Mamestra brassicae*. J Comp Physiol 154:13–21

Hitzemann K (1979) Untersuchungen über den Energiestoffwechsel in der Sprungmuskulatur von *Locusta migratoria* (L.) Diplomarbeit, Universität Münster

Hochachka PW, Somero GN (1973) Strategies of biochemical adaptation. Saunders, Philadelphia

Hodkova M, Kubista V (1971) Anaerobic formation of alanine in the metathoric musculature of *Periplaneta americana*. Insect Biochem 1:461–466

Hodson AC, Alrawy AA (1958) Temperature in relation to developmental thresholds of insects. Proc 10th Int Congr Entomol 2:61–65

Hoffmann KH (1973) Der Einfluß der Temperatur auf die chemische Zusammensetzung von Grillen (*Gryllus*, Orthopt.). Oecologia (Berl) 13:147–175

Hoffmann KH (1974) Wirkung von konstanten und tagesperiodisch alternierenden Temperaturen auf Lebensdauer, Nahrungsverwertung und Fertilität adulter *Gryllus bimaculatus*. Oecologia (Berl) 17:39–54

Hoffmann KH (1975) Pyruvate kinase from muscle and fat body of the house cricket *Acheta domesticus* L.: Purification and catalytic studies. J Comp Physiol 104:59–69

Hoffmann KH (1976a) Catalytic efficiency and structural properties of invertebrate muscle pyruvate kinases: Correlation with body temperature and oxygen consumption rates. J Comp Physiol 110:185–195

Hoffmann KH (1976b) Organic body constituents of *Protophormia terrae-novae* (Dipt.) from Spitsbergen compared with flies from a laboratory stock. Oecologia (Berl) 23:13–16

Hoffmann KH (1978) Thermoregulation bei Insekten. Biol unserer Zeit 8:17–26

Hoffmann KH (1980) Anpassungen im Stoffwechsel von Insekten an tagesperiodische Wechseltemperaturen. Verh Dtsch Zool Ges 1980: 214–227

Hoffmann KH (1981) Leuchtende Tiere: Chemie und biologische Bedeutung. Biol Unserer Zeit 11:97–106

Hoffmann KH, Krüll F (1979) Körpertemperaturmessungen bei Insekten. Biol Unserer Zeit 9: 153–156

Hoffmann KH, Marstatt H (1977) The influence of temperature on catalytic efficiency of pyruvate kinase of crickets (Orthoptera: Gryllidae). J Therm Biol 2:203–207

Hoffmann KH, Stockmeier E (1975) Wirkung von vollsynthetischer Nahrung und Temperatur auf Gesamtfettsäurespektrum und Neutralfette von *Gryllus bimaculatus* (Orthopt.). Verh Dtsch Zool Ges 1974: 286–291

Hoffmann KH, Behrens W, Ressin W (1981) Effect of a daily temperature cycle on ecdysteroid titres and cyclic nucleotide titres in adult femal crickets, *Gryllus bimaculatus*. Physiol Entomol 6:375–385

Hofsvang L (1976) Development of *Anthocoris sibiricus* Reuter (Het., Anthocoridae) at constant and fluctuating temperatures with the green peach aphid *Myzus persicae* (Sulzer) as prey. Norw J Entomol 23:29–34

Holzapfel CM, Bradshaw WE (1981) Geography of larval dormancy in the tree hole mosquito, *Aedes triseriatus*. Can J Zool 59:1014–1021

Hondka K (1981) Environmental factors affecting the pupal coloration in *Papilio protenor demetrius* (Lepidoptera, Papilionidae). 2. Effect of physical stimuli. Appl Entomol Zool 16: 467–471

Honěk A (1973) Induction of a winter coloration in *Chrysopa carnea*. Steph. (Neuroptera, Chrysopidae). Vestn Cesk Spol Zool 37:253–257

Honěk A (1975) Color polymorphism in *Adalia bipunctata* in Bohemia, Czechoslovakia (Coleoptera, Coccinellidae). Entomol Ger 1:293–299

Horwath KL, Duman JG (1983a) Preparatory adaptations for winter survival in the cold hardy beetles, *Dendroides canadensis* and *Dendroides concolor*. J Comp Physiol 151B:225–232

Horwath KL, Duman JG (1983b) Induction of antifreeze protein production by juvenile hormone in larvae of the beetle, *Dendroides canadensis*. J Comp Physiol 151B:233–240

House HL (1974a) In: Rockstein M (ed) The physiology of insects. Academic, New York London, pp 1–62

House HL (1974b) Digestion. In: Rockstein M (ed) The physiology of insects. Academic, New York London, pp 63–117

Howe RW (1967) Temperature effects on embryonic development in insects. Annu Rev Entomol 12:15–42

Huffaker CB (1944) The temperature relations of the immature stages of the malarial mosquito, *Anopheles quadrimaculatus* Say, with a comparison of the developmental power of constant and variable temperatures in insect metabolism. Ann Entomol Soc Am 37:1–27

Hughes RD (1960) Induction of diapause in *Erioishia brassicae* Bouché (Diptera: Anthomyidae). J Exp Biol 37:218–223

Hunter AS (1968) Effects of temperature on *Drosophila* – IV. Adaptation of *D. immigrans*. Comp Biochem Physiol 24:327–333

Hunter AS, Cediel N (1970) Effects of temperature on *Drosophila* – VI. Respiratory enzymes. Comp Biochem Physiol 37:243–249

Hustert R (1974) Morphologie und Atmungsbewegungen des 5. Abdominalsegments von *Locusta migratoria migratorioides*. Zool Jahrb Physiol 78:157–174

Hutten H (1971) Untersuchungen nichtstationärer Austauschvorgänge in gekoppelten Konvektions-Diffusions-Systemen. Abhandlungen der Mathematisch-Naturwissenschaftlichen Klasse vol 1970, nr 4:329–386. Akademie der Wissenschaften und der Literatur Mainz, Steiner, Wiesbaden

Ibrahim MM (1974) Environmental effects on colour variation in *Acrida pellucida* Klug. Z Angew Entomol 77:133–136

Ikemoto H (1980) Hormonal and neural control of the body color change in the larvae of the larger pellucid hawk moth, *Cephonodes hylas*. New Entomol 29:33–40

Ikemoto H (1983) The role of juvenile hormone in the density-related color variation in larvae of *Cephonodes hylas* L. (Lepidoptera: Sphingidae). Appl Entomol Zool 18:57–61

Isobe M, Goto T (1980) Diapause hormones. In: Miller TA (ed) Neurohormonal techniques in insects. Springer, Berlin Heidelberg New York, pp 216–243

Ivanović JP, Janković-Hladni MI, Milanović MP (1975) Effect of constant temperature on survival rate, neurosecretion and endocrine cells, and digestive enzymes in *Morimus funereus* larvae (Cerambycidae: Coleoptera). Comp Biochem Physiol 50A;125–130

Ivanović JP, Janković-Hladni MI, Stanić V, Milanović MP, Nenadović V (1979) Possible role of neurohormones in the process of acclimatization and acclimation in *Morimus funereus* larvae (Insecta) – I. Changes in the neuroendocrine system and target organs (midgut, haemolymph) during the annual cycle. Comp Biochem Physiol 63A:95–102

Ivanović JP, Janković-Hladni MI, Milanović MP, Stanić V, Božidarac M (1982) Possible role of neurohormones in the process of acclimatization and acclimation in *Morimus funereus* (Insecta) – II. Changes in the haemolymph free amino acid content and digestive enzyme activity. Comp Biochem Physiol 718:695–701

Iwao S (1962) Studies on the phase variation and related phenomena in some lepidopterous insects. Mem Coll Agric Kyoto 84:1–80

Jacobs MH (1935) Diffusion processes. Ergeb Biol 12:1–160

Jacobs MH (1967) Diffusion processes. Springer, Berlin Heidelberg New York

Jacovlev V, Krüger F (1953) Vergleichende Untersuchungen zur Physiologie der Transpiration der Orthopteren. Zool Jahrb Physiol 64:391–428

Janda V (1936) Über den Farbwechsel transplantierter Hautstücke und künstlich verbundener Körperfragmente bei *Dixippus morosus* (Br. et Redt.). Zool Anz 115:177–185

Janisch E (1925) Über die Temperaturabhängigkeit biologischer Vorgänge und ihre kurvenmäßige Analyse. Pflügers Archiv 209: 414–436

Janković-Hladni MI, Stanić V, Ivanović JP (1976) The effect of temperature and crowding on midgut amylotic activity in *Tenebrio molitor* adults. J Insect Physiol 22:851–853

Johnston JM (1970) Intestinal absorption of fats. In: Florkin M, Stotz EH (eds) Comprehensive biochemistry, vol 18, Elsevier, Amsterdam, pp 1–18

Joly L, Joly P (1974) Comparaison de la phase grégaire et de la phase solitaire de *Locusta migratoria migratorioides* (Orthoptére) du point de vue de la teneur de leur hemolymphe en hormone juvenile. C R Acad Sci (Paris) Ser D 279:1007–1009

Joly P (1945) La fonction ovarienne et son contrôle humorale chez les dytiscides. Arch Zool Exp Gen 84:47–164

Jones BM (1956) Endocrine activity during insect embryogenesis. Control of events in development following the embryonic moult *(Locusta migratoria* and *Locustana pardalina,* Orthoptera). J Exp Biol 33:685–696

Jones D, Jones G, Hammock BDD (1981) Growth parameters associated with endocrine events in larval *Trichoplusia ni* (Hübner) and timing of these events with developmental markers. J Insect Physiol 27:779–788

Joosse ENG, Veltcamp E (1970) Some aspects of growth, moulting and reproduction in five species of surface dwelling Collembola. Neth J Zool 20:315–328

Jordan RG (1980) Embryonic diapause in three populations of the western tree hole mosquito, *Aedes sierrensis.* Ann Entomol Soc Am 73:357–359

Jost W, Hauffe K (1972) Diffusion – Methoden der Messung und Auswertung, 2nd edn. Dr. Dietrich Steinkopf Verlag, Darmstadt, pp 1–327 (Fortschritte der Physikalischen Chemie, vol 1, Diffusion)

Jungreis AM (1979) Physiology of moulting in insects. Adv Insect Physiol 14:109–183

Jungreis AM (1980) Hemolymph as a dynamic tissue. In: Locke M, Smith DS (eds) Insect biology in the future "VBW 80". Academic, New York London, pp 273–294

Kaars C (1981) Insects – spiracle control. In: Herreid II Cf, Fourtner CR (eds) Locomotion and energetics in arthropods. Plenum, New York London, pp 337–366

Kageyama T (1976) Pathways of carbohydrate metabolism in the eggs of the silkworm, *Bomyx mori.* Insect Biochem 6:507–511

Kageyama T, Ohnishi E (1971) Carbohydrate metabolism in the eggs of the silkworm, *Bombyx mori.* I. Absence of phosphofructokinase in the diapausing egg. Dev Growth Differ 13: 97–106

Kallapur VL, Downer RGH, George JC, Thompson JE (1982) Effect of environmental temperature on the phase properties and lipid composition of flight muscle mitochondria of *Schistocerca gregaria.* Insect Biochem 12:115–121

Kammer AE (1981) Physiological mechanisms of thermoregulation. In: Heinrich B (ed) Insect thermoregulation. Wiley, New York, pp 115–158

Kammer AE, Heinrich B (1978) Insect flight metabolism. Adv Insect Physiol 13:133–228

Kanwisher JW (1966) Tracheal gas dynamics in pupae of the *Cecropia* silkworm. Biol Bull (Woods Hole) 130:96–105

Karlson P, Sekeris CE (1962) Zum Tyrosinstoffwechsel der Insekten. IX. Kontrolle des Tyrosinstoffwechsels durch Ecdyson. Biochim Biophys Acta 63:489–495

Kaufmann O (1932) Einige Bemerkungen über den Einfluß von Temperaturschwankungen auf die Entwicklungsdauer und Streuung bei Insekten und seine graphische Darstellung durch Kettenlinie und Hyperbel. Z Morphol Oekol Tiere 25:353–361

Kaufmann T (1965) Ecological and biological studies on the West African firefly *Luciola discicollis* (Coleoptera: Lampyridae). Ann Entomol Soc Am 58:414–426

Kayser H (1982) Carotenoids in insects. In: Britton G, Goodwin TW (eds) Carotenoid chemistry and biochemistry. Pergamon, Oxford New York, pp 195–210

Kayser H (1984) Pigments. In: Kerkut GA, Gilbert LI (eds) Comprehensive insect physiology, biochemistry and pharmacology. Pergamon, Oxford (to be published)

Kayser H, Angersbach D (1974) Action spectra for light-controlled pupal pigmentation in *Pieris brassicae:* melanization and level of bile pigment. J Insect Physiol 20:2277–2285

Kayser H, Dettner K (1984) Biliverdin IX_g in beetles (Dytiscidae: Laccophilinae). Comp Biochem Physiol 77B:639–643

Kayser-Wegmann I (1976) Differences in black pigmentation in lepidopteran cuticles as revealed by light and electron microscopy. Cell Tissue Res 171:513–521

Keister M, Buck J (1974) Respiration: Some exogeneous and endogeneous effects on rate of respiration. In: Rockstein M (ed) The physiology of insecta, vol VI, 2nd edn. Academic, New York London, pp 470–509

Kenagy GJ, Stevenson RD (1982) Role of body temperature in the seasonality of daily activity in tenebrionid beetles of eastern Washington. Ecology 63:1491–1503

Kerkut GA, Taylor JR (1957) A temperature receptor in the tarsus of the cockroach, *Periplaneta americana.* J Exp Biol 34:486–493

Kestler P (1971) Die diskontinuierliche Ventilation bei *Periplaneta americana* L. und anderen Insekten. Thesis, Julius-Maximilians-Universität, Würzburg

Kestler P (1978a) Gas balance of external respiration by electronic weighing in water saturated air. Eur J Physiol Suppl 373: Abstract 36

Kestler P (1978b) Atembewegungen und Gasaustausch bei der Ruheatmung adulter terrestrischer Insekten. Verh Dtsch Zool Ges 1978: 269

Kestler P (1980) Saugventilation verhindert bei Insekten die Wasserabgabe aus dem Tracheensystem. Verh Dtsch Zool Ges 1980: 306

Kestler P (1982) How do insect spiracles prevent water loss? Verh Dtsch Zool Ges 1982: 262

Kestler P (1983) Weshalb gerade ganz kleine Insekten im Flug einen Gasaustausch durch Diffusion verhindern müssen. In: Nachtigall W (ed) BIONA-report 2:11–20 Akad Wiss, Mainz. Fischer, Stuttgart New York

Kettlewell HBD (1961) The phenomenon of industrial melanism in Lepidoptera. Annu Rev Entomol 6:245–262

Key KHL (1957) Kentromorphic phases in three species of Phasmatodea. Aust J Zool 5:247–284

Key KHL, Day MF (1954) A temperature-controlled physiological colour response in the grasshopper *Kosciuscola tristis* Sjöst. (Orthoptera: Acrididae). Aust J Zool 2:309–339

Khan MA (1983) Control of corpus allatum activity in the adult colorado potato beetle. Thesis, Agricultural University, Wageningen

Khan MA, Koopmanschap AB, de Kort CAD (1982a) The effect of juvenile hormone, 20-hydroxyedysone and precocene II on activity of corpora allata and the mode of negative-feedback regulation of these glands in the adult Colorado potato beetle. J Insect Physiol 28:995–1001

Khan MA, Koopmanshap AB, Privee H, de Kort CAD (1982b) The mode of regulation of the corpus allatum activity during starvation in adult females of the Colorado potato beetle, *Leptinotarsa decemlineata* (Say). J Insect Physiol 28:791–796

Khan MA, Koopmanschap AB, de Kort CAD (1983) The relative importance of nervous and humoral pathways for control of corpus allatum activity in the adult Colorado potato beetle, *Leptinotarsa decemlineata* (Say). Gen Comp Endocrinol 52:214–221

Kieffer S, Wegener C (1983) Wirkungen von Anoxie auf den Gehirnstoffwechsel der Honigbiene *(Apis mellifica)* und der Schmeißfliege *(Calliphora erythrocephala)*. Hoppe Seyler's Z Physiol Chem 364:1156

Kindler SC, Staples R (1970) The influence of fluctuating and constant temperatures, photoperiod, and soil moisture on the resistance of alfalfa to the spotted alfalfa aphid. J Econ Entomol 63:1198–1201

Kingsolver JG, Watt WB (1983) Thermoregulatory strategies in *Colias* butterflies: Thermal stress and the limits to adaptation in temporally varying environments. Am Nat 121:32–55

King EG, Martin DF (1975) *Lixophaga diatraeae:* development at different constant temperatures. Environ Entomol 4:329–332

Kloft WJ, Lloyd HE, Bhatkar AP (1975) Nachweis von Reflex-Bluten bei Lampyriden (Coleoptera) mit Hilfe der "Tracer"-Methode. Experientia (Basel) 31:450–451

Kloot WG van der (1963) The electrophysiology and nervous control of the spiracular muscle of pupae of the giant silkmoths. Comp Biochem Physiol 9:317–333

Knapp E, Wieser W (1980) Looking for "metabolic thermometers" in poikilothermic animals. Naturwissenschaften 67:313–314

Koch K (1982) Biologischer Farbwechsel bei Metriona trisignata – einer tropischen Casside. Entomol Blätter 78:187–191

Koch PB, Bückmann D (1984) Vergleichende Untersuchungen der Farbmuster und der Farbanpassung von Nymphalidenpuppen (Lepidoptera). Zool Beitr (to be published)

Koch PB, Hoffmann KH (1984) Juvenile hormones and reproduction in crickets, Gryllus bimaculatus de Geer. Physiol Entomol (to be published)

Koch UT (1983) A method for recording respiratory movements in an unrestrained insect. In: Nachtigall W (ed) BIONA-report 1:35–40 Akad Wiss, Mainz. Fischer, Stuttgart New York

Koeppe JK, Brantley SG, Nijhout MM (1980) Structure, haemolymph titre, and regulatory effects of juvenile hormone during ovarian maturation in *Leucophaea maderae.* J Insect Physiol 26:749–753

Kogure M (1933) The influence of light and temperature on certain characters of the silkworm, *Bombyx mori.* J Dep Agric Kyushu Imp Univ Fukuoka Jpn 4:1–93

Koidsumi K (1934) Experimentelle Studien über die Transpiration und den Wärmehaushalt bei Insekten. Mem Fac Sci Taihoku 1–12:1–380

Komatsu A (1977) Change of respiratory movement after emergence in the cockroach, *Periplaneta australasiae.* Jpn J Appl Entomol Zool 21:179–183

Kort CAD de, Granger NA (1981) Regulation of the juvenile hormone titer. Annu Rev Entomol 26:1–28

Kricka LJ, DeLuca M (1982) Effect of solvents on the catalytic activity of firefly *Photinus pyralis* luciferase. Arch Biochem Biophys 217:674–681

Krieger F (1954) Untersuchungen über den Farbwechsel der Libellenlarven. Z Vgl Physiol 36: 352–366

Krogh A (1920a) Studien über Tracheenrespiration. II. Über Gasdiffusion in den Tracheen. Pflüger's Arch Gesamte Physiol Menschen Tiere 179:95–112

Krogh A (1920b) Studien über Tracheenrespiration. III. Die Kombination von mechanischer Ventilation mit Gasdiffusion nach Versuchen an *Dytiscus*larven. Pflüger's Arch Gesamte Physiol Menschen Tiere 179:113–120

Krogh A (1941) The comparative physiology of respiratory mechanisms. University of Pennsylvania Press. Reprint (1968) by Dover Publications, New York

Krogh A, Weis-Fogh T (1951) The respiratory exchange of the desert locust *(Schistocerca gregaria)* before, during and after flight. J Exp Biol 28:344–357

Kubista V (1957) Accumulation of a stable phosphorous compound in glycolysing insect muscle. Nature (Lond) 180:549

Kubista V (1958) Anaerobe Glykolyse in den Insektenmuskeln. Biochem Z 330:315–323

Küppers J, Thurm U (1980) Water transport by electroosmosis. In: Locke M, Smith DS (eds) Insect biology in the future "VBW 80". Academic, New York London, pp 125–144

Kurland CG, Schneiderman HA (1959) The respiratory enzymes of diapausing silkworm pupae: A new interpretation of carbon monoxide insensitive respiration. Biol Bull (Woods Hole) 116: 136–161

Laak S van der (1982) Physiological adaptations to low temperature in freezing tolerant *Phyllodecta laticollis* beetles. Comp Biochem Physiol 73A:613–620

Lall AB (1981) Vision tuned to species bioluminescence emission in firefly *Photinus pyralis.* J Exp Zool 216:317–320

Lall AB, Seliger HH, Biggley WH (1980) Ecology of colors of firefly bioluminescence. Science (Wash DC) 210:560–562

Lall AB, Lord ET, Trouth CO (1982) Vision in the firefly *Photuris lucicrescens* (Coleoptera: Lampyridae): Spectra sensitivity and selective adaptation in the compound eye. J Comp Physiol 147A:195–200

Laquet LM (1976) Convection and diffusion in the human airways. Thesis, Katholieke Universiteit Leuven. SINTAL, Leuven

Laudien H (1973) Changing reaction systems. In: Precht H, Christophersen J, Hensel J, Larcher W. Temperature and life. Springer, Berlin Heidelberg New York, pp 355–399

Laugé G, Launois M (1980) Effets de deux conditions photothérmoperiodique ou apériodique sur le criquet migrateur Malgache *Locusta migratoria capito* (Saussure) (Orthopt., Acrididae). Ann Soc Entomol Fr 16:221–231

Lee J (1976) Bioluminescence of the Australian glow-worm, *Arachnocampa richardsae* Harrison. Photochem Photobiol 24:279–285

Lee RE, Baust JG (1982) Absence of metabolic cold adaptation and compensatory acclimation in the Antarctic fly, *Belgica antarctica.* J Insect Physiol 28:725–729

Lees AD (1955) The physiology of diapause in arthropods. Cambridge University Press, London Cambridge

Lees AD (1964) The location of the photoperiodic receptors in the aphid *Megoura viciae* Buckton. J Exp Biol 41:119–133

Levy RI, Schneiderman HA (1957) The direct measurement and significance of changes in intratracheal gas composition during the respiratory cycle of the *Cecropia* moth (Abstract 205). Anat Rec 128:583

Levy RI, Schneiderman HA (1958) An experimental solution to the paradox of discontinuous respiration in insects. Nature (Lond) 182:491–493

Levy RI, Schneiderman HA (1966a) Discontinuous respiration in insects. II. The direct measurement and significance of changes in tracheal gas composition during the respiratory cycle of silkworm pupae. J Insect Physiol 12:83–104

Levy RI, Schneiderman HA (1966b) Discontinuous respiration in insects. III. The effect of temperature and ambient oxygen tension on the gaseous composition of the tracheal system of silkworm pupae. J Insect Physiol 12:105–121

Levy RI, Schneiderman HA (1966c) Discontinuous respiration in insects. IV. Changes in intratracheal pressure during the respiratory cycle of silkworm pupae. J Insect Physiol 12:465–492

L'Honoré MJ (1971) Données cytophysiologiques sur les tubes de Malpighi de *Gryllotalpa gryllotalpa* Latr. (Orthoptères, Gryllotalpidae). C R Acad Sci (Paris) Ser D 272:2788–2791

Lin S, Hodson AC, Richards AG (1954) An analysis of threshold temperature for development of *Oncopeltus* and *Tribolium* eggs. Physiol Zool 27:287–311

Linzen B (1974) The tryptophan –> ommochrome pathway in insects. Adv Insect Physiol 10: 117–246

Livingstone DA (1963) Chemical composition of rivers and lakes. In: Fleischer M (ed) Data of geochemistry, 6th, edn. Geological Survey Professional, Paper 440, Chapter G. GPO, Washington

Lloyd JE (1968) Illumination, another function of firefly flashes? Entomol News 79:265–268
Lloyd JE (1971) Bioluminescent communication in insects. Annu Rev Entomol 16:97–122
Lloyd JE (1973) Fireflies of Melanesia: Bioluminescence, mating behavior, and synchronous flashing (Coleoptera: Lampyridae). Environ Entomol 2:991–1008
Lloyd JE (1978) Insect bioluminescence. In: Herring PJ (ed) Bioluminescence in action. Academic, New York, pp 241–272
Lloyd JE (1981) Firefly mate-rivals mimic their predators and *vice versa.* Nature (Lond) 290: 498–500
Lloyd JE (1983) Bioluminescence and communication in insects. Annu Rev Entomol 28:131–160
Locke M (1958) The coordination of growth in the tracheal system of insects. Q J Microsc Sci 99:373–391
Loeb MJ (1982) Diapause and development in the tobacco budworm, *Heliothis virescens:* A comparison of haemolymph ecdysteroid titres. J Insect Physiol 28:667–673
Logan JA, Wollkind DJ, Hoyt SC, Tanigoshi LK (1976) An analytic model for description of temperature dependent rate phenomena in arthropods. Environ Entomol 5:1133–1140
Loof A de, de Wilde J (1970) Hormonal control of synthesis of vitellogenic female proteins in the Colorado potato beetle, *Leptinotarsa decemlineata.* J Insect Physiol 16:1455–1466
Lounibos LP, Bradshaw WE (1975) A second diapause in *Wyeomyia smithii:* seasonal incidence and maintenance by photoperiod. Can J Zool 53:215–221
Loveridge JP (1968) Control of water loss in *Locusta migratoria migratorioides* R. and F. II. Water loss through the spiracles. J Exp Biol 49:15–29
Loveridge JP (1975) Studies on the water balance of adult locusts. III. The water balance of non-flying locusts. Zoologica Africana 10:1–28
Ludwig D (1928) The effects of temperature on the development of an insect (*Popillia japonica* Newman). Physiol Zool 1:358–389
Ludwig D, Cable RM (1933) The effect of alternating temperatures on the pupal development of *Drosophila melanogaster* Meigen. Physiol Zool 6:493–508
Lüscher M (1965) The influence of hormones on tissue respiration in the insect, *Leucophaea maderae.* Gen Comp Endocrinol 5:669–670
Maddrell SHP (1962) A diuretic hormone in *Rhodnius prolixus* Stal. Nature (Lond) 194:605–606
Maddrell SHP (1963) Excretion in the blood-sucking bug, *Rhodnius prolixus* Stal. I. The control of diuresis. J Exp Biol 40:247–256
Maddrell SHP (1977) Insect Malpighian tubules. In: Gupta BL, Moreton RB, Oschman JL, Wall BJ (eds) Transport of ions and water in animals. Academic, New York London, pp 541–569
Maddrell SHP (1980) The control of water relations in insects. In: Locke M, Smith DS (eds) Insect biology in the future "VBW 80". Academic, New York London, pp 179–200
Madubunyi LC, Koehler CS (1974) Effects of photoperiod and temperature on development in *Hypera brunneipennis.* Environ Entomol 3:1017–1021
Mandl K (1931) Künstliche Veränderungen der Farben an *Cicindela nitida* Licht. und an anderen *Cicindela*-Arten. Z Morphol Oekol Tiere 22:110–120
Mansingh A (1971) Physiological classification of dormancies in insects. Can Entomol 103:983–1009
Marrewijk WJA van, van den Broek ATM, Beenakkers AMT (1980) Regulation of glycogenolysis in the locust fat body during flight. Insect Biochem 10:675–679
Martin MM (1983) Cellulose digestion in insects. Comp Biochem Physiol 75A:313–324
Martin MM, Boyd ND, Gieselmann MJ, Silver RG (1975) Activity of faecal fluid of a leaf-cutting ant toward plant cell wall polysaccharides. J Insect Physiol 21:1887–1892
Marzusch K (1952) Untersuchungen über die Temperaturabhängigkeit von Lebensprozessen bei Insekten unter besonderer Berücksichtigung winterschlafender Kartoffelkäfer. Z Vgl Physiol 34:75–92
Masaki S (1956) The effect of temperature on the termination of diapause in the egg of *Lymantria dispar* L.. Jpn J Appl Entomol Zool 21:148–157
Masaki S (1980) Summer diapause. Annu Rev Entomol 25:1–26
Matsumoto S, Isogai A, Suzuki A, Ogura N, Sonobe H (1981) Purification and properties of the melanization and reddish colouration hormone (MRCH) in the armyworm, *Leucania separata* (Lepidoptera). Insect Biochem 11:725–733

Matthee JJ (1978) The induction of diapause in eggs of *Locustana pardalina* (Acrididae) by high temperatures. J Entomol Soc S Afr 41:25–30

Matthews EG (1976) Insect Ecology. University of Queensland Press, St. Lucia

Matthews JR, Downer RGH (1973) Hyperglycemia induced by anesthesia in the American cockroach, *Periplaneta americana* (L.). Can J Zool 51:395–397

May ML (1976) Physiological color change in new-world damselflies (Zygoptera). Odonatologica (Utr) 5:165–171

May ML (1979) Insect thermoregulation. Annu Rev Entomol 24:313–349

McDaniel CM (1979) Haemolymph edcysone concentration in *Hyalophora cecropia* pupae, dauerpupae and adults. J Insect Physiol 25:143–145

McCapra F (1978) The chemistry of bioluminescence. In: Herring PJ (ed) Bioluminescence in action. Academic, New York, pp 49–74

McCapra F (1982) The chemistry of bioluminescence. Proc R Soc Lond Ser B 215:247–272

McCutcheon FH (1940) The respiratory mechanism in the grasshopper. Ann Entomol Soc Am 33:35–55

McElroy WD, DeLuca M (1978) Chemistry of firefly luminescence. In: Herring PJ (ed) Bioluminescence in action. Academic, New York, pp 109–128

McElroy WD, DeLuca M (1981) The chemistry and applications of firefly luminescence. In: DeLuca MA, McElroy WD (eds) Bioluminescence and chemiluminescence. Academic, New York, pp 179–186

McElroy ED, Seliger HH, DeLuca M (1965) Enzyme catalysis and color of light in bioluminescent reactions. In: Bryson V, Volge H (eds) Evolving genes and proteins. Academic, New York, pp 319–340

McElroy WD, Seliger HH, DeLuca M (1974) Insect bioluminescence. In: Rockstein M (ed) The physiology of insects, vol II. Academic, New York, pp 411–460

McFarlane A, McLusky DS (1972) The oxygen consumption of chironomid larvae from Loch Leven in relation to temperature. Comp Biochem Physiol 43A:991–1001

McLaren IA (1963) Effects of temperature on growth of zooplankton and the adaptive value of vertical migration. J Fish Res Board Can 20:685–727

Meats A (1973) Rapid acclimatization to low temperature in the Queensland fruit fly, *Dacus tryoni*. J Insect Physiol 19:1903–1911

Mellors WK, Bassow FE (1983) Temperature-dependent development of Mexican bean beetle (Coleoptera: Coccinellidae) immatures on snap bean and soybean foliage. Ann Entomol Soc Am 76:692–698

Merkel G (1977) The effects of temperature and food quality on the larval development of *Gryllus bimaculatus* (Orthoptera, Gryllidae). Oecologia (Berl) 30:129–140

Messenger PS (1964) The influence of rhythmically fluctuating temperatures on the development and reproduction of the spotted alfalfa aphid, *Therioaphis maculata*. J Econ Entomol 57: 71–76

Messenger PS, Flitters NE (1958) Effect of constant temperature environments on the egg stage of three species of Hawaiian fruit flies. Ann Entomol Soc Am 51:109–119

Messenger PS, Flitters NE (1959) Effect of variable temperature environments on egg development of three species of fruit flies. Ann Entomol Soc Am 52:191–204

Meyer SGE (1971) Der respiratorische Stoffwechsel von *Callitroga macellaria* (Fabricius) (Diptera, Calliphoridae) in Abhängigkeit von Entwicklung und Umwelt (mit einem Rechenprogramm zur mathematisch-statistischen Auswertung von WARBURG-Versuchen). PHD Thesis, Universität Bonn

Meyer SGE (1974) Variation von Glycerin, Sorbit und anderen Metabolitgehalten infolge von Temperaturstreß bei Larven von *Callitroga macellaria*. J Insect Physiol 20:395–403

Meyer SGE (1977) Concentrations of some glycolytic and other intermediates in larvae of *Callitroga macellaria* (F.) (Diptera, Calliphoridae) during anaerobiosis. Comp Biochem Physiol 58B:49–55

Meyer SGE (1978) Effects of heat, cold, anaerobiosis and inhibitors on metabolite concentrations in larvae of *Callitroga macellaria*. Insect Biochem 8:471–477

Meyer SGE (1980) Studies on anaerobic glucose and glutamate metabolism in larvae of *Callitroga macellaria*. Insect Biochem 10:449–455

Meyer SGE, Müller R (1977) Änderungen der Aminosäurekonzentration bei Anaerobiose in Larven von *Callitroga macellaria.* Insect Biochem 7:459–461

Meyer SGE, Schaub G (1973) Der respiratorische Stoffwechsel von Calliphoridenlarven. J Insect Physiol 19:2183–2198

Michel R, Wegener R (1982) Metabolic effects of anoxia in the central nervous system of the locust *(Locusta migratoria).* Hoppe Seyler's Z Physiol Chem 363:1310

Michieli S, Zener B (1968) Der Sauerstoffverbrauch verschiedener Farbstadien bei der Wanze *Nezara viridula* (L.). Z Vgl Physiol 58:223–224

Mill PJ (1972) Respiration in the invertebrates. Macmillan, London

Mill PJ (1974) Respiration: Aquatic insects. In: Rockstein M (ed) The physiology of insecta, 2nd edn, vol VI. Academic, New York London, pp 403–467

Miller LK (1978) Freezing tolerance in relation to cooling rate in an adult insect. Cryobiology 15:345–349

Miller LK (1982) Cold-hardiness strategies of some adult and immature insects overwintering in interior Alaska. Comp Biochem Physiol 73A:595–604

Miller PL (1964) Factors altering spiracle control in adult dragonflies: water balance. J Exp Biol 41:331–343

Miller PL (1966) The regulation of breathing in insects. Adv Insect Physiol 3:279–354

Miller PL (1973) Spatial and temporal changes in coupling of cockroach spiracles to ventilation. J Exp Biol 59:137–148

Miller PL (1974) Respiration – Aerial gas transport. In: Rockstein M (ed) The physiology of insecta, 2nd, vol VI. Academic, New York London, pp 345–402

Miller PL (1980) Pacemaker neurons and rhythmic behavior. In: Locke M, Smith DS (eds) Insect biology in the future "VBW 80". Academic, New York London, pp 819–846

Miller PL (1981) Ventilation in active and inactive insects. In: Herreid II CF, Fourtner CR (eds) Locomotion and energetics in arthropods. Plenum, New York London, pp 367–390

Miller PL (1982) Respiration. In: Bell WJ, Adiyodi KG (eds) The American cockroach. Chapman and Hall, London New York, pp 87–116

Mills RR, Cochran DG (1967) Adenosinetriphosphatases from thoracic muscle mitochondria of the American cockroach. Comp Biochem Physiol 20:919–923

Mishima J (1976) The estimation of the life cycles of *Pieris rapae* (Lepidoptera, Pieridae) from the effective accumulative temperature. Sci Rep Tokyo K D Sect B 16:159–170

Miyashita K (1971) Effects of constant and alternating temperatures on the development of *Spodoptera litura* F. (Lepidoptera, Noctuidae). Appl Entomol Zool 6:105–111

Moon RD (1983) Simulating developmental time of preadult face flies (Diptera: Muscidae) from air temperatures records. Environ Entomol 12:943–948

Mordue W (1972) Hormones and excretion in locusts. Gen Comp Endocrinol Suppl 3:289–298

Moreau R (1975) Relations entre les modifications du métabolisme respiratoire et l'utilisation du cycle des pentoses chez *Bombyx mori* L. somuis à différentes températures. Soc Biol Bordeaux 28:953–958

Morgan KR, Bartholomew GA (1982) Homeothermic response to reduced ambient temperature in a scarab beetle. Science (Wash DC) 216:1409–1410

Morgan TD, Chippendale GM (1983) Free amino acids of the haemolymph of the southwestern corn borer in relation to their diapause. J Insect Physiol 29:735–740

Moscardi F, Barfield CS, Allen GE (1981) Effects of temperature on adult velvetbean caterpillar oviposition, egg hatch, and longevity. Ann Entomol Soc Am 74:167–171

Mothes G (1960) Weitere Untersuchungen über den physiologischen Farbwechsel von *Carausius morosus* (Br.). Zool Jahrb Physiol 69:133–162

Mourik J, Raeven, Steuer K, Addink ADF (1982) Anaerobic metabolism of red skeletal muscle of goldfish, *Carassius auratus* (L.). Mitochondrial produced acetaldehyde as an anaerobic electron acceptor. FEBS Lett 137:111–114

Müller HJ (1970) Formen der Dormanz bei Insekten. Nove Acta Leopold N F 35 191:7–27

Müller HJ (1974) Farb-Polymorphismus bei Larven der Jasside *Mocydia crocera* H. S. (Homoptera, Auchenorrhyncha). Zool Anz 192:303–315

Müller HJ (1976) Formen der Dormanz bei Insekten als Mechanismen ökologischer Anpassung. Verh Dtsch Zool Ges 1976:46–58

Mulligan FS (1980) Direct induction of embryonic diapause in colonized *Acedes dorsalis*. Ann Entomol Soc Am 73:589–592

Mullins DE (1982) Osmoregulation and excretion. In: Bell WJ, Adiyodi KG (eds) The American cockroach. Chapman and Hall, London New York, pp 117–149

Murad H, Ahmad J (1977) Histomorphology of the male reproductive organs of the red flour beetle, *Tribolium castaneum* (Coleoptera: Tenebrionidae). J Anim Morphol Physiol 24: 35–41

Murray DRP (1968) The importance of water in the normal growth of larvae of *Tenebrio molitor.* Entomol Exp Appl 11:149–168

Mutchmor JA (1967) Temperature adaptation in insects. In: Prosser CL (ed) Molecular mechanisms of temperature adaptation. American Association for the Advancement of Science, Washington DC, pp 165–176

Myers TB (1958) Some aspects of tracheal ventilation in the cockroach, *Byrsotria fumigata* (Guerin). Thesis, Ohio State University, Columbus

Myers TB, Fisk FW (1962) Breathing movements of the Cuban burrowing cockroach. Ohio J Sci 62:253–257

Myers TB, Retzlaff E (1963) Localization and action of the respiratory centre of the Cuban borrowing cockroach. J Insect Physiol 9:607–614

Needham AE (1974) The significance of zoochromes. Springer, Berlin Heidelberg New York

Needham AE (1978) Insect biochromes: Their chemistry and role. In: Rockstein M (ed) Biochemistry of insects. Academic, New York, pp 233–305

Needham JE (1964) The growth process in animals. Sir Isaac Pitman, London

Neumann D (1962) Die Anaerobiose-Toleranz der Larven zweier Subspezies von *Chironomus thummi.* Z Vgl Physiol 46:150–162

Neumann D (1976) Adaptations of chironomids to intertidal environments. Annu Rev Entomol 21:387–414

Neumann D, Heimbach F (1975) Das Wachstum des Kohlweißlings bei konstanten und tagesperiodisch wechselnden Temperaturen. Oecologia (Berl) 20:135–141

Neville AC (1965) Circadian organization of chitin in some insect skeletons. Q J Microsc Sci 106:315–325

Neville AC (1977) Metallic gold and silver colours in some insect cuticles. J Insect Physiol 23: 1267–1274

Newell RC (1967) Oxidative activity of poikilotherm mitochondria as a function of temperature. J Zool 151:299–311

Newsholme EA, Start C (1973) Regulation in metabolism. Wiley, London

Newsholme EA, Crabtree B, Higgins SJ, Thornton SD, Start C (1972) The activities of fructose diphosphatase in flight muscles from the bumblebee and the role of this enzyme in heat generation. Biochem J 128:89–97

Nicholson SW (1976) The hormonal control of diuresis in the Cabbage white butterfly, *Pieris brassicae.* J Exp Biol 65:565–575

Nicol JAC (1978) Bioluminescence and vision. In: Herring PJ (ed) Bioluminescence in action. Academic, New York, pp 367–398

Nicolas G, Fuzeau-Braesch S, Djeddour K (1979) Influence de la suppression du thermo-photoperiodisme sur les charactéristiques morphométriques du criquet migrateur, *Locusta migratoria* L. élevé en groupe. C R Acad Sci (Paris) Ser D 288:343–346

Nicolaus RA (1968) Melanins. Hermann, Paris

Nijhout HF (1975) A treshold size for metamorphosis in the tobacco hornworm, *Manduca sexta* (L.). Biol Bull (Woods Hole) 149:214–225

Nijhout HF (1979) Stretch-induced moulting in *Oncopeltus fasciatus.* J Insect Physiol 25: 277–281

Nijhout HF (1981) Physiological control of moulting in insects. Am Zool 21:631–640

Nijhout HF, Wheeler DE (1982) Juvenile hormone and the physiological basis of insect polymorphisms. Q Rev Biol 57:109–133

Nuñez JA (1956) Untersuchungen über die Regelung des Wasserhaushaltes bei *Anisotarsus cupripennis* Germ. Z Vgl Physiol 38:341–354

Nunome Z (1944a) Studies on the respiration of the silkworm. I. Diffusion of oxygen in the respiratory system of the silkworm. Bull Seric Exp Stn Japan 12:17–39

Nunome Z (1944b) Studies on the respiration of the silkworm. II. Several measurements in the respiratory organ. Bull Seric Exp Stn Japan 12:41–90

Nunome Z (1951) Studies on the respiration of the silkworm. III. On the air current of respiration. J Seric Sci Tokyo 20:111–127 (Translations loaned by courtesy of Dr. John Buck)

O'Farrell (1963) Temperature-controlled physiological colour change in some Australian damselflies (Odonata: Zygoptera). Aust J Sci 25:437–438

Okuda T, Hodek I (1983) Response to constant photoperiods in *Coccinella septempunctata bruckii* populations from central Japan (Coleoptera: Coccinellidae). Acta Entomol Bohemoslov 80:74–75

Oltmer A (1968) Die Steuerung des Melanineinbaus in das Farbmuster der Kohlweißlingspuppe *Pieris brassicae* L. Wilhelm Roux Arch Dev Biol 160:401-427

Opalka B (1977) Untersuchungen zum Anaerobiosestoffwechsel der Larven von *Chaoborus crystallinus* de Geer. Diplomarbeit, Universität Münster

Orchard J, Lange AB (1983) The hormonal control of hemolymph lipid during flight in *Locusta migratoria*. J Insect Physiol 29:639–642

Otte D (1980) Notes and comments on theories of flash synchronization in fireflies. Am Nat 116:587–590

Otto D, Weber T (1982) Interneurons descending from the cricket cephalic ganglia that discharge in the pattern of two motor rhythms J Comp Physiol 148A:209–219

Owen OE, Reichard GA Jr, Patel MS, Boden G (1979) Energy metabolism in feasting and fasting. Adv Exp Med Biol 11:169–188

Pagel R, Wulff K (1980) On the mechanism of *Phausis splendidula* bioluminescence. Photochem Photobiol 32:111–112

Papillon M, Cassier P, Girardie J, Lafon-Cazal M (1976) L' influence de la température d'élevage sur l'activité endocrine de *Schistocerca gregaria*. Arch Biol 87:103–127

Papillon M, Porcheron P, Baehr JC (1980) Effects de la température d'élevage sur la croissance et l'équilibre hormonal de *Schistocerca gregaria* au cours des deux derniers stades larvaires. Experientia (Basel) 36:419–422

Patterson JL, Duman JG (1979) Composition of a protein antifreeze from larvae of the beetle, *Tenebrio molitor*. J Exp Zool 210:361–367

Patterson JL, Duman JG (1982) Purification and composition of protein antifreeze with high cysteine contents from larvae of the beetle, *Tenebrio molitor*. J Exp Zool 219:381–384

Patterson JL, Kelly TJ, Duman JG (1981) Purification and composition of a thermal hysteresis producing protein from the milkweeg bug, *Oncopeltus fasciatus*. J Comp Physiol 142B: 539–542

Paulpandian A (1964) Cyclic ventilation movement in the common cockroach, *Periplaneta americana*. Curr Sci (Bangalore) 33:404–405

Pavlidis T, Zimmerman WF, Osborn J (1968) A mathematical model for the temperature effects on circadian rhythms. J Theor Biol 18:210–221

Peacock AJ, Anstee JH, Bowler K (1977) The effect of temperature on development and the histology of the neuroendocrine system of the tettigoniid, *Jamaicana flava* (Caudell). J Therm Biol 2:35–40

Peairs LM (1914) The relation of temperature to insect development. J Econ Entomol 7:1974–1981

Peferoen M, Stynen D, de Loof A (1982) A re-examination of the protein pattern of the haemolymph of *Leptinotarsa decemlineata*, with special reference to vitellogenins and diapause proteins. Comp Biochem Physiol 72B:345–351

Pener MP, Orshan L (1980) Reversible reproductive diapause and intermediate states between diapause and full reproductive activity in male *Oedipodia miniata* grasshoppers. Physiol Entomol 5:417–426

Peterson MK (1970) The fine structure of the larval firefly light organ. J Morphol 131:103–115

Pette D (1965) Plan und Muster im zellulären Stoffwechsel. Naturwissenschaften 52:597–616

Pfadt RE, Lloyd JE, Sharafi G (1975) Pupal development of cattle grubs at constant and alternating temperatures. J Econ Entomol 68:325–328

Pflüger W (1973) Die Sanduhrsteuerung der gezeitensynchronen Schlüpfrhythmik der Mücke *Clunio marinus* im arktischen Mittsommer. Oecologia (Berl) 11:113–150

Phillip JS, Watson TF (1971) Influence of temperature on population growth of the pink bollworm, *Pectinophora gossypiella* (Lepidoptera: Gelechiidae). Ann Entomol Soc Am 64: 334–340

Phillips JE (1964) Rectal absorption in the desert locust *Schistocerca gregaria* Forskal. II. Sodium, potassium and chloride. J Exp Biol 41:39–67

Phillips JE (1977) Excretion in insects. Function of gut and rectum in concentrating and diluting urine. Fed Proc 36:2480–2486

Phillips JE (1980) Epithelial transport and control in recta of terrestrial insects. In: Locke M, Smith DS (eds) Insect biology in the future "VBW 80". Academic, New York London, pp 145–177

Phillips JE (1983) Cellular mechanisms and control of excretion in insects: recent advances. Verh Dtsch Zool Ges 1983:55–68

Phillips JE, Bradley TJ, Maddrell SHP (1977) Mechanisms of ionic and osmotic regulation in saline-water mosquito larvae. In: Schmidt-Nielsen K, Bolis L, Maddrell SHP (eds). Comparative physiology: Water, ions and fluid mechanisms. Cambridge University Press, London, pp 151–171

Phillips JE, Spring J, Hanrahan J, Mordue W, Meredith J (1982a) Hormonal control of salt reabsorption by the excretory system of an insect: isolation of a new protein. In: Farmer DS, Lederis K (eds) Neurosecretion: molecules, cells, systems. Plenum,New York, pp 371–380

Phillips JE, Meredith J, Spring J, Chamberlin M (1982b) Control of ion reabsorption in locust rectum: Implication for fluid transport. J Exp Zool 222:297–308

Piccione W, Baust JG (1977) The effects of low temperature acclimation of neural (sodium-potassium dependent) ATPase in *Periplaneta americana.* Insect Biochem 7:185–189

Pickfort R (1953) A two-year cycle in grasshoppers (Orthoptera: Acrididae) overwintering as eggs and nymphs. Can Entomol 85:9–14

Piiper J (1982) Physiologie der Atmung. Urban und Schwarzenberg, München Berlin Wien (Physiologie des Menschen, vol 6)

Piiper J, Dejours P, Haab P, Rahn H (1971) Concepts and basic quantities in gas exchange physiology. Respir Physiol 13:292–304

Popescu C (1979) Natural selection in the industrial melanic psocid *Mesopsocus unipunctatus* (Insecta: Psocoptera) in northern England UK. Heredity 42:133–142

Poras M, Baehr JC, Cassier P (1983) Control of corpus allatum activity during the imaginal diapause in females of *Locusta migratoria* L.. Int J Invertebr Reprod 6:111–122

Porcheron P, Papillon M, Baehr JC (1982) Hormonal levels and protein variations during sexual maturation of *Schistocerca gregaria;* effect of rearing temperature. Experientia (Basel) 38: 970–972

Powell JA (1974) Occurrence of prolonged diapause in ethiniid moths. Pan-Pac Entomol 50: 220–225

Precht H (1949) Die Temperaturabhängigkeit von Lebensprozessen. Z Naturforsch 46:26–35

Precht H, Christophersen J, Hensel H, Larcher W (1973) Temperature and life. Springer, Berlin Heidelberg New York

Prosser CL (1958) General Summary: The nature of physiological adaptation. In: Prosser CL (ed) Physiological adaptation. Am Physiol Soc (Wash DC), pp 167–180

Prosser CL, Nelson DO (1981) The role of nervous systems in temperature adaptation of poikilotherms. Annu Rev Physiol 43:281–300

Provansal A, Baudry-Partiaoglou N, Slama K (1977) Haemolymph pressure pulses in the metamorphosis of *Tenebrio molitor.* Acta Entomol Bohemoslov 74:362–374

Pryor MGM (1940) On the hardening of the cuticle of insects. Proc R Soc Lond Ser B 128: 393–407

Punt A (1944) De gaswisseling van enkele bloedzuigende parasieten van warmbloedige dieren *(Cimex, Rhodnius, Triatoma)*. Onderzoekingen Physiol Lab Rijks-Universiteit Utrecht Ser 8 Pt 3:122–141

Punt A (1950) The respiration of insects. Physiol Comp Oecol 2:59–74

Punt A (1955) New diaferometric investigations on the respiration in insects. Acta Physiol Pharmacol Neerl 4:114

Punt A (1956) The influence of carbon dioxide on the respiration of *Carabus nemoralis* Müll. Physiol Comp Oecol 4:132–141

Punt A, Parser WJ, Kuchlein J (1957) Oxygen uptake in insects with cyclic CO_2 release. Biol Bull (Woods Hole) 112:108–119

Raabe M (1966) Recherches sur la neurosécrétion dans la chaine nerveuse ventrale du phasme *Carausius morosus:* Liaison entre l'activité des cellules B 1 et la pigmentation. C R Acad Sci (Paris) Ser D 263:408–411

Raabe M (1982) Insect neurohormones. Plenum, New York London

Raabe M (1983) Pigment metabolism. In: Downer RGH, Laufer H (eds) Endocrinology of insects. Alan R. Liss, New York, pp 493–500 (Invertebrate endocrinology, vol 1)

Rahn H, Paganelli CV (1968) Gas exchange in gas gills of diving insects. Respir Physiol 5:145–164

Ramsay JA (1935) The evaporation of water from the cockroach. J Exp Biol 12:373–383

Ramsay JA (1953) Active transport of potassium by the Malpighian tubules of insects. J Exp Biol 30:358–369

Ramsay JA (1954) Active transport of water by the Malpighian tubules of the stick insect *Dixippus morosus* (Orthoptera, Phasmidae). J Exp Biol 31:104–113

Ramsay JA (1955) The excretion of sodium, potassium and water by the Malpighian tubules of the stick insect, *Dixippus morosus* (Orthoptera, Phasmidae). J Exp Biol 32:200–216

Ramsay JA (1958) Excretion by the Malpighian tubules of the stick insect *Dixippus morosus* (Orthoptera, Phasmidae). Amino acids, sugar, urea. J Exp Biol 35:871–891

Ratte HT (1979) Tagesperiodische Vertikalwanderung in thermisch geschichteten Gewässern: Einfluß von Temperatur- und Photoperiode-Zyklen auf *Chaoborus crystallinus* de Geer (Diptera: Chaoboridae). Arch Hydrobiol Suppl 57:1–37

Ratte HT (1983) Nachweis einer tagezeitlichen Temperatursensitivität für Diapause-Induktion und Gewichtszunahme bei *Chaoborus crystallinus* (Diptera, Chaoboridae). Verh Dtsch Zool Ges 1983:313

Redfern CPF (1982) 20-hydroxy-ecdysone and ovarian development in *Anopheles stephensi.* J Insect Physiol 28:97–109

Reinecke P, Buckner JS, Grugel SR (1980) Life cycle of laboratory-reared tobacco hornworms, *Manduca sexta,* a study of development and behavior, using time lapse cinematography. Biol Bull (Woods Hole) 158:129–140

Remmert H (1960) Über tagesperiodische Änderungen des Licht- und Temperaturpräferendums bei Insekten (Untersuchungen an *Cicindela campestris* und *Gryllus domesticus*). Biol Zentralbl 79:577–584

Remmert H (1980) Arctic animal ecology. Springer, Berlin Heidelberg New York

Remmert H, Wünderling K (1970) Temperature differences between arctic and alpine meadows and their ecological significance. Oecologia (Berl) 4:208–210

Rence B, Loher W (1975) Arhythmically singing crickets: Thermoperiodic reentrainment after bilobectomy. Science (Wash DC) 190:385–387

Rensing L (1973) Biologische Rhythmen und Regulationen. Fischer, Stuttgart (Grundbegriffe der modernen Biologie, Bd 10)

Ressin WJ (1980) The effect of juvenile hormone on pupal pigmentation of *Pieris brassicae* L. J Insect Physiol 26:295–302

Reynolds WW (1979) Thermoregulation in ectotherms. Am Zool 19:191–384

Richards AM (1960) Observations on the New Zealand glowworm *Arachnocampa luminosa* (Skuse) 1890. Trans R Soc NZ 88:559–574

Richards OW, Davies RG (1977) Imms' general textbook of entomology. 10th edn, vol 1. Chapman and Hall, London

Richter H (1976) The water status in the plant – experimental evidence. In: Lange OL, Kappen L, Schulze ED (eds) Water and plant life. Springer, Berlin Heidelberg New York, pp 42–58

Riddiford LM, Truman JW (1978) Biochemistry of insect hormones and insect growth regulators. In: Rockstein M (ed) The physiology of insects. Academic, New York, pp 307–357

Ring RA (1982) Freezing-tolerant insects with low supercooling points. Comp Biochem Physiol 73A:605–612

Ring DR, Calcote VR, Harris MK (1983) Verification and generalization of a degree-day model predicting pecan nut casebearer (Lepidoptera: Pyralidae) activity. Environ Entomol 12:487–489

Robb R, Hammond R, Bieber LL (1972) Temperature-dependent changes in sphingosine composition of fatty acids of glycerophosphatides from *Musca domestica.* larvae. Insect Biochem 2:131–136

Robbins WE, Kaplanis JN, Svoboda JA, Thompson MJ (1971) Steroid metabolism in insects. Annu Rev Entomol 16:53–72

Roberts JL (1974) Temperature acclimation and behavioural thermoregulation in cold-blooded animals. Fed Proc 3:2155–2161

Roberts SKdeF (1962) Circadian activity in cockroaches. II. Entrainment and phase-shifting. J Cell Comp Physiol 59:175–186

Robertson HA, Carlson AD (1976) Octopamine: presence in firefly lantern suggests a transmitter role. J Exp Zool 195:159–164

Rockstein M, Miquel J (1973) Ageing in insects. In: Rockstein M (ed) The physiology of insecta, vol I, 2nd edn. Academic, New York London, pp 371–478

Rock GC, Shaffer PL (1983) Developmental rates of codling moth (Lepidoptera: Olethreutidae) reared on apple at four constant temperatures. Environ Entomol 12:831–834

Roe RM, Clifford CW, Woodring JP (1980) The effect of temperature on feeding, growth, and metabolism during the last larval stadium of the female house cricket, *Acheta domestica.* J Insect Physiol 26:639–644

Roe RM, Hammond AM Jr, Sparks TC (1982) Growth of larval *Diatraea saccharalis* (Lepidoptera: Pyralidae) on an artificial diet and synchronization of the last larval stadium. Ann Entomol Soc Am 75:421–429

Rojas RR, Lee RE, Luu TA, Baust JG (1983) Temperature dependence-independence of antifreeze turnover in *Eurosta solidaginis* (Fitch). J Insect Physiol 29:865–869

Roland J (1978) Variation in spectral reflectance of alpine and arctic *Colias* (Lepidoptera: Pieridae). Can J Zool 56:1447–1453

Rothe U (1983) Preliminary results of metabolic measurements in resting and flying honeybees. In: Nachtigall W (ed) BIONA-report 1:21–28 Akad Wiss, Mainz. Fischer, Stuttgart New York

Rothschild M (1973) Remarks on carotenoids in the evolution of signals. In: Gilbert LE, Raven PH (eds) Coevolution of animals and plants. University of Texas Press, Austin London, pp 20–50

Rowell CHF (1971) The variable coloration of the acridoid grasshoppers. Adv Insect Physiol 8: 145–198

Rüdiger W, Klose W (1970) Über die Pigmente der Florfliege *Chrysopa carnea.* Experientia (Basel) 26:498

Ruiter L de (1956) Countershading in caterpillars. Arch Neerl Zool 11:285–341

Sacktor B (1965) Energetics and respiratory metabolism of muscular contraction. In: Rockstein M (ed) The physiology of insecta, vol II. Academic, London, pp 483–580

Sacktor B (1970) Regulation of intermediary metabolism, with special reference to the control mechanisms in insect flight muscle. Adv Insect Physiol 7:267–347

Sacktor B (1975) Biochemistry of insect flight. Part 1 – Utilization of fuels by muscle. In: Candy DJ, Kilby BA (eds) Insect biochemistry and function. Champan and Hall, London, pp 1–88

Safranek L, Cymborowski B, Williams CM (1980) Effects of juvenile hormone on ecdysone-dependent development in the tobacco hornworm, *Manduca sexta.* Biol Bull (Woods Hole) 158: 248–256

Sahota TS, Ruth DS, Ibaraki A, Farris SH, Peet FG (1982) Diapause in the pharate adult stage of insect development. Can Entomol 114:1179–1184

Saunders DS (1971) The temperature-compensated photoperiodic clock "programming" development and pupal diapause in the flesh fly, *Sarcophaga argyrostoma*. J Insect Physiol 17: 801–812

Saunders DS (1974) Circadian rhythms and photoperiodism in insects. In: Rockstein M (ed) The physiology of insecta, vol II, 2nd edn. Academic, New York London, pp 461–533

Saunders DS (1976) Insect clocks. Pergamon, Oxford

Saunders DS (1982) Insect clocks, 2nd edn, Pergamon, Oxford

Scheid P, Hook C, Bridges CR (1982) Diffusion in gas exchange of insects. Fed Proc 41:2143–2145

Schleip W (1910) Der Farbwechsel von *Dixippus morosus*. Zool Jb Allg Zool 30:45–132

Schmidt G (1956) Der Stoffwechsel der Caraben (Ins., Coleoptera) und seine Beziehung zum Wasserhaushalt. Zool Jahrb Physiol 66:273–294

Schmidt-Nielsen K (1979) Animal physiology. Adaptation and environment. Cambridge University Press, Cambridge

Schneiderman HA (1960) Discontinuous respiration in insects: role of the spiracles. Biol Bull (Woods Hole) 119:494–528

Schneiderman HA, Schechter AN (1966) Discontinuous respiration in insects. V. Pressure and volume changes in the tracheal system of silkworm pupae. J Insect Physiol 12:1143–1170

Schneiderman JA, Williams CM (1953a) The physiology of insect diapause. VII. The respiratory metabolism of the cecropia silkworm during diapause and development. Biol Bull (Woods Hole) 105:320–334

Schneiderman HA, Williams CM (1953b) Discontinuous carbon dioxide output by diapausing pupae of the giant silkworm, *Platysamia cecropia*. Biol Bull (Woods Hole) 105:382

Schneiderman HA, Williams CM (1955) An experimental analysis of the discontinuous respiration of the cecropia silkworm. Biol Bull (Woods Hole) 109:123–143

Schneppenheim R, Theede H (1980) Isolation and characterization of freezing-point depressing peptides form larvae of *Tenebrio molitor*. Comp Biochem Physiol 67B:561–568

Schöttler U (1980) Der Energiestoffwechsel bei biotopbedingter Anaerobiose: Untersuchungen an Anneliden. Verh Dtsch Zool Ges 1980:228–240

Schopf A (1980) Diapause of the pupal parasitoid *Pimpla turionella* (Hymenoptera: Ichneumonidae). Zool Jahrb Abt Syst Oekol Geogr Tiere 107:537–567

Schroff G, Schöttler U (1977) Anaerobic reduction of fumarate in the body wall musculature of *Arenicola marina* (Polychaeta). J Comp Physiol 116B:325–336

Schulz M (1984) Untersuchungen des Lipid- und Kohlenhydratstoffwechsels während des Fluges von *Manduca sexta*. Masters Thesis, Freie Universität Berlin

Schulz M, Ziegler R (1983) Untersuchungen zur Regulation des Kohlenhydratstoffwechsels während des Fluges bei *Manduca sexta*. Verh Dtsch Zool Ges 1983:322

Schwalb HH (1961) Beiträge zur Biologie der einheimischen Lampyriden *Lampyris noctiluca* und *Phausis splendidula*, und experimentelle Analysen ihres Beutefang- und Sexualverhaltens. Zool Jahrb Abt Syst Oekol Geogr Tiere 88:399–550

Seliger HH, Lall AB, Lloyd JE, Biggley WH (1982a) The colors of firefly bioluminescence – II. Experimental evidence for the optimization model. Photochem Photobiol 36:681–688

Seliger HH, Lall AB, Lloyd JR, Biggley WH (1982b) Colors of fireflies – I. Optimization model. Photochem Photobiol 36:673–680

Shapiro JP, Low JH (1983) Locust adipokinetic hormone stimulates lipid mobilization in *Manduca sexta*. Biochem Biophys Res Commun 115:924–931

Shappirio DG (1974) Comparative studies of oxidative enzyme systems in epidermis and fat body of diapausing and nondiapausing silkmoths. J Insect Physiol 20:291–300

Shappirio DG, Williams CM (1957a) The cytochrome system of the cecropia silkworm. I. Spectroscopic studies of individual tissues. Proc R Soc Lond Ser B 147:218–232

Shappirio DG, Williams CM (1957b) The cytochrome system of the cecropia silkworm. II. Spectrophotometric studies of oxidative enzyme system in the wing epithelium. Proc R Soc Lond Ser B 147:233–246

Sharma U, Sahni SL, Sinha DP (1980) Effect of unfavourable environmental temperature on the neuroendocrine system and oviposition in *Dysdercus koenigii* (Hemiptera: Pyrrhocoridae). Indian J Exp Biol 18:1192–1194

Sharpe PJH, De Michele DW (1977) Reaction kinestics of poikilotherm development. J Theor Biol 64:649–670

Sharpe PJH, Curry GL, De Michele DW, Cole CL (1977) Distribution model of organism development times. J Theor Biol 66:21–38

Shaw J (1955) Ionic regulation and water balance in the aquatic larva of *Sialis lutaria*. J Exp Biol 32:330–352

Shelford VW (1927) An experimental investigation of the relations of the codling moth to weather and climate. Ill St Nat Hist Surv Bull 16:311–440

Shimada K (1983) Photoperiodic induction of diapause in normal and allatectomized precocious pupae of *Papilio machaon*. J Insect Physiol 29:801–806

Shimizu J, Kitabatake S, Kato M (1981) Effect of carotenoid deficiency on photosensitivities in the silkworm, *Bombyx mori*. J Insect Physiol 27:593–599

Shoubridge EA, Hochachka PW (1980) Ethanol: Novel endproduct of vertebrate anaerobic metabolism. Science (Wash DC) 209:308–309

Shroyer DA, Craig GB (1980) Egg hatchability and diapause in *Aedes triseriatus* (Diptera: Culicidae). Temperature induced and photoperiod induced latencies. Ann Entomol Soc Am 73: 39–43

Siddiqui WH, Barlow CA (1972) Population growth of *Drosophila melanogaster* (Diptera: Drosophilidae) at constant and alternating temperatures. Ann Entomol Soc Am 65:993–1001

Siddiqui WH, Barlow CA (1973) Population growth of *Anagasta kuehniella* (Lepidoptera:Pyralidae) at constant and alternating temperatures. Ann Entomol Soc Am 66:579–585

Siddiqui WH, Barlow CA, Randolph DA (1973) Effect of some constant and alternating temperatures on population growth of the pea aphid *Acyrthosiphon pisum* (Homoptera: Aphididae). Can Entomol 105:145–156

Sidibé B, Laugé G (1977) Incidence de thermopériodes et de températures constantes sur quelques critères biologiques de *Spodoptera* littoralis Boisduval (Lepidoptera, Noctuidae). Ann Soc Entomol Fr 13:369–379

Sieber R, Benz G (1977) Juvenile hormone in larval diapause of the codling moth, *Laspeyresia pomonella* L. (Leptidoptera: Tortricidae). Experientia (Basel) 33:1598–1599

Sieber R, Benz G (1980a) Hormonal regulation of pupation in the codling moth, *Laspeyresia pomonella*. Physiol Entomol 5:283–290

Sieber R, Benz G (1980b) The hormonal regulation of the larval diapause in the codling moth, *Laspeyresia pomonella* (Lepidoptera: Tortricidae). J Insect Physiol 26:213–218

Siegert K (1983) Regulation des Kohlenhydrat- und des Lipidstoffwechsels während der späten Larvalentwicklung und des Hungerns von *Manduca sexta*. PHD Thesis, Freie Universität Berlin

Siegert K, Ziegler R (1982) Sauerstoffverbrauch und respiratorische Quotienten bei *Manduca sexta* (Lepidoptera: Sphingidae). Verh Dtsch Zool Ges 1982:332

Siegert K, Ziegler R (1983a) A hormone from the corpora cardiaca controls fat body glycogen phosphorylase during starvation in tobacco hornworm larvae. Nature (Lond) 301:526–527

Siegert K, Ziegler R (1983b) Regulation of release of glycogen phosphoylase activating hormone from the corpora cardiaca of larval *Manduca sexta* during starvation. Acta Endocrinol Suppl 253:64–65

Siegert K, Krippeit P, Ziegler R (1982) Regulation of fat body glycogen phosphorylase during starvation in larvae of *Manduca sexta* (Lepidoptera: Sphingidae). Acta Endocrinol Suppl 246:33–34

Simonet DE, Pienkowski RL (1980) Temperature effect on development and morphometrics of the potato leafhopper. Environ Entomol 9:798–800

Sims SR (1983) The genetic and environmental basis of pupal color dimorphism in *Papilio zelicaon* (Lepidoptera: Papilionidae). Heredity 50:159–168

Singh SP, Das AB (1977) Thermal acclimatory responses of salivary amylase of *Periplaneta americana* (Linn.). Experientia (Basel) 33:168–169

Singh SP, Das AB (1980) Biochemical changes in tissue composition of *Periplaneta americana* (Linn.) acclimated to high and low temperatures. J Therm Biol 5:211–218

Sivinsky J (1981) The nature and possible functions of luminescence in Coleoptera larvae. Coleopt Bull 35:167–179

Slama K (1976) Insect haemolymph pressure and its determination. Acta Entomol Bohemoslov 73:65–75

Slama K, Baudry-Partiaglou N, Provansal-Bauder A (1979) Control of extracardiac haemolymph pressure pulses in *Tenebrio molitor.* J Insect Physiol 25:825–831

Smalley KN, Tarwater DE, Davidson T (1977) Localization of fluorescent compounds in the firefly light organ. Cited in: Herring PJ (ed) Bioluminescence in action. Academic, New York, p 540

Smith AG (1978) Environmental factors influencing pupal colour determination in Lepidoptera. I. Experiments with *Papilio polytes, Papilio demoleus* and *Papilio polyxenes.* Proc R Soc Lond Ser B 200:295–329

Smith AG (1980) Environmental factors influencing pupal colour determination in Lepidoptera. II. Experiments with *Pieris rapae, Pieris napi* and *Pieris brassicae.* Proc R Soc Lond Ser B 207: 163–186

Snell PA (1932) The control of luminescence in the male lampyrid firefly, *Photuris pennsylvanica,* with special references to the effect of oxygen tension on flashing. J Cell Comp Physiol 1:37–51

Sømme L (1964) Effects of glycerol on cold-hardiness in insects. Can J Zool 42:87–101

Sømme L (1967) The effect of temperature and anoxia on haemolymph composition and supercooling in three overwintering insects. J Insect Physiol 13:805–814

Sømme L (1968) Acclimation to low temperatures in *Tribolium confusum* Duvat (Coleopt., Tenebrionidae). Nor Entomol Tidsskr 15:134–136

Sømme L (1972) The effects of acclimation and low temperatures on enzyme activites in larvae of *Ephestia kuehniella* Zell. (Lep., Pyralidae). Entomol Scand 3:12–18

Sømme L (1982) Supercooling and winter survival in terrestrial arthropods. Comp Biochem Physiol 73A:519–543

Sømme L, Block W (1982) Cold hardiness of Collembola at Signy Island, maritime Antarctic, Oikos 38:168–176

Spencer-Davis P, Tribe MA (1969) Temperature dependence of metabolic rate in animals. Nature (Lond) 224:723–724

Scrivastava BBL, Hopkins TL (1975) Bursicon release and activity in haemolymph during metamorphosis of the cockroach, *Leucophaea maderae.* J Insect Physiol 21:1985–1993

Staal GB (1961) Studies on the physiology of phase induction in *Locusta migratoria migratorioides* R. & F. Meded Lab Entomol Wageningen 72:1–125

Steel CHG, Lees AD (1977) The role of neurosecretion in photoperiodic control of polymorphism in the aphid *Megoura viciae.* J Exp Biol 67:117–135

Steele JE (1976) Hormonal control of metabolism in insects. Adv Insect Physiol 12:239–323

Steele JE (1981) The role of carbohydrate metabolism in physiological function. In: Downer RGH (ed) Energy metabolism in insects. Plenum, New York London, pp 101–133

Steele JE (1982) Glycogen phosphorylase in insects. Insect Biochem 12:131–147

Stell W, Marshak D, Yamada T, Brechna N, Karten H (1980) Peptides in the eye of the beholder. Trends Neurosci 3:292–295

Stobbart RH (1967) The effect of some anions and cations upon the fluxes and net uptake of chloride in the larvae of *Aedes aegypti* (L.), and the nature of the uptake mechanisms for sodium and chloride. J Exp Biol 47:35–57

Stone JV, Mordue W, Batley KE, Morris HR (1976) Structure of locust adipokinetic hormone, a neurohormone that regulates lipid utilization during flight. Nature (Lond) 263:207–211

Storey KB (1982) Phosphofructokinase from the overwintering gall fly larva, *Eurosta solidaginis:* Control of cryoprotectant polyol synthesis. Insect Biochem 12:501–505

Storey KB (1983) Metabolism and bound water in overwintering insects. Cryobiology 20: 365–379

Storey KB, Storey JM (1983) Biochemistry of freeze tolerance in terrestrial insects. Trends Biochem Sci 8:242–245

Storey KB, Park IRA, Storey JM (1981) Isozyme composition and low temperature acclimation in the overwintering gall fly larva, *Eurosta solidaginis.* Cryo Lett 2:279–284

Strause LG, DeLuca M (1981) Characteristics of luciferases from a variety of firefly species: Evidence for the presence of luciferase isozymes. Insect Biochem 11:417–422

Strong FE, Sheldahl JA (1970) The influence of temperature on longevity and fecundity in the bug *Lygus hesperus* (Hemiptera: Miridae). Ann Entomol Soc Am 63:1509–1515

Strong FE, Dunkle RL, Dunn RL (1970) Low-temperature physiology of Antarctic arthropods. Antarct J US 5:123

Surholt B, Newsholme EA (1983) The rate of substrate cycle between glucose and glucose-6-phosphate in muscle and fat-body of the hawk moth *(Acherontia atropos)* at rest and during flight. Biochem J 210:49–54

Tager HS, Markese J, Kramer KJ, Spiers RD, Childs CN (1976) Glucagon-like and insulin-like hormones of the insect neurosecretory system. Biochem J 156:515–520

Takami T (1959) Induced growth of diapausing silkworm embryos *in vitro*. Science (Wash DC) 130:98–99

Takeda M, Masaki S (1979) Asymmetric perception of twilight affecting diapause induction by the fall webworm, *Hyphantria cunea*. Entomol Exp Appl 25:317–327

Tanaka H (1982) The migratory locust, *Locusta migratoria* (Orthoptera: Acrididae) in Japan. 1. Experiments on crowding effects. Appl Entomol Zool 17:467–479

Tauber MJ, Tauber CA (1976) Insect seasonality: diapause maintenance, termination, and post-diapause development. Annu Rev Entomol 21:81–107

Tauber MJ, Tauber CA (1978) Evolution of phenological strategies in insects: a comparative approach with eco-physiological and genetic considerations. In: Dingle H (ed) Evolution of insect migration and diapause. Springer, Berlin Heidelberg New York, pp 53–71

Tauber MJ, Tauber CA Nechols JR, Helgesen RG (1982) A new role for temperature in insect dormancy: cold maintains diapause in temperate zone Diptera. Science (Wash DC) 218: 690–691

Thauer RK, Jungermann K, Decker K (1977) Energy conservation in chemotrophic anaerobic bacteria. Bacteriol Rev 41:100–180

Thiele HU (1973) Remarks about Mansingh's and Müller's classifications of dormancies in insects. Can Entomol 105:925–928

Thiessen CI, Mutchmor JA (1967) Some effects of thermal acclimation on muscle apyrase activity and mitochondrial number in *Periplaneta americana* and *Musca domestica*. J Insect Physiol 13:1837–1842

Thillart G van den, Kesbeke F (1978) Anaerobic production of carbon dioxide and ammonia by goldfish, *Carassius auratus* (L.). Comp Biochem Physiol 59A:393–400

Thorpe WH, Crisp DJ (1947a) Studies on plastron respiration. I. The biology of *Aphelocheirus* [Hemiptera, Aphelocheiridae (Naucoridae)] and the mechanism of plastron retentinon. J Exp Biol 24:227–269

Thorpe WH, Crisp DJ (1947b) Studies on plaston respiration. II. The respiratory efficiency of the plastron in *Aphelocheirus*. J Exp Biol 24:270–303

Thorpe WH, Crisp DJ (1947c) Studies on plastron respiration. III. The orientation responses of *Aphelocheirus* [Hemiptera, Apheloheiridae (Naucoridae)] in relation to plastron respiration; together with an account of specialized pressure receptors in aquatic insects. J Exp Biol 24: 310–328

Tilbrook PJ, Block W (1972) Oxygen uptake in an Antarctic collembole *Cryptopygus antarcticus*. Oikos 23:313–317

Tobe SS, Chapman CS (1979) The effects of starvation and subsequent feeding on juvenile hormone synthesis and oocyte growth in *Schistocerca gregaria*. J Insect Physiol 25:701–708

Tombes AS (1966) Aestivation (summer diapause) in *Hypera postica* (Coleoptera: Curculionidae). I. Effect of aestivation, photoperiods, and diet on total fatty acids. Ann Entomol Soc Am 59: 376–380

Tomescu N, Chis V, Coroiu I, Jeleriu S (1982) The influence of temperature upon the response of the males of *Mamestra brassicae* (Lepidoptera, Noctuidae) to the sexual pheromone. Stud Univ Babes-Bolyai Biol 27:49–52

Treherne JE (1967) Gut absorption. Annu Rev Entomol 12:43–58

Treviranus (1814) Biologie, oder Philosophie der lebenden Natur. Vol IV

Tribe MA, Bowler K (1968) Temperature dependence of "standard metabolic rate" in a poikilotherm. Comp Biochem Physiol 25:427–436

Trpis M, Horsfall WR (1969) Development of *Aedes sticticus* (Meigen) in relation to the temperature, diet, density and depth. Ann Zool Fenn 6:156–160

Tsitsipis JA (1980) Effect of constant temperatures on larval and pupal development of olive fruit flies reared on artificial diet. Environ Entomol 9:764–768

Turunen S (1979) Digestion and absorption of lipids in insects. Comp Biochem Physiol 63A: 455–460

Turunen S, Chippendale GM (1979) Possible function of juvenile hormone-dependent protein in larval insect diapause. Nature (Lond) 280:836–838

Turunen S, Chippendale GM (1980a) Fat body protein associated with the larval diapause of the southwestern corn borer, *Diatraea grandiosella:* synthesis and characteristics. Comp Biochem Physiol 65B:595–603

Turunen S, Chippendale GM (1980b) Proteins of the fat body of non-diapausing and diapausing larvae of the southwestern corn borer, *Diatraea grandiosella:* effect of juvenile hormone. J Insect Physiol 26:163–169

Ugarova NN, Filippova NYu, Berezin IV (1981) Inhibition of luciferase of the firefly *Luciola mingrelica* by inorganic salts. Biokhimiya 46:851–858

Vejbjerg K, Normann TC (1974) Secretion of hyperlgycaemic hormone from the corpus cardiacum of flying blowflies, *Calliphora erythrocephala.* J Insect Physiol 20:1189–1192

Veron JEN (1973) Physiological control of the chromatophores of *Austrolestes annulosus* (Odonata). J Insect Physiol 19:1689–1703

Wade NH, West DA (1979) Environmental control of pupal color in swallowtail butterflies (Lepidoptera: Papilioninae) *Battus philenor* and *Papilio polyxenes.* Ecol Entomol 4:393–400

Wagner TL, Gagne JA, Cover JD, Coulson RN, Pulley PE (1981) Comparison of gallery construction oviposition, and reemergence by *Dendroctonus frontalis* females producing first and second broods. Ann Entomol Soc Am 74:570–575

Wall BJ, Oschman (1975) Structure and function of the rectum in insects. In: Wessing A (ed) Excretion. Fortschr Zool 23:2/3, Fischer, Stuttgart, pp 193–222

Wampler JE (1978) Measurements and physical characteristics of luminescence. In: Herring PJ (ed) Bioluminescence in action. Academic, New York, pp 1–48

Wasserthal LT (1981) Oscillating hemolymph 'circulation' and discontinuous tracheal ventilation in the giant silk moth *Attacus atlas* L. J Comp Physiol 145:1–15

Watt WB (1969) Adaptive significance of pigment polymorphisms in *Colias* butterflies. II. Thermoregulation and photoperiodically controlled melanin variation in *Colias eurytheme.* Proc Nat Acad Sci USA 63:767–774

Watts DT (1951) Intratracheal pressure in insect respiration. Ann Entomol Soc Am 44:527–538

Way MJ,Hopkins BA (1950) The influence of photoperiod and temperature on the induction of diapause in *Diataraxia oleracea* L. (Lepidoptera). J Exp Biol 27:365–376

Weaver RJ, Pratt GE (1981) Effects of starvation and feeding upon corpus allatum activity and oocyte growth in adult female *Periplaneta americana.* J Insect Physiol 27:75–83

Weeda E (1981) Hormonal regulation of proline synthesis and glucose release in the fat body of the colorado potato beetle, *Leptinotarsa decemlineata.* J Insect Physiol 27:411–417

Weeda E, de Kort CAD, Beenakkers AMT (1979) Fuels for energy metabolism in the Colorado potato beetle, *Leptinotarsa decemlineata* Say. J Insect Physiol 25:951–955

Wegener G (1981) Comparative aspects of energy metabolism in nonmammalian brains under normoxic and hypoxic conditions. In: Stefanovich V (ed) Animal models and hypoxia. Pergamon, Oxford, pp 87–109

Wegener G (1983) Brains burning fat: Different forms of energy metabolism in the CNS of insects. Naturwissenschaften 70:43–45

Weis-Fogh T (1952) Fat combustion and metabolic rate of flying locusts *(Schistocerca gregaria* Forskal). Philos Trans R Soc Lond B 237:1–36

Weis-Fogh T (1964) Diffusion in insect wing muscle, the most active tissue known. J Exp Biol 41:229–256

Weis-Fogh T (1967) Respiration and tracheal ventilation in locusts and other flying insects. J. Exp Biol 47:561–587

Welbers P (1975a) Der Einfluß von tagesperiodischen Wechseltemperaturen bei der Motte *Pectinophora*. I. Entwicklungsdauer, Larvengewicht, Reproduktionsrate. Oecologia (Berl) 21: 31–42

Welbers P (1975b) Der Einfluß von tagesperiodischen Wechseltemperaturen bei der Motte *Pectinophora* II. Der Sauerstoffverbrauch. Oecologia (Berl) 21:43–56

Welbers P (1976) Analyse des Sauerstoffverbrauchs von *Periplaneta americana* bei konstanten und tagesperiodisch wechselnden Temperaturbedingungen. Oecologia (Berl) 24:175–192

Wells RMG (1980) Invertebrate Respiration. Edward Arnold, London (The institute of biology's studies in biology, no 127)

Wessing A, Eichelberg D (1978) Malpighian tubules, rectal papillae and excretion. In: Ashburner M, Wright TRF (eds) The genetics and biology of *Drosophila,* vol 2c. Academic, New York London, pp 1–42

Wessing A, Hevert F, Rönnau K (1978) Harnproduktion der Malpighi-Gefäße *in vitro.* Verh Dtsch Zool Ges 1978:300

White EH, McCapra F, Field GF, McElroy WD (1961) The structure and synthesis of firefly luciferin. J Am Chem Soc 83:2402–2403

White EH, McCapra F, Field GF (1963) The structure and synthesis of firefly luciferin. J Am Chem Soc 85:337–343

Wieczorek H (1982) A biochemical approach to the electrogenic potassium pump of insect sensilla: Potassium sensitive ATPase in labellum of the fly. J Comp Physiol 148:303–311

Wieser W (ed) (1973) Effects of temperature on ecotothermic organisms. Springer, Berlin Heidelberg New York

Wigglesworth VB (1930) A theory of tracheal respiration in insects. Proc R Soc Lond Ser B 106: 229–250

Wigglesworth VB (1931) The physiology of excretion in a bloodsucking insect, *Rhodnius prolixus* (Hemiptera, Reduviidae). I. Composition of the urine. J Exp Biol 8:411–427

Wigglesworth VB (1934) Insect physiology. Methuen, London

Wigglesworth VB (1935) The regulation of respiration in the flea, *Xenopsylla cheopis,* Roths. (Pulicidae). Proc R Soc Lond Ser B 118:397–419

Wigglesworth VB (1953a) Surface forces in the tracheal system of insects. Q J Microsc Sci 94: 507–522

Wigglesworth VB (1953b) The principles of insect physiology. Methuen, London

Wigglesworth VB (1954) Growth and regeneration in the tracheal system of an insect, *Rhodnius prolixus* (Hemiptera). Q J Microsc Sci 95:115–137

Wigglesworth VB (1959) The control of growth and form: a study of the epidermal cell in an insect. Cornell University Press, Ithaca

Wigglesworth VB (1964) The hormonal regulation of growth and reproduction in insects. Adv Insect Physiol 2:247–336

Wigglesworth VB (1972) The principles of insect physiology. Chapman and Hall, London

Wigglesworth VB (1981) The natural history of insect tracheoles. Physiol Entomol 6:121–128

Wigglesworth VB, Lee WM (1982) The supply of oxygen to the flight muscles of insects: A theory of tracheole physiology. Tissue & Cell 14:501–518

Wilde J de (1962) Photoperiodism in insects and mites. Annu Rev Entomol 7:1–26

Wilde J de (1983) Endocrine aspects of diapause in the adult stage. In: Downer RGH, Laufer H (eds) Endocrinology of insects. Alan R. Liss, New York, pp 357–367 (Invertebrate endocrinology, vol 1)

Wilhelm RC, Schneiderman HA, Daniel DJ (1961) The effects of anaerobiosis on the giant silkworm *Hyalophora cecropia* and *Samia cynthia* with special reference to the accumulation of glycerol and lactic acid. J Insect Physiol 7:273–288

Wilkins MB (1960) A temperature-dependent endogenous rhythm in the rate of carbon dioxide output of *Periplaneta americana.* Nature (Lond) 185:481–482

Williams CM (1946) Physiology of insect diapause: the role of the brain in the production and termination of pupal dormancy in the giant silkworm, *Plathysamia cecropia.* Biol Bull (Woods Hole) 90:234–243

Williams CM (1947) Physiology of insect diapause. II. Interaction between the pupal brain and the prothoracic glands in the metamorphosis of the giant silkworm, *Platysamia cecropia*. Biol Bull (Woods Hole) 93:89–98

Williams CM (1952) Physiology of insect diapause. IV. The brain and prothoracic glands as an endocrine system in the cecropia silkworm. Biol Bull (Woods Hole) 103:120–138

Williams CM (1980) Growth in insects. In: Locke M, Smith DS (eds) Insect biology in the future "VBW 80". Academic, New York London, pp 369–385

Willig A (1969) Die Carotinoide und der Gallenfarbstoff der Stabheuschrecke, *Carausius morosus* und ihre Beteiligung an der Entstehung der Farbmodifikationen. J Insect Physiol 15:1907–1927

Willis JH (1974) Morphogenetic action of insect hormones. Annu Rev Entomol 19:97–115

Willmer PG (1982) Microclimate and the environmental physiology of insects. Adv Insect Physiol 16:1–57

Willmer PG, Unwin DM (1981) Field analyses of insect budgets: reflectance, size and heating rates. Oecologia (Berl) 50:250–255

Wilps H (1976) Glykogenabbau und Endprodukte des anaeroben Stoffwechsel der aquatischen Dipterenlarve *Chironomus thummi thummi* (Diptera). Diplomarbeit, Universität Münster

Wilps H (1978) Untersuchungen zum Anaerobiosestoffwechsel der Dipterenlarve *Chironomus thummi thummi*. PHD Thesis, Universität Münster

Wilps H, Schöttler U (1980) *In vitro* studies on the anaerobic formation of ethanol by the larvae of *Chironomus thummi thummi* (Diptera). Comp Biochem Physiol 67B:239–242

Wilps H, Zebe (1976) The end-products of anaerobic carhohydrate metabolism in the larvae of *Chironomus thummi thummi*. J Comp Physiol 112:263–272

Winteringham FPW (1960) Phosphorylated compounds in the head and thoracic tissues of the adult housefly, *Musca domestica* L., during flight, rest, anoxia and starvation. Biochem J 87: 35–38

Witsack W (1973) Experimentell-ökologische Untersuchungen über Dormanz-Formen von Zikaden (Homoptera: Anchenorrhyncha). II. Zur Ovarial-Parapause und obligatorischen Embryonal-Diapause von *Philaenus spumarius* (L.) (Aphrophoridae). Zool Jahrb Syst 100:517–562

Wood FE, Nordin JH (1976) Studies on the low temperature induced biogenesis of glycerol by adult *Protophormia terranovae*. J Insect Physiol 22:1665–1673

Wood FE, Nordin JH (1980) Temperature effects on mitochondrial respiration of *Protophormia terranovae* and *Musca domestica*. Insect Biochem 10:95–99

Wood FE, Mahar P, Nordin JH (1977) Metabolite levels and enzyme activities in *Protophormia terranovae* during low temperature induced glycerol accumulation. Insect Biochem 7:141–149

Woodring JP (1983) Control of moulting in the house cricket, *Acheta domesticus*. J Insect Physiol 29:461–464

Wynne-Edwards VC (1962) Animal dispersion in relation to social behaviour. Oliver and Boyd, Edinburgh London

Yagi S (1981) Physiological aspects of diapause in rice stem borers and the effect of juvenile hormone (Lepidoptera: Pyralidae). Entomol Gen 7:213–221

Yamashita O (1983) Egg diapause. In: Downer RGH, Laufer H (eds) Endocrinology of insects. Alan R, Liss, New York, pp 337–342 (Invertebrate endocrinology, vol 1)

Yamashita O, Yaginuma T, Hasegawa K (1981) Hormonal and metabolic control of egg diapause of the silkworm, *Bombyx mori* (Lepidoptera: Bombycidae). Entomol Gen 7:195–211

Yeargan KV (1980) Effects of temperature on developmental rate of *Telenomus podisi* (Hymenoptera: Scelionidae). Ann Entomol Soc Am 73:339–342

Yeargan KV (1983) Effects of temperature on developmental rate of *Trissolcus euschisti* (Hymenoptera: Scelionidae), a parasite of stink bug eggs. Ann Entomol Soc Am 76:757–760

Young SR (1979) Effect of temperature change on the metabolic rate of an Antarctic mite. J Comp Physiol 131B:341–346

Zachariassen KE (1980) The role of polyols and nucleating agents in cold-hardy beetles. J Comp Physiol 140:227–234

Zalucki MP (1982) Temperature and rate of development in *Danaus plexippus* L. and *D. chrysippus* L. (Lepidoptera: Nymphalidae). J Aust Entomol Soc 21:241–246

Zebe E (1961) Vergleichende Physiologie des Energiestoffwechsels von Muskeln. Ergeb Biol 24: 247–286

Zerahn K (1978) Transport across insect gut epithelium. In: Giebisch G, Tosteson DC, Ussing HH (eds) Membrane transport in biology III: Transport across multimembrane systems. Springer, Berlin Heidelberg New York, pp 273–306

Ziegler H (1977) Flüssigkeitsströme in Pflanzen. In: Hoppe W, Lohmann W, Markl H, Ziegler H (eds) Biophysik. Springer, Berlin Heidelberg New York, pp 561–577

Ziegler R (1979) Hyperglycaemic factor from the corpora cardiaca of *Manduca sexta*. Gen Comp Endocrinol 39:350–357

Ziegler R (1980) Hypertrehalokämische und adipokinetische Aktivität bei *Manduca sexta*. Verh Dtsch Zool Ges 1980:309

Ziegler R (1984) Developmental changes in the response of the fat body of *Manduca sexta* to injections of corpora cardiaca extracts. Gen Comp Endocrinol 54:51–88

Ziegler R, Gäde G (1984) Preliminary characterization of glycogen phosphylase activating hormone and adipokinetic hormone from *Manduca sexta* corpora cardiaca. Physiol Entomol 9:229–236

Ziegler R, Schulz M (1982) Hormonal regulation of lipid mobilization in the tobacco hornworm, *Manduca sexta* (Lepidoptera: Sphingidae). Acta Endocrinol Suppl 246:32–33

Ziegler R, Siegert K, Rohde M (1978) Veränderungen im Kohlenhydratstoffwechsel während der Larvalentwicklung von *Manduca sexta* (Lepidoptera: Sphingidae). Verh Dtsch Zool Ges 1978: 303

Zwölfer W (1935) Die Temperaturabhängigkeit der Entwicklung der Nonne (*Lymantria monacha* L.) und ihre Bevölkerungswissenschaftliche Auswertung. Z Angew Entomol 21:333–384

Subject Index

Taxonomic Index